Technologies in Cell Culture - A Journey From Basics to Advanced Applications

Edited by Soumya Basu,
Amit Ranjan and Subhayan Sur

Published in London, United Kingdom

Technologies in Cell Culture - A Journey From Basics to Advanced Applications
http://dx.doi.org/10.5772/intechopen.107775
Edited by Soumya Basu, Amit Ranjan and Subhayan Sur

Contributors
Amit Ranjan, Anja Rodenbrock, Anthony Leone, Arturo Agilar-Rojas, Aziza Manceur, Bianka Grunow, Brijesh Kumar Singh, Burcu Ayhan-Sahin, Chris Chumsae, Dayan A. Carrión-Estrada, Dimitrios Bouzianas, Enes Bal, Estelle Leclerc, Gloria Li, Jillian Cornish, Julien Robitaille, Karen Callon, Letha Chemmalil, Madhulika Tripathi, Manish Kumar Verma, Marcus Ground, Martin Loignon, Miguel Vargas, Ozge Rencuzogullari, Paget Milsom, Paola Briseño-Diaz, Pawel Borowicz, Pelin Ozfiliz-Kilbas, Qiang Li, Rob Walker, Sandra Delfin-Azuara, Sangeeta Ballav, Subhayan Sur, Soumya Basu, Stefan W. Vetter, Valeria Di Leonardo, Yousuf Alam

First published in London, United Kingdom, 2024 by IntechOpen
IntechOpen is the global imprint of INTECHOPEN LIMITED, registered in England and Wales, registration number: 11086078, 167-169 Great Portland Street, London, W1W 5PF, United Kingdom

British Library Cataloguing-in-Publication Data
A catalogue record for this book is available from the British Library

Additional hard and PDF copies can be obtained from orders@intechopen.com

Technologies in Cell Culture - A Journey From Basics to Advanced Applications
Edited by Soumya Basu, Amit Ranjan and Subhayan Sur
p. cm.

This title is part of the Biochemistry Book Series, Volume 54
Topic: Cell and Molecular Biology
Series Editor: Andrei Surguchov
Topic Editor: Mary C. Maj
Associate Topic Editor: Felicia Ikolo

Print ISBN 978-1-83768-291-1
Online ISBN 978-1-83768-328-4
eBook (PDF) ISBN 978-1-83768-329-1
ISSN 2632-0983

For EU product safety concerns:
IN TECH d.o.o., Prolaz Marije Krucifikse Kozulić 3, 51000 Rijeka, Croatia,
info@intechopen.com or visit our website at intechopen.com.

IntechOpen Book Series
Biochemistry
Volume 54

Aims and Scope of the Series

Biochemistry, the study of chemical transformations occurring within living organisms, impacts all of the life sciences, from molecular crystallography and genetics, to ecology, medicine and population biology. Biochemistry studies macromolecules - proteins, nucleic acids, carbohydrates and lipids –their building blocks, structures, functions and interactions. Much of biochemistry is devoted to enzymes, proteins that catalyze chemical reactions, enzyme structures, mechanisms of action and their roles within cells. Biochemistry also studies small signaling molecules, coenzymes, inhibitors, vitamins and hormones, which play roles in the life process. Biochemical experimentation, besides coopting the methods of classical chemistry, e.g., chromatography, adopted new techniques, e.g., X-ray diffraction, electron microscopy, NMR, radioisotopes, and developed sophisticated microbial genetic tools, e.g., auxotroph mutants and their revertants, fermentation, etc. More recently, biochemistry embraced the 'big data' omics systems. Initial biochemical studies have been exclusively analytic: dissecting, purifying and examining individual components of a biological system; in exemplary words of Efraim Racker, (1913 –1991) "Don't waste clean thinking on dirty enzymes." Today, however, biochemistry is becoming more agglomerative and comprehensive, setting out to integrate and describe fully a particular biological system. The 'big data' metabolomics can define the complement of small molecules, e.g., in a soil or biofilm sample; proteomics can distinguish all the proteins comprising e.g., serum; metagenomics can identify all the genes in a complex environment e.g., the bovine rumen.

This Biochemistry Series will address both the current research on biomolecules, and the emerging trends with great promise.

Meet the Series Editor

Andrei Surguchov, Ph.D., joined Baylor College of Medicine, Houston, TX, as a faculty member in 1992, where he studied the mechanisms of the genetic control of lipid metabolism. At the University of Utah, his research interests focused on cloning new genes encoding retinal proteins. He studied molecular and cellular mechanisms of neurodegenerative diseases and retinal degeneration at Washington University, St. Louis. Currently, his research focuses on the structure-function relationship of proteins involved in neurodegeneration and ocular diseases. Andrei Surguchov is an Editor-in-Chief at Biochemistry Research International and Associate Editor in several biomedical journals.

Meet the Volume Editors

Dr. Soumya Basu is an associate professor at Dr. D. Y. Patil Biotechnology and Bioinformatics Institute, Dr. D. Y. Patil Vidyapeeth University (DPU), Pune, India. With more than 17 years of research experience in drug design, development, and preclinical validation, she has published more than forty peer-reviewed articles. She has taught post-graduate students for more than 10 years. Dr. Basu's research focuses on designing small-molecule partial agonists of PPAR-γ and using artificial intelligence to understand exosomal miRNA's role in cancer metastasis. She is the recipient of many awards and recognitions, including the Early Career Research Award from the Department of Science and Technology, Government of India. She is a reviewer for several international journals and has received many awards and recognitions. She is also a member of statutory boards and the Board of Studies at Dr. D. Y. Patil Vidyapeeth University, and a life member of professional bodies The Cytometry Society and All India Cell Biology Society in India. Dr. Basu qualified for several national competitive examinations, including the National Eligibility Test (NET) and Graduate Aptitude Test in Engineering (GATE). She is also recognized as an Innovation Ambassador of the Ministry of Education's Innovation Cell (MIC), Ministry of Education of India (MHRD). Additionally, she plays a key role in fostering industry-academia collaborations and shaping curriculum strategies.

Dr. Amit Ranjan is currently an assistant professor at Dr. D.Y. Patil Biotechnology and Bioinformatics Institute, Dr. D. Y. Patil Vidyapeeth University (DPU), Pune, India. He is a research professional with more than 12 years of teaching and research experience. He obtained an MSc in Molecular Biology and Biotechnology from Tezpur University, India, and a Ph.D. from the Advanced Centre for Treatment, Research and Education in Cancer (ACTREC), Tata Memorial Center, India. He qualified for several national competitive examinations, including the National Eligibility Test (NET), the Graduate Aptitude Test in Engineering (GATE), and the Biotechnology Eligibility Test (BET). He is a recipient of a travel award and an Innovation Ambassador of the Ministry of Education's Innovation Cell (MIC), Ministry of Education of India (MHRD). His areas of research include the role of glycans in cancer metastasis. He has published thirteen journal articles and four book chapters.

Dr. Subhayan Sur is currently a Ramalingaswami Re-entry Fellow, Department of Biotechnology (DBT), India, and an assistant professor at Dr. D. Y. Patil Biotechnology and Bioinformatics Institute, Dr. D. Y. Patil Vidyapeeth University (DPU), Pune, India. Dr. Sur has a keen interest in the exploration of non-coding RNA and natural compounds for cancer prevention and therapy. Dr. Sur obtained his Ph.D. in Biotechnology from the University of Calcutta, India, and completed a post-doctoral fellowship at Saint Louis University, MO, USA, focusing on cancer biology, non-coding RNA, and the pathogenesis of hepatitis C virus and SARS-CoV-2. Dr. Sur has been actively involved in research and teaching. He is a journal reviewer and has many publications in international peer-reviewed journals to his credit. He is also a member of the Society of Biological Chemists (SBC) (India).

Contents

Preface

Welcome to *Technologies in Cell Culture – A Journey from Basics to Advanced Applications*. In the ever-evolving landscape of biotechnology, cell culture is a cornerstone that enables innovations and amazing discoveries in fields ranging from basic cell biology to medicine, with current and upcoming research on biocomputing. This book is a comprehensive exploration of this vital discipline, offering a roadmap that navigates from fundamental principles to cutting-edge methodologies.

The goal of this book is to give a clear picture of the current state-of-the-art technological and application updates in cell culture. It is a useful handbook for students, researchers, and practitioners in the field. We hope that it will inspire future research ideas and directions.

In a time when scientific discoveries have the potential to both lessen human suffering and advance our knowledge of intricate biological systems, the chapters in this book act as guiding lights toward advancement. This book unravels the complexities of cell culture and provides both beginners and experienced researchers with a clear understanding of the underlying principles, techniques, and applications. Each chapter marks a significant step in the quest for knowledge and understanding, ranging from the fundamentals of cell culture to the cutting edge of sophisticated applications.

This book includes eleven chapters organized into six sections based on topics related to cell culture technological advancements and applications.

Section 1 consists of four chapters that narrate methods of two-dimensional (2D) and three-dimensional (3D) cell cultures. Chapter 1 describes valvular interstitial cellular (VIC) physiology, as well as the isolation and culture of these types of cells. In our aging world, valve pathology is becoming more common than ever, and there is very little information available regarding VIC pathophysiology. VIC stable cell populations in the lab are necessary for the scientific investigation of valve disease. These populations are frequently isolated from tissue from pigs and humans. For experimental integrity to be maintained, it is crucial to comprehend the many VIC phenotypes as well as the molecular cues that control the transition between phenotypes. Thus, this chapter presents an outline of VIC physiology and a tried-and-true technique for isolating and cultivating these cells. Chapter 2 provides a comprehensive review of culturing techniques for human pluripotent stem cells (hPSCs), covering 2D and 3D adherent culture, suspension culture, and the utilization of hydrogel scaffolds in 3D hPSC culture. To meet the demands of cutting-edge biomedical applications, the chapter also investigates the use of sophisticated 3D cell manufacturing techniques to simplify the creation of vast quantities of high-quality hPSCs. By covering these subjects, this chapter provides a thorough summary of various culturing techniques and the many uses they have in hPSC research, including basic research and cutting-edge biomedical studies.

The study of interactions between tumor microenvironment cells, including adipocytes, endothelial cells, fibroblasts, macrophages, and cancer cells, is commonly conducted by *in vitro* coculture techniques. Indirect and direct methods are the two primary categories into which coculture techniques are divided. Chapter 3 discusses the general techniques for coculture research, focusing on the outcomes of coculturing cancer cells and adipocytes. Chapter 4 describes the formation of spontaneously contracting heart primary cultures obtained from fish embryos and larvae. These cultures are useful models for several purposes, including studying heart-related viral infections and conducting research in biomedicine, pharmacology, and cardiology. Chapter 4 also presents a summary of the findings regarding the *in vitro* production of 3D heart-like cell aggregates from fully digested fish embryos and larvae.

To evaluate and validate the qualities of plant-based natural substances for use as possible therapies, Chapter 5 in Section 2 addresses the latest technological advancements in cell-free and cell-based approaches for determining properties like induction of oxidative DNA damage, inflammatory factors, apoptotic/anti-apoptotic factors, and so on.

In Section 3, Chapters 6 and 7 discuss the application of cell culture in enhancing drug discovery. Chapter 6 offers a thorough analysis of the traits and attributes of *in vitro* techniques that bolster the efficacy of two compounds. Cytotoxicity assays in cell lines and primary cell cultures of pancreatic and breast cancer grown in 2D and 3D conditions are used to determine the efficacy of both molecules, specifically against mutant forms of K-Ras as antitumor agents. Recent technologies for cultivating cells have recently been developed to mimic the biophysical characteristics of tumors. With the use of these new technologies, it is possible to create 3D tumor-like structures, or spheroids, in which cancer cells connect in a manner akin to that seen in actual tumors. As a result of these developments in cell culture technology, cell-based assays must be modified and improved. Chapter 7 focuses on certain problems encountered and the solutions found during the assessment of the effects of two anticancer medications (temozolomide and vemurafenib) on spheroids-grown melanoma cell lines.

In Section 3, Chapter 8 discusses autologous adipose tissue stem cell-derived terminally differentiated retinal pigment epithelium monolayer cell development on a biocompatible scaffold in the restoration of vision in age-related macular degeneration patients.

Section 5 includes Chapters 9 and 10, which discuss the application of cell culture in bioprocessing. Chapter 9 highlights the recent advancements in process analytical technology-based tool development, specifically for biopharmaceutical applications, along with the techniques inherited from the traditional pharmaceutical sector to monitor and control critical process parameters and critical quality attributes. Considering current and upcoming trends to address bottlenecks and problems, Chapter 10 discusses techniques for manufacturing lentiviral vectors in HEK293 cells grown in serum-free media and the concepts of optimizing along with scaling-up of bioprocesses.

Finally, Chapter 11 in Section 6 discusses the most recent development and application of cell culture technology, the intersection of biology and machine learning and organoid intelligence (OI), which can reshape our understanding of life and health. The goal of the OI concept is to interface brain organoids with computer technology in order to utilize their inherent biological capabilities for biocomputing and synthetic intelligence memorization.

We hope this book will suffice its purpose of disseminating knowledge to the students and scientific community.

We would like to acknowledge the authors and reviewers for their excellent contributions. We also wish to thank the staff at IntechOpen for their enormous support.

Dr. Soumya Basu
Associate Professor,
Cancer and Translational Research Centre,
Dr. D.Y. Patil Biotechnology and Bioinformatics Institute,
Dr. D. Y. Patil Vidyapeeth,
Pune, Maharashtra, India

Dr. Amit Ranjan
Assistant Professor,
Cancer and Translational Research Centre,
Dr. D.Y. Patil Biotechnology and Bioinformatics Institute,
Dr. D. Y. Patil Vidyapeeth,
Pune, Maharashtra, India

Dr. Subhayan Sur
DBT-Ramalingaswami Re-Entry Fellow,
Assistant Professor,
Cancer and Translational Research Centre,
Dr. D.Y. Patil Biotechnology and Bioinformatics Institute,
Dr. D. Y. Patil Vidyapeeth,
Pune, Maharashtra, India

Dedication

To the past, current, and future scientists whose unwavering effort has advanced and will continue to progress this field of study.

To Henrietta Lacks, and to all the nameless patients and individuals who donated their cells for the advancement of cell culture technology and biomedical research.

Section 1

Advances in Cell Culture Technologies

Chapter 1

Perspective Chapter: Valvular Interstitial Cells – Physiology, Isolation, and Culture

Marcus Ground, Karen Callon, Rob Walker, Paget Milsom and Jillian Cornish

Abstract

Valvular interstitial cells (VICs) are the primary cellular component of the heart valve. Their function is to maintain the structure of the valve leaflets as they endure some three billion beats in the course of a human lifespan. Valvular pathology is becoming ever more prevalent in our ageing world, and there has never been a greater need for understanding of the pathological processes that underpin these diseases. Despite this, our knowledge of VIC pathology is limited. The scientific enquiry of valve disease necessitates stable populations of VICs in the laboratory. Such populations are commonly isolated from porcine and human tissue. This is achieved by digesting valve tissue from healthy or diseased sources. Understanding of the many VIC phenotypes, and the biochemical cues that govern the transition between phenotypes is essential for experimental integrity. Here we present an overview of VIC physiology, and a tried-and-true method for their isolation and culture. We make mention of several biochemical cues that the researcher may use in their culture media to ensure high quality and stable VIC populations.

Keywords: valvular interstitial cell, heart valve, cell culture, myofibroblasts, aortic stenosis

1. Introduction

The four valves of the mammalian heart are delicate, complex structures that ensure unidirectional flow of blood through the heart's four chambers. Two atrioventricular valves, the tricuspid and mitral, sit between the atria and ventricles. During systole, these valves are closed to prevent the backflow of blood into the atria. The semilunar valves—so named for their half-moon shaped leaflets—sit between the ventricles are their outflows: the pulmonary trunk and the aorta. During systole, these valves are open, allowing blood to exit the heart. During diastole, they slam shut, preventing the blood from flowing backwards. The valves themselves comprise two main structural features: the valve leaflets (also called 'cusps') and the fibrocartilaginous ring onto which they are mounted. The leaflets are beautifully thin sheets of tissue organised into three distinct layers each with a definite function. The upstream layer, the 'ventricularis' of the semilunar valves and the 'atrialis' of the atrioventricular valves is rich in radially oriented elastin, allowing the leaflet to hinge open and closed. The middle layer, the 'spongiosa', is rich in glycosaminoglycans, forming the bulk of the matrix and structurally linking the superficial layers. Finally, the downstream surface, the 'fibrosa', is rich in circumferentially oriented collagen that confers strength against the mechanical load of the blood pressure when the valve is closed [1, 2].

Two key cell types are responsible for maintaining the structure and function of the heart valve leaflet. Valvular endothelial cells (VECs) form a confluent monolayer on the surface of the valve. These cells are phenotypically distinct from endothelial cells elsewhere in the vasculature.

The second key cell type, and the object of this chapter, is the valvular interstitial cell (VIC). These are effectively fibroblasts that populate the extracellular matrix of the valve leaflets.

Functionally, VICs are responsible for maintaining the matrix of heart valve over the course of growth and development. This is no mean feat—the heart valves must endure some three billion beats over their lifetime.

2. Valvular interstitial phenotypes and their physiology

There are five recognised phenotypes of valvular interstitial cell: embryonic, quiescent, progenitor, activated, and osteoblastic.

2.1 Embryonic VICs

Embryonic VICs are responsible for valvulogenesis—the formation of the valves from the endocardial cushion during fetal development. The precise detail of how this occurs in not well understood, but several key features of these embryonic VICs are known. Firstly, they are highly proliferative, a feature which distinguishes them from other VIC phenotypes [3]. Second, is that these cells are endothelial in origin, having undergone the endothelial-to-mesenchymal transition coordinated by a complex sequence of cell signallers. Notch and TGF family ligands (such as BMP-2 and VEGF) induce TGF-β and Wnt/β-catenin signalling which supresses endothelial-type gene expression in the endothelial cells, and instead encourages the cell to become embryonic VICs, a mesenchymal cell type [4–6]. Once this transition is complete, the maturing valve and its embryonic VICs 'striate' the matrix into the three aforementioned layers. Precisely how this occurs is unknown, but recent evidence suggests that the differentiation is aided by the haemodynamic forces of the fetal circulation [7].

2.2 Progenitor VICs

The VIC population is not a static one. Rather, a subpopulation of VICs present in the valve tissue in adulthood express surface markers of progenitor cells. Studies on healthy porcine valves show that approximately 5% of VICs express ABCG2, NG2, or SSEA-4 [8]. The origin of these progenitor cells in not entirely clear. Evidence suggests that some of these progenitor cells migrate to the valve from the blood stream—studies in mice have shown that bone marrow progenitor cells marked with green fluorescent protein appear in the valve, and assume the role of qVICs by secreting collagen 1 [9]. Other researchers suggest that another source of pVICs are VICs already residing within the valve, and are 'activated' in response to injury [3].

2.3 Quiescent VICs and activated VICs

Quiescent VICs (qVICs) maintain the matrix throughout adult life [3]. They are neither very metabolically nor mitotically active. Instead, their role in maintaining the valve matrix is by sensing mechanical forces, and responding by differentiating into activated VICs (aVICs). The activated VIC is a myofibroblastic phenotype—characterised by the synthesis of ECM proteins and α-smooth muscle actin (a contractile protein) [10]. The differentiation from qVIC to aVIC is the key feature of mechano-regulated growth of this tissue from fetus to adult. As the heart grows, the pressures across the valve leaflet increase, introducing undue stress to the ECM. qVICs resident in the ECM can sense this stress, and differentiate into aVICs that then secrete matrix proteins in order to strengthen the matrix, until such a time when the matrix is able to support the increased pressure. At this time the 'stress' stimulus is lost and the aVIC then dedifferentiates back into a qVIC. It should be no surprise that the fetal heart valve contains a high proportion of aVICs, and that the healthy adult valve contains nearly no aVICs [11].

This mechano-regulation of heart valve growth and development relies on the ability of the qVIC's mechano-sensing ability. This is achieved through three surface protein classes: integrins, cadherins, and mechano-sensitive ion channels [12]. Integrins bind the VIC to the matrix, forming a continuation of the cytoskeleton intracellularly with the ECM extracellularly. Cadherins are transmembrane proteins that bind other cells—in the case of VICs, cadherin 2 is upregulated in qVICs, and cadherin 11 in aVICs [12]. Mechano-sensitive ion channels, particularly calcium channels, also enable mechano-sensation by altering ion concentration within the cell in response to stress-induced deformation [13]. Collectively, these three mechano-sensing modalities induce a transcriptional change within a qVIC in response to stress, which causes aVIC differentiation, upregulation of ECM component synthesis, and the production of α-SMA [10].

2.4 Osteoblastic VICs

Osteoblastic VICs (obVICs) are, as the name suggests, bone-producing cells that arise from VICs in the context of heart valve calcification. Up to 13% of the cellular content of end-stage diseased valve leaflets are positive for osteoblast markers RUNX2 and ALP, and histology of diseased leaflets shows evidence of lamellar bone [14]. Exactly where the obVICs come from is not certain. Some authors suggest that obVICs arise exclusively from qVICs in the context of age-related valve disease [10], while others suggest that obVICs come from a restricted range of susceptible VICs or perhaps pVICs [14].

3. Valvular interstitial cells in health and disease

3.1 From qVIC to aVIC and back again: dynamic reversibility of VIC activation

The key feature of VIC physiology that underpins normal growth and development, and the pathogenesis of heart valve disease, is the reversibility of the transition between qVIC and aVIC [15]. This reversibility (depicted in **Figure 1**), or the failure thereof, is perhaps best illustrated by two examples: growth and development by fetal heart valves, and the development of calcific aortic valve disease. Each involves substantial activation of VICs and the subsequent production of matrix.

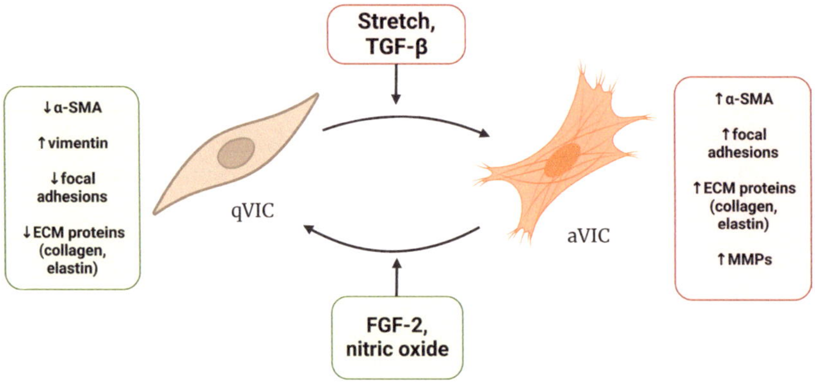

Figure 1.
Shows the 'dynamic reversibility' of VICs as they transition between their quiescent and activated states, as well as the key protein expression markers associated with each phenotype.

The fetal heart valve is an immature structure—not yet required to bear the forces of systolic blood pressure while the fetal circulation, with its shunts, remains active. Once a baby is born, however, those shunts close, and the semilunar valves must immediately assume their role in preventing blood from flowing backwards into the ventricles from their respective outflows. The transition from immature fetal valve to fully competent valve requires several histological changes. First, the immature, loosely organised ECM network of the early valve must become more structured and load-bearing. Secondly, the resident VICs must proliferate in order to adequately populate the valve matrix, and will need to change their phenotype to meet the increased mechanical needs of the growing valve [11].

The immature valve matrix has a lower proportion of elastin than the adult valve, and a much greater proportion of glycosaminoglycan [11]. Interestingly, the proportion of collagen—the primary ECM component of the valve by weight—appears to be fairly stable from the first trimester through to adulthood. What does change markedly, however, is the degree of 'organisation', assessed by various authors using methods such as picrosirius red staining viewed under polarised light [11], or by quantifying the degree of hydroxyproline cross-linking [16]. In the first instance, collagen fibre thickness and alignment is significantly increased from fetus to child, and in the second, hydroxyproline crosslinking is significantly elevated. The net effect is substantial increase in the strength of the valve matrix.

How do the resident VICs achieve this? The answer lies in the presence of aVICs. The cell population of the fetal valve is up to 30% activated VIC, as determined by α-SMA staining [11]. These cells contribute to the changes in valve ECM makeup by simultaneous upregulation of matrix proteins (collagen, for example) and matrix metalloproteases (MMPs) [17]. The combination of upregulated ECM structural proteins and the enzymes that break them down is 'remodelling'—i.e., the overall proportion of collagen remains static, but the maturation into strong, organised bundles proceeds.

In the fetal valve, the laydown of collagen in the correct orientation is driven by the mechano-sensing ability of the resident VICs. But as we age, there are situations where the dynamic reversibility of qVIC and aVIC phenotypes is upset, and the resulting products of VIC activation lead to disease. The prescient example is calcific aortic valve disease (CAVD)—the most common valvular pathology of the developed world, with prevalence rates of up to 15% in those over 75 years old [18]. In the early phase of this disease, aVIC activation and the accumulation of structural

ECM components mimics the changes seen in fetal valves, only in this circumstance, the interruption of the 'dynamic reversibility' of aVICs is lost, and the valves become stiff and non-compliant [19, 20]. Valves affected by CAVD are characterised by the overproduction of collagen [21] and the dysregulation of MMP expression [22].

The 'dynamic reversibility' of VIC phenotypes and the mechanical properties of the valve matrix are a true 'chicken and egg' problem. VIC activation invariably leads to stiffening and matrix remodelling (either physiological in the case of the fetal valve, or pathological in the case of CAVD), which, in turn, results in a stiffer matrix, which then serves a signal for qVICs to activate [23].

3.2 VIC calcification: murky origins

The origin if the calcification in heart valve disease is not perfectly understood. What is known is that VICs can calcify via two distinct mechanisms: calcific nodule formation and obVIC differentiation (depicted in **Figure 2**) [10]. The relative contributions of these two mechanisms to heart valve disease, or if the two mechanisms are independent of one another is not known.

Current evidence suggests that calcific nodules arise from aVICs, and are in fact calcified apoptotic bodies left after the cell dies. This type of calcification is termed 'dystrophic' to distinguish it from the matrix calcification caused by obVICs [24]. Where the obVICs arise from, is not clear. Some have suggested that obVICs arise from the qVIC population independently, and in response to changes in matrix mechanics or the accumulation of calcium [25], while others suggest that obVICs arise from aVICs in a manner that depends 'activating' signalling [26].

Support for the idea that the obVIC and aVIC calcification are distinct pathways is seen in the apparent changes in α-SMA expression, which is markedly reduced in presence of osteogenic medium and markedly increased in the presence of TGF-β.

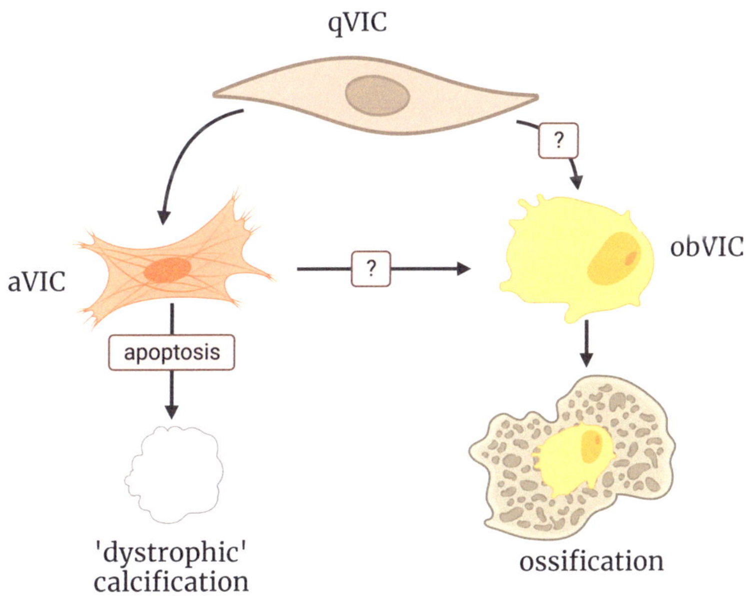

Figure 2.
Shows the two phenotypes of VIC calcification: the dystrophic calcific nodules that occur after aVIC apoptosis, and the ossification of obVICs. What is unclear is if the obVIC phenotype arises directly from qVICs or aVICs.

Conversely, qVICs cultured in osteogenic medium will demonstrate upregulation in key osteoblast markers: ALP and OCN [25]. To further confuse the true origin of obVICs, the same pattern of gene expression occurs when valvular endothelial cells are cultured in osteogenic media [27].

3.3 VIC signalling molecules

The various signalling molecules that induced change in VICs has been extensively reviewed elsewhere [10]. There are, however, three molecules that commonly feature in culture medium preparations, and the physiology that underpins their use is included here.

3.3.1 TGF-β

Perhaps the most well characterised stimulus for VIC activation is transforming growth factor β (TGF-β) [28–34]. TGF-β, or more specifically the isoform TGF-β1 is one member of a much larger 'TGF-β superfamily', which includes 30 species such as bone morphogenic proteins (BMPs), growth and differentiation factors, and various other peptides key to growth and development [35]. In adulthood, TGF-β1 is involved in a myriad of cell functions, from cell fate determination, cell migration, response to injury, and ECM regulation [34, 36, 37]. In the context of VICs, the key function of TGF-β is its ability to regulate activation of VICs and maintain homeostasis of the valve ECM by induction of the aVIC phenotype. As well as structural proteins, ECM contains large amounts of 'latent' signalling molecules—signallers bound up with latency-associated peptides. These peptide complexes are hidden from cells, and so do not enact their functions until they are cleaved from the matrix during proteolysis following injury [36, 38]. TGF-β, as a promoter of ECM production, serves as a 'switch' in heart valves: when the valve is unable to cope with a tension force, microscopic injury releases TGF-β from the matrix, which in turn signals to cells to secrete more ECM, until such time that the matrix can withhold the tension and the TGF-β switch is turned off. It is through delicate control of TGF-β stimulated ECM secretion that a valve grows and remodels throughout the changing haemodynamic environment from fetal development to adulthood [6, 39].

TGF-β acts on cells via several biochemical pathways. The most well characterised pathway with respect to VIC physiology is the 'canonical TGF-β' pathway, or Smad pathway [34, 38, 40]. TGF-β forms a homodimer in the extracellular space and binds two type 2 TGF-β receptors (Tgfbr2), which then recruit two type 1 TGF-β receptors (Tgfbr1) forming an activated receptor-ligand complex [39]. The activated Tgfbr1 is then able to phosphorylate intracellular messenger proteins Smad2 and Smad3, which form a complex with Smad4 and move to the nucleus as transcription factors to activate the associated genes [39, 41]. It is this process that is responsible for the activation of qVICs to myofibroblasts and for the development of dystrophic nodules [3].

Non-Smad pathways initiated by TGF-β are termed 'non canonical'. These include the ERK/MAPK pathway [41], which has been shown to regulate key aspects of aVIC expression in a manner independent of the canonical TGF-β pathway [42].

3.3.2 FGFs

The fibroblast growth factor (FGF) family comprises 23 distinct proteins, 18 of which are ligands to various FGF receptors, effecting a wide range of cellular functions: from migration to proliferation to cell survival [43, 44]. Of interest to the VIC researcher is FGF-2, also called basic FGF (bFGF), so named for its discovery in the alkaline fraction of pituitary extracts [45]. FGF-2 has been implicated in numerous biological processes, from cell migration and fate determination to pathological signalling in cancer metastasis [46]. Most interesting is that while this molecule is present and active throughout the body, a distinct physiological role has not been identified. Homozygous FGF-2 knockout mice survive perfectly well and give birth to healthy young.

FGF-2's signalling pathway first involves heparin binding [46]. Two heparin-FGF2 complexes then bind two FGF receptors (FGFR) to form a stable dimer, which then activates intracellular tyrosine kinase domains and effects changes in gene expression through a number of signalling cascades, among them MAPK and AKT pathways [43].

FGF-2 has been shown to be particularly important in regulating the quiescent VIC phenotype [34, 47–49]. The mechanism of FGF-2's effect appears to be multifaceted: both repressing the canonical TGF-β pathway by blocking Smad nuclear translocation [48], and by activating of the Akt/mTOR pathway [34, 49]. The downstream effects of this activation have been shown to oppose the effect of TGF-β. FGF-2 appears to promote the qVIC phenotype: VICs treated with FGF-2 have a spindle-shaped morphology, and express more vimentin and less α-SMA [17].

3.3.3 Nitric oxide

The function of nitric oxide (NO) in the vasculature is understood to be a local control mechanism for maintaining blood pressure and perfusion. First, NO is produced in endothelial cells by endothelial nitric oxide synthase (eNOS) in response to stretch or biochemical stimuli [50]. Then, NO is able to rapidly diffuse into the surrounding tissue by virtue of its small size, where it binds guanylyl cyclase (a cytoplasmic receptor) [51] and leads to a myriad of downstream effects including relaxation of the vascular smooth muscle [50]. With respect to heart valves, NO production by VECs and its subsequent action of VICs is an important paracrine regulator of valve cell function [3, 10]. More importantly, dysregulation of NO signalling is implicated in the development of heart valve disease [52–55]. In the initial stages of CAVD, eNOS is upregulated on the aortic-side endothelium [56], where it contributes to pathological neovascularisation of the valve tissue [57]. The early eNOS response is a protective mechanism to temper the proliferation of VICs, but as overexpression of eNOS continues, resistance to NO develops—a phenomenon termed 'NOS decoupling' [58].

With respect to VIC activation, exogenous NO inhibits the formation of calcific nodules in porcine VIC cultures under TGF-β stimulation [59]. Culturing VICs in media supplemented with DETA-NONOate—a biologically inert molecule that spontaneously releases two NO molecules in a predictable manner [60]—limits the formation of new nodules.

This effect is not purely induced by exogenous NO, rather it can be replicated by co-culturing VICs with NO-producing VECs [61]. Under these conditions, VECs can inhibit the formation of aVIC-related calcific nodules in a manner that is reversed by addition of a NO scavenger. This suggests that of the multitude of paracrine signals that travel between the endothelium and the interstitium, NO is paramount in maintaining the valve in its quiescent state.

4. Sources of valvular interstitial cells

There are a number of VIC sources for a researcher's consideration. No one option is without issue. Perhaps the easiest source of VICs is from animals. Pigs are by far the most common source of VICs for research [61–81], as they are very easily obtainable from meat processing facilities. As a biproduct of industry, obtaining and using porcine VICs requires little or no ethical oversight [82]. Another notable benefit is the anatomical similarity between the pig and human heart. This is of particular importance as the mechanical and structural forces regulate VIC physiology and gene expression [23, 68], and so choosing a structurally similar heart approximates a physiologically similar VIC. In a similar vein, pigs housed for research can be induced into diseased states for investigation of a particular pathology. For example, pigs fed on high-cholesterol diets experience atherosclerotic lesions, though no porcine model of calcific valve disease has been trialled to date [52, 83]. It should be noted that VICs derived from meat industry pigs can be considered healthy qVICs by definition—their valves are not experiencing rapid growth, nor has age-related disease begun to influence their phenotype. This fact is either a boon or an issue for the cell culturist. Those interested in studying the development of valve disease might be eager to use a healthy population of qVICs, while those searching for therapeutic options for persistent aVIC or obVIC presence will likely have to choose another source. In our experience, porcine VICs are easily obtained and fairly forgiving when it comes to cell culture. Cultures derived from digested porcine valves are adequately proliferative and less sensitive to aVIC formation when cultured on hard plastic.

Sheep may represent a promising animal alternative for obtaining VICs liable to calcify. Sheep exhibit an elevated calcium metabolism compared to humans, and this has led to their use as an experimental model of heart valve calcification [84]. Several authors have made use ovine VICs for experimentation [85, 86].

Human VICs are the best option for studying heart valve disease, as none of the animal models approximate our valvular pathologies perfectly. The most ideal source is a young, healthy adult human heart valve. Cells from this source, in theory, are almost entirely qVICs. However, there are a number of constraints that make sourcing healthy human tissue difficult.

First is availability—unlike other cell types, VICs are not sourced from a healthy tissue that is routinely removed from people undergoing surgery (as it the case in, for example, adipose tissue removed during breast cancer surgery), and suitable heart valve tissue from younger donors is contained within a heart that is likely more suitable for heart transplantation. To complicate matters further, in the instance where a young donor heart is not suitable for transplant, the heart valve may be harvested as a 'homograft' (a heart valve transplant).

The second constraint is time. While VICs are relatively robust in the absence of nutrient supply [10], good laboratory practice dictates that cells should be isolated and incubated as quickly as possible. Very rarely are healthy heart valves available within temporal or spatial proximity to a laboratory. Despite these constraints, various researchers have made use of healthy human VICs, typically from donor hearts not suitable for either transplantation or homograft use [47, 87].

A much more commonplace source of human VICs is those isolated from patients undergoing valve replacement surgery. Valve replacement surgeries are very common—our experience at Auckland City Hospital saw an average of 5 valves available per week. The obvious problem with this source is the inherent disease state of the cells within the valve. Though our protocol above recommends removal of the more calcified portion of the valve, the pathology is not localised at the calcification, and altered stiffness characteristics across the whole leaflet ensures that all cells present will have some degree of pathological programming [53].

The diseased state of these cells may actually be useful for researchers concerned with the study of aortic calcification—the most common valvular pathology in the developed world. Many examples of diseased valves used for this purpose are available in the literature [47, 87–102].

The key consideration to address with diseased human VICs is ethics. Patients must consent to their tissue being harvested and used in research, and protection of the tissue and any data generated must be ensured by the researcher. In addition, many cultures have unique requirements for the disposal or return of human tissue [103].

5. Isolation protocol

This isolation protocol was developed by our laboratory for the isolation of human VICs from patients undergoing valve replacement surgery [104]. An abbreviated flow chart is included in **Figure 3**.

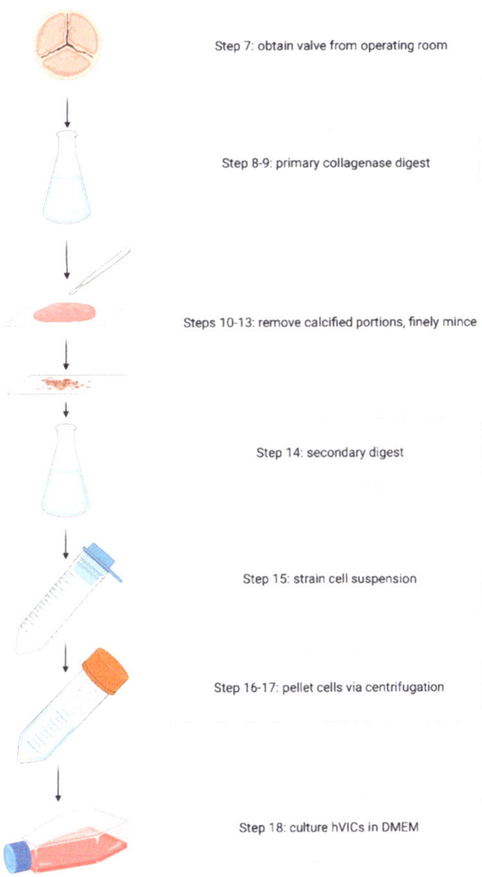

Figure 3.
Shows and abbreviated flow chart of steps 7–18.

5.1 Consumables

- Collagenase II (Sigma, #C6885)
- Sterile phosphate buffered saline
- Sterile DMEM (Sigma, #D5649) supplemented with sodium bicarbonate per manufacturer's instructions
- Fetal bovine serum (FBS)
- 50 mL containers
- 15 mL centrifuge tubes
- 50 mL centrifuge tube
- 90 mm petri dishes
- Sterile guards
- #10 Scalpel blades and handle
- Sterile forceps
- 70 µm cell strainer (Bio-Strategy, #BDAA352350)
- 0.2 µm sterile filter (Bio-Strategy, #SART16534-K)
- 30 mL syringe
- T25 cell culture flasks (Sigma, #C6356)
- 25 mL serological pipettes
- Disposable 'squeezy' pipettes

5.2 Hardware

- Tissue culture hood
- Incubator, 37°C at 5% CO_2
- Orbital shaker-incubator, 37°C
- Sterilisation services (i.e., autoclaving service, sterile filtration capabilities)
- Balance
- Centrifuge

- Electric pipette filler
- Microscope with haemocytometer slide
- Tube rack

5.3 Method

Prior to tissue collection:

1. In the tissue culture hood, prepare at least 100 mL of DMEM supplemented with 10% FBS, and warm to 37°C
2. Prepare a > 20 mL aliquot of sterile PBS for tissue collection, and warm to 37°C
3. Turn on the shaker-incubator to equilibrate to 37°C
4. Weigh out 30kU collagenase II. Note that the mass of the collagenase is lot-number specific - refer to the specification sheet for the number of units per milligramme. Add the collagenase II to a 50 mL container containing 30 mL PBS, gently agitate to dissolve
5. In the tissue culture hood, aspirate the collagenase II into the 30 mL syringe. Attach the 0.2 μL filter to the syringe and gently push the collagenase through the filter into a sterile container. Collagenase II loses its activity within hours, and should be prepared as close to the time of digest as possible. Divide the collagenase into 2 portions of 15 mL each in 50 mL containers. Label them 'Digest 1' and 'Digest 2'
6. Once sealed, remove 'Digest 2' from the hood and weigh the container, recording the mass to the nearest milligramme. This will allow weighing of the usable valve tissue in a sterile manner

In the operating room:

7. Make sure to bring the warmed aliquot of PBS for tissue collection. When the surgeon removes the valves, ensure the scrub nurse or assisting surgeon offers the container in a sterile manner, and that the tissue is added to the container and sealed shut all while in the sterile field. Return to the laboratory as quickly as possible

In the laboratory:

8. Using sterile forceps, add the leaflets to the 'Digest 1' container
9. Add to the shaker-incubator for 10 min at 37°C with gentle agitation
10. Once complete, arrange the leaflets on a 90 mm petri dish, and scrape down the outside using a scalpel blade. If the calcifications make this too difficult, then consider removing the calcified sections before scraping. Rinse the valves using a squeezy pipette and sterile PBS

11. If hVECs are desired, collect the rinsing media and pool this with suspension in 'Digest 1', centrifuge, and resuspend the cell pellet in 1 mL of DMEM +10% FBS. Note that the expected pellet is small, and the suspension should be plated in a 6-well plate or smaller

12. If not already done so, remove calcified sections of the valve. Finely mince the tissue using the scalpel, and transfer to the tissue to 'Digest 2'

13. Weigh 'Digest 2' and calculate the difference from recorded mass in step 7. If the mass is less than 500 mg, then there is a low likelihood that the culture will generate enough cells to succeed. We recommend proceeding only if the tissue mass exceeds 500 mg

14. Add 'Digest 2' to the shaker-incubator at 37°C for 2 h with gentle agitation

15. In a tube rack, assemble the 50 mL tube with the 70 μm cell strainer. Pour the digestate through the strainer, and agitate the contents using a squeezy pipette

16. Centrifuge at 1200 RPM for 2 min

17. Discard the supernatant, and, without disturbing the pellet, rinse the tube with 15 mL of DMEM +10% FBS to remove any residual collagenase. Spin briefly to re-pellet any cells that may have been disturbed

18. Discard the supernatant. Resuspend the cell pellet in 2 mL of DMEM +10% FBS. Take and 30 μL sample and pipette directly onto the haemocytometer. Do not perform cell count immediately - instead, get the cells into the incubator as quickly as possible: immediately add the suspension to T25 culture flask and make up to final volume with 8 mL of DMEM +10% FBS. Add to the incubator at 37°C

19. Perform a cell count, and record. Note that digests that yield <100,000 cells are unlikely to be successful

20. After the digest:

21. Monitor the culture daily for confluency and signs of myofibroblastic activation. hVIC cultures generally reach 80% confluency in 8 days. Monitor for signs of myofibroblast differentiation, including pleiomorphism and non-fusiform morphologies

22. Replace entire medium every four days

23. When using hVICs in experiments, make sure to seed at a density of 10,000 cells/cm^2, and anticipate that growth will plateau within 5 days

6. Quality control considerations

Regardless of cell source, the goal for VIC culture is a population of phenotypically stable cells (i.e. few aVICs) with as little endothelial contamination as possible.

To assess phenotypic stability, we encourage researchers to plate out a sample of their VIC culture and assess it for features of myofibroblast activation in aVICs. Typically, we >5000 cells from our digestate and plated these in a 24-well plate, and left to grow for up to 6 days. We then assessed for the presence of aVICs first by visual inspection on phase microscopy, and then by α-SMA staining. aVICs differ from pVICs in their morphology. On phase microscopy, qVICs are slender, spindle-shaped cells with a central body surrounded by two or three thin processes. Their area rarely exceeds $5000\mu m^2$, and their aspect ratio (the ratio between their major and minor axes) should be greater than 3 [17]. aVICs on the other hand are much larger, with areas often exceeding $10{,}000\mu m^2$. Their shape is much more circular, with a large cell body and loss of the thin processes. Similar cell area measurements were found by Porras and colleagues, who used FIB media to prevent (or reverse) aVIC formation in porcine cultures [72].

We suggest staining for α-SMA and vimentin to assess the degree of aVIC presence (an example of which can be seen in **Figure 4**). aVICs have dense bands of α-SMA running across the cell, terminating in adhesion complexes that link the intracellular force generated by this protein to the ECM [105]. True qVICs will have very little α-SMA staining, if at all.

Many researchers will quantify aVIC numbers by using α-SMA positivity, either by fluorescent intensity measurement or thresholding numbers of positive cells [47, 48]. In our experience, there is little utility in attempting to quantify the degree of VIC activation using α-SMA staining or phase microscopy, as all VICs will show some α-SMA expression and variation in morphology. More intensive methods for quantifying α-SMA include western blotting and real time PCR [47, 106]. These are comparative methods (i.e., they require two groups), and so are of little use in the 'quality control' stage. Experimentally, we have found that expression of ACTA2 (α-SMA's gene) and a number of other aVIC-associated genes correlate with the

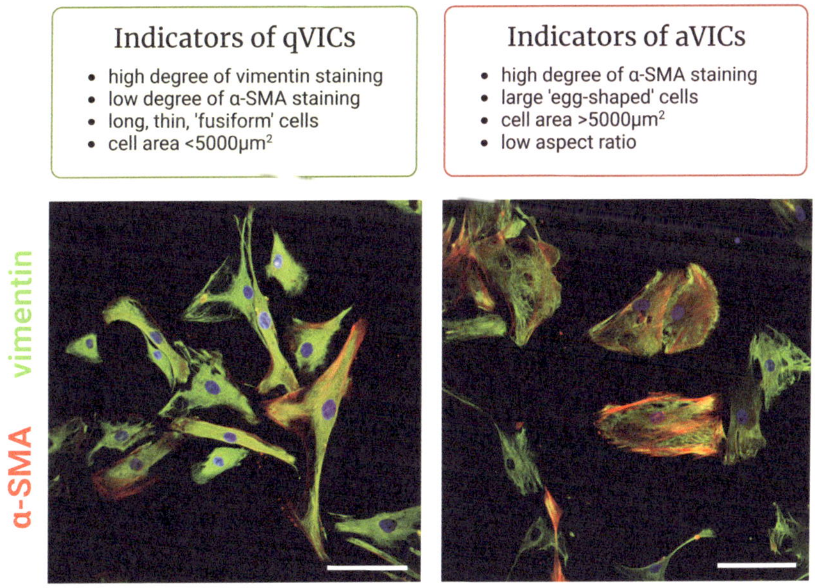

Figure 4.
Shows two cultures of VICs: one rich in qVICs on the left, and another rich in aVICs. Cells are stained for vimentin (green) and α-SMA (red). Cells were cultured for four days on plastic. Scale bars represent 100 μm.

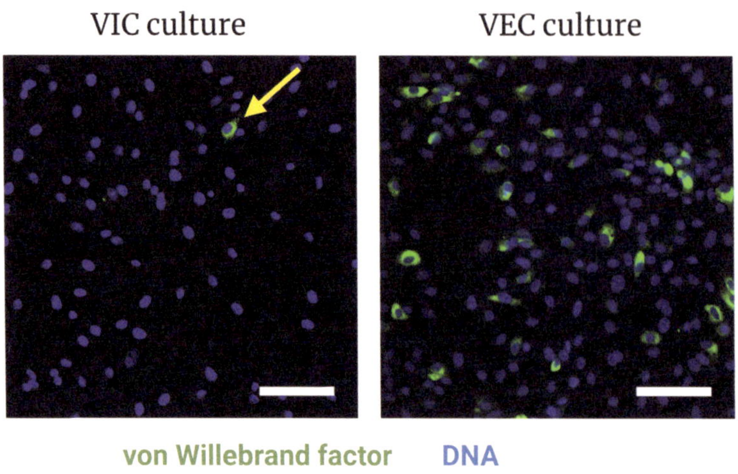

Figure 5.
Shows a quality control procedure performed on a VIC culture (left). The yellow arrow indicates VEC contamination. This was performed alongside a VEC culture as a positive control. This culture was grown from 'Digest 1' aspirate. Both cultures were grown for four days on plastic. Scale bars represent 100 μm.

presence of 'aVIC-looking' cells on phase microscopy and α-SMA staining. We suggest that researchers subjectively assess the appearance of their VIC cultures on day 6 post-isolation.

The second main feature of VIC quality control is endothelial contamination (**Figure 5**). Despite the initial digest and scraping we suggest in the protocol, it is impossible to remove all VECs. Endothelial contamination can run as high as 10% [10]. In our experience, VIC cultures generated by the means outlined above will produce cultures with less than 1% von Willebrand factor (vWF) positivity. To assess VEC contamination, we suggest first saving and plating the digestate from digest 1 in a 6-well plate or smaller (see step 11) for use as a positive control. In the same plate, we suggest reserving a small volume of the VIC suspension from digest 2, and culturing both for up to 6 days. We then suggest staining the two wells for an endothelial marker. The choice of marker is not especially important: VECs express vWF, CD31 (PECAM) and CD34, and are negative for α-SMA [74, 107].

7. Culture conditions

The majority of the literature has VICs cultured on flat plastic in Dublecco's modified Eagle medium (DMEM) supplemented with 10% fetal bovine serum (FBS). We have had good results with this formulation. There are, however, a number of other culture media that should be considered. Alternatives to DMEM include M199 medium [29, 108] and DMEM-F12 [73, 109].

Good laboratory practice stipulates that isolated cells should be cultured in such a way that preserves their *in vivo* phenotype as faithfully as possible. And so, it befalls the VIC researcher to consider the inherent aVIC promoting properties of chosen media. The supplementation of media with 10% FBS is standard practice for the culture of any fibroblast [110]. However, careful thought should be put into the inherent content of peptides that stimulate VIC activation. TGF-β, for example, is present in FBS in concentrations ranging from 10 to 20 ng/mL. TGF-β is a strong promoter of VIC activation, and several authors have mitigated its effect by using a 2% FBS medium supplemented with insulin as a mitotic agent. This so-called 'FIB' media, first used by Latif and colleagues [47, 72], is also supplemented with 10 ng/mL FGF-2—an agent shown to reduce VIC activation in favour of the quiescent phenotype. We have demonstrated that VICs cultured in FIB media show significantly downregulated expression of aVIC-associated genes (α-SMA in particular), and a morphology more typical of qVICs (greater aspect ratio, and reduced total cell area) compared to VICs cultured in the standard DMEM preparation [17]. The inherent ability of FIB medium to suppress aVIC formation can be potentiated further by culturing VICs on collagen-coated plates [72].

Induction of the obVIC phenotype requires osteogenic medium. This invariably contains ascorbic acid, dexamethasone, and β-glycerophosphate [28, 109, 111], supplemented in some instances with TGF-β or BMP-2 [89].

Seeding densities vary widely in the literature, and will differ depending on the surface on which the VICs are grown. On hard plastic or glass, we have found VICs to be rather sensitive to contact inhibition and myofibroblastic activation. We suggest that VICs are seeded at a density of 10,000 cells/cm^2 for experiments running for less than five days. At this density, cell growth reaches a plateau on day 4 [104].

8. Conclusion

VICs are the specialised fibroblasts that populate the human valve and maintain its structure and function throughout our lives. The increasing incidence of valvular pathology means that there is no better time to gain understanding about the physiological and pathological processes that affect these cells. Despite this, our understanding of VIC health and disease is in its infancy. Culture of VICs is an important part of the scientific exploration of heart valve disease. Several key methodological considerations need to be addressed for the jobbing scientist:

- VICs come in a variety of phenotypes. Some are implicated in embryogenesis, while other in disease

- Biochemical signals govern the phenotypic transition of VICs between these phenotypes

- A normal, physiological dynamism exists between the qVIC and the aVIC. This is the key process that underpins normal valve growth and development

- Upsetting this dynamic transition is implicated in the development of valvular disease. Chief among them is calcific aortic valve disease

- Phenotypically stable VIC cultures can come from pigs and humans

- Diseased VICs are the most readily attainable human cell source, and with careful consideration of culture medium preparations, stable cultures of qVICs can be readily and easily obtained

- qVICs are liable to become aVICs in culture. The researcher is therefore tasked with ensuring the culture conditions avoid errant VIC activation

Acknowledgements

This work was supported by the Green Lane Research and Educational Fund and the Health Research Council of New Zealand.

Author details

Marcus Ground[1,2]*, Karen Callon[2], Rob Walker[1], Paget Milsom[3] and Jillian Cornish[2]

1 University of Otago, Dunedin, New Zealand

2 University of Auckland, Auckland, New Zealand

3 Auckland City Hospital, Auckland, New Zealand

*Address all correspondence to: marcus.ground@auckland.ac.nz

References

[1] Chester AH, El-Hamamsy I, Butcher JT, Latif N, Bertazzo S, Yacoub MH. The living aortic valve: From molecules to function. Global Cardiology Science Practise. 2014;**2014**(1):11

[2] Loukas M, Bilinsky E, Bilinsky S, Blaak C, Tubbs RS, Anderson RH. The anatomy of the aortic root. Clinical Anatomy. 2014;**27**(5):748-756

[3] Liu AC, Joag VR, Gotlieb AI. The emerging role of valve interstitial cell phenotypes in regulating heart valve pathobiology. American Journal of Pathology [Internet]. 2007;**171**(5):1407-1418. DOI: 10.2353/ajpath.2007.070251

[4] Bosada FM, Devasthali V, Jones KA, Stankunas K. Wnt/β-catenin signaling enables developmental transitions during valvulogenesis. Development. 2016;**143**(6):1041-1054

[5] Neri T, Hiriart E, van Vliet PP, Faure E, Norris RA, Farhat B, et al. Human pre-valvular endocardial cells derived from pluripotent stem cells recapitulate cardiac pathophysiological valvulogenesis. Nature Communication [Internet]. 2019;**10**(1):1-14. DOI: 10.1038/s41467-019-09459-5

[6] Butcher JT, Markwald RR. Valvulogenesis: The moving target. Philosophical Transaction on Royal Society B Biology Science. 2007;**362**(1484):1489-1503

[7] Lindsey SE, Butcher JT. The cycle of form and function in cardiac valvulogenesis. Aswan Heart Center Science Practise Series. 2011;**2011**(2)

[8] Wang H, Sridhar B, Leinwand LA, Anseth KS. Characterization of cell subpopulations expressing progenitor cell markers in porcine cardiac valves. PLoS One. 2013;**8**(7):1-11

[9] Hajdu Z, Romeo SJ, Fleming PA, Markwald RR, Visconti RP, Drake CJ. Recruitment of bone marrow-derived valve interstitial cells is a normal homeostatic process. Journal of Molecular Cell Cardiology [Internet]. 2011;**51**(6):955-965. DOI: 10.1016/j.yjmcc.2011.08.006

[10] Rutkovskiy A, Malashicheva A, Sullivan G, Bogdanova M, Kostareva A, Stensløkken KO, et al. Valve interstitial cells: The key to understanding the pathophysiology of heart valve calcification. Journal of American Heart Association [Internet]. 2017;**6**(9). Available from: http://www.ncbi.nlm.nih.gov/pubmed/28912209

[11] Aikawa E, Whittaker P, Farber M, Mendelson K, Padera RF, Aikawa M, et al. Human semilunar cardiac valve remodeling by activated cells from fetus to adult: Implications for postnatal adaptation, pathology, and tissue engineering. Circulation. 2006;**113**(10):1344-1352

[12] Howsmon DP, Sacks MS. On valve interstitial cell signaling: The link between multiscale mechanics and mechanobiology. Cardiovascular Engineering and Technology. 2021;**12**(1):15-27

[13] Simard C, Aize M, Chaigne S, Mpweme Bangando H, Guinamard R. Ion channels in the development and remodeling of the aortic valve. International Journal of Molecular Sciences. 2023;**24**(6):5860

[14] Liu X, Xu Z. Osteogenesis in calcified aortic valve disease: From

histopathological observation towards molecular understanding. Progress in Biophysics Molecular Biology [Internet]. 2016;**122**(2):156-161. DOI: 10.1016/j.pbiomolbio.2016.02.002

[15] Rabkin-Aikawa E, Farber M, Aikawa M, Schoen FJ. Dynamic and reversible changes of interstitial cell phenotype during remodeling of cardiac valves. The Journal of Heart Valve Disease. 2004;**13**(5):841-847

[16] Van Geemen D, Soares ALF, Oomen PJA, Driessen-Mol A, Janssen-Van Den Broek MWJT, Van Den Bogaerdt AJ, et al. Age-dependent changes in geometry, tissue composition and mechanical properties of fetal to adult cryopreserved human heart valves. PLoS One. 2016;**11**(2)

[17] Ground M, Ground M, Yutzey K, Waqanivavalagi S, Park YE, Callon K, et al. Fibroblast growth factor 2 inhibits myofibroblastic activation of valvular interstitial cells. PLoS One [Internet]. 2022;**17**(6):e0270227. DOI: 10.1371/journal.pone.0270227

[18] Coffey S, Cairns BJ, Iung B. The modern epidemiology of heart valve disease. Heart [Internet]. 2016;**102**(1):75-85. Available from: http://heart.bmj.com/lookup/doi/10.1136/heartjnl-2014-307020

[19] Mrsic Z, Hopkins SP, Antevil JL, Mullenix PS. Valvular heart disease. Primary Care – Clinical Official Practise [Internet]. 2018;**45**(1):81-94. DOI: 10.1016/j.pop.2017.10.002

[20] Helms AS, Bach DS. Valvular heart disease. Primary Care – Clinical Official Practise [Internet]. 2013;**40**(1):91-108. DOI: 10.1016/j.pop.2012.11.005

[21] Gomez-Stallons MV, Tretter JT, Hassel K, Gonzalez-Ramos O, Amofa D, Ollberding NJ, et al. Calcification and extracellular matrix dysregulation in human postmortem and surgical aortic valves. Heart. 2019;**105**(21):1616-1621

[22] Ohukainen P, Ruskoaho H, Rysa J. Cellular mechanisms of valvular thickening in early and intermediate calcific aortic valve disease. Current Cardiology Reviews. 2018;**14**(4):264-271

[23] Butcher JT, Nerem RM. Valvular endothelial cells and the mechanoregulation of valvular pathology. Philosophical Transaction Royal Society B Biological Science. 2007;**362**(1484):1445-1457

[24] Merryman WD, Schoen FJ. Mechanisms of calcification in aortic valve disease: Role of mechanokinetics and mechanodynamics. Current Cardiology Reports. 2013;**15**(5):355

[25] Monzack EL, Masters KS. Can valvular interstitial cells become true osteoblasts? A side-by-side comparison. Journal of Heart Valve Diseases [Internet]. 2011;**20**(4):449-463. Available from: https://www.ncbi.nlm.nih.gov/pmc/articles/PMC3624763/pdf/nihms412728.pdf

[26] Galeone A, Paparella D, Colucci S, Grano M, Brunetti G. The role of TNF- α and TNF superfamily members in the pathogenesis of calcific aortic valvular disease. Scientific World Journal. 2013;**2013**(Ldl):1-10

[27] Hjortnaes J, Shapero K, Goettsch C, Hutcheson JD, Keegan J, Kluin J, et al. Valvular interstitial cells suppress calcification of valvular endothelial cells. Atherosclerosis [Internet]. 2015;**242**(1):251-260. DOI: 10.1016/j.atherosclerosis.2015.07.008

[28] Cloyd KL, El-Hamamsy I, Boonrungsiman S, Hedegaard M,

Gentleman E, Sarathchandra P, et al. Characterization of porcine aortic valvular interstitial cell "calcified" nodules. PLoS One. 2012;**7**(10):1-9

[29] Song R, Fullerton DA, Ao L, Zheng D, Zhao K, seng, Meng X. BMP-2 and TGF-β1 mediate biglycan-induced pro-osteogenic reprogramming in aortic valve interstitial cells. Journal of Molecular Medicine. 2015;**93**(4):403-412

[30] Das D, Holmes A, Murphy GA, Mishra K, Rosenkranz AC, Horowitz JD, et al. TGF-beta1-induced MAPK activation promotes collagen synthesis, nodule formation, redox stress and cellular senescence in porcine aortic valve interstitial cells. Journal of Heart Valve Disease [Internet]. 2013;**22**(5):621-630. Available from: http://www.ncbi.nlm.nih.gov/pubmed/24383372

[31] Walker GA, Masters KS, Shah DN, Anseth KS, Leinwand LA. Valvular myofibroblast activation by transforming growth factor-β: Implications for pathological extracellular matrix remodeling in heart valve disease. Circulation Research. 2004;**95**(3):253-260

[32] Cushing MC, Liao JT, Anseth KS. Activation of valvular interstitial cells is mediated by transforming growth factor-β1 interactions with matrix molecules. Matrix Biology. 2005;**24**(6):428-437

[33] Wang F, Zhang C, Kwagh J, Strassle B, Li J, Huang M, et al. TGFβ2 and TGFβ3 mediate appropriate context-dependent phenotype of rat valvular interstitial cells. iScience. 2021;**24**(3):102133

[34] Han L, Gotlieb AI. Fibroblast growth factor-2 promotes in vitro heart valve interstitial cell repair through the Akt1 pathway. Cardiovascular Pathology [Internet]. 2012;**21**(5):382-389. DOI: 10.1016/j.carpath.2011.12.001

[35] Weiss A, Attisano L. The TGFbeta superfamily signaling pathway. Wiley Interdisciplinary Review Developmental Biology [Internet]. 2013;**2**(1):47-63. Available from: https://onlinelibrary.wiley.com/doi/10.1002/wdev.86

[36] Miyazono K. Positive and negative regulation of TGF-β signaling. Journal of Cell Science. 2000;**113**(7):1101-1109

[37] Meng X, Nikolic-Paterson DJ, Lan HY. TGF-β: The master regulator of fibrosis. Nature Review Nephrology [Internet]. 2016;**12**(6):325-338. DOI: 10.1038/nrneph.2016.48

[38] Piersma B, Bank RA, Boersema M. Signaling in fibrosis: TGF-β, WNT, and YAP/TAZ converge. Frontiers in Medicine. 2015;**2**(SEP):1-14

[39] Saitoh M. Regulation of EMT by TGF-$β$ signaling in cancer cells. In: Shinomiya N, Kataoka H, Xie Q, editors. Regulation of Signal Transduction in Human Cell Research [Internet]. Singapore: Springer Singapore; 2018. pp. 71-84. DOI: 10.1007/978-981-10-7296-3_4

[40] Leopold JA. Cellular mechanisms of aortic valve calcification. Circulation. Cardiovascular Interventions. 2012;**5**(4):605-614

[41] Clayton SW, Ban GI, Liu C, Serra R. Canonical and noncanonical TGF-β signaling regulate fibrous tissue differentiation in the axial skeleton. Scientific Report [Internet]. 2020;**10**(1):1-11. DOI: 10.1038/s41598-020-78206-4

[42] Hutcheson JD, Chen J, Sewell-Loftin MK, Ryzhova LM, Fisher CI, Su YR, et al. Cadherin-11 regulates cell-cell tension necessary for calcific nodule

formation by valvular myofibroblasts. Arteriosclerosis, Thrombosis, and Vascular Biology. 2013;**33**(1):114-120

[43] Turner N, Grose R. Fibroblast growth factor signalling: From development to cancer. Nature Reviews. Cancer. 2010;**10**(February):116-129

[44] Li X. Design and discovery of FGF/FGFR inhibitors. In: Li X, editor. Fibroblast Growth Factors [Internet]. Amsterdam: Elsevier; 2018. pp. 339-383. Available from: https://www.sciencedirect.com/science/article/pii/B9780128161425000060

[45] Gospodarowicz D. Localisation of a fibroblast growth factor and its effect alone and with hydrocortisone on 3T3 cell growth. Nature [Internet]. 1974;**249**(5453):123-127. DOI: 10.1038/249123a0

[46] Beenken A, Mohammadi M. The FGF family: Biology, pathophysiology and therapy. Nature Reviews. Drug Discovery. 2009;**8**(3):235-253

[47] Latif N, Quillon A, Sarathchandra P, McCormack A, Lozanoski A, Yacoub MH, et al. Modulation of human valve interstitial cell phenotype and function using a fibroblast growth factor 2 formulation. PLoS One [Internet]. 2015;**10**(6):e0127844. Available from: https://dx.plos.org/10.1371/journal.pone.0127844

[48] Cushing MC, Mariner PD, Liao J, Sims EA, Anseth KS. Fibroblast growth factor represses Smad-mediated myofibroblast activation in aortic valvular interstitial cells. FASEB Journal [Internet]. 2008;**22**(6):1769-1777. Available from: https://onlinelibrary.wiley.com/doi/10.1096/fj.07-087627

[49] Lam NT, Tandon I, Balachandran K. The role of fibroblast growth factor 1 and 2 on the pathological behavior of valve interstitial cells in a three-dimensional mechanically-conditioned model. Journal of Biological Engineering. 2019;**13**(1):1-14

[50] Knowles RG, Moncada S. Nitric oxide as a signal in blood vessels. Trends in Biochemical Science [Internet]. 1992;**17**(10):399-402. Available from: https://linkinghub.elsevier.com/retrieve/pii/096800049290008W

[51] Martin E, Berka V, Tsai AL, Murad F. Soluble guanylyl cyclase: The nitric oxide receptor. Methods in Enzymology [Internet]. 2005;**396**:478-492. Available from: http://www.ncbi.nlm.nih.gov/pubmed/16291255

[52] Miller JD, Weiss RM, Heistad DD. Calcific aortic valve stenosis: Methods, models, and mechanisms. Circulation Research. 2011;**108**(11):1392-1412

[53] Sverdlov AL, Ngo DT, Chapman MJ, Ali OA, Chirkov YY, Horowitz JD. Pathogenesis of aortic stenosis: Not just a matter of wear and tear. American Journal of Cardiovascular Diseases [Internet]. 2011;**1**(2):185-199. Available from: http://www.ncbi.nlm.nih.gov/pubmed/22254198%0Ahttp://www.pubmedcentral.nih.gov/articlerender.fcgi?artid=PMC3253493

[54] Nightingale AK, Horowitz JD. Aortic sclerosis: Not an innocent murmur but a marker of increased cardiovascular risk. Heart. 2005;**91**(11):1389-1393

[55] Lindman BR, Clavel MA, Mathieu P, Iung B, Lancellotti P, Otto CM, et al. Calcific aortic stenosis. Nature Review in Disease Primer. 2016;**2**:16006

[56] Guerraty MA, Grant GR, Karanian JW, Chiesa OA, Pritchard WF, Davies PF. Hypercholesterolemia induces side-specific phenotypic changes and peroxisome proliferator-activated receptor-gamma pathway activation

in swine aortic valve endothelium. Arteriosclerosis Thrombosis Vascular Biology [Internet]. 2010;**30**(2):225-231. Available from: http://www.ncbi.nlm.nih.gov/pubmed/19926833

[57] Soini Y, Salo T, Satta J. Angiogenesis is involved in the pathogenesis of nonrheumatic aortic valve stenosis. Human Pathology. 2003;**34**(8):756-763

[58] Miller JD, Chu Y, Brooks RM, Richenbacher WE, Peña-Silva R, Heistad DD. Dysregulation of antioxidant mechanisms contributes to increased oxidative stress in calcific aortic valvular stenosis in humans. Journal of American College Cardiology [Internet]. 2008;**52**(10):843-850. Available from: https://linkinghub.elsevier.com/retrieve/pii/S0735109708021165

[59] Kennedy JA, Hua X, Mishra K, Murphy GA, Rosenkranz AC, Horowitz JD. Inhibition of calcifying nodule formation in cultured porcine aortic valve cells by nitric oxide donors. European Journal of Pharmacology [Internet]. 2009;**602**(1):28-35. DOI: 10.1016/j.ejphar.2008.11.029

[60] Pulfer SK. Nitric Oxide Releasing Polymers and their Application to Vascular Devices [Internet]. Ann Arbor: The University of Akron; 1995. Available from: http://ezproxy.auckland.ac.nz/login?url=https://www.proquest.com/dissertations-theses/nitric-oxide-releasing-polymers-their-application/docview/304166110/se-2?accountid=8424

[61] Gould ST, Matherly EE, Smith JN, Heistad DD, Anseth KS. The role of valvular endothelial cell paracrine signaling and matrix elasticity on valvular interstitial cell activation. Biomaterials [Internet]. 2014;**35**(11):3596-3606. DOI: 10.1016/j.biomaterials.2014.01.005

[62] Blevins TL, Carroll JL, Raza AM, Grande-Allen KJ. Phenotypic characterization of isolated valvular interstitial cell subpopulations. Journal of Heart Valve Disease [Internet]. 2006;**15**(6):815-822. Available from: http://europepmc.org/abstract/MED/17152790

[63] Benton JA, Fairbanks BD, Anseth KS. Characterization of valvular interstitial cell function in three dimensional matrix metalloproteinase degradable PEG hydrogels. Biomaterials [Internet]. 2009;**30**(34):6593-6603. DOI: 10.1016/j.biomaterials.2009.08.031

[64] Benton JA, Deforest CA, Vivekanandan V, Anseth KS. Photocrosslinking of gelatin macromers to synthesize porous hydrogels that promote valvular interstitial cell function. Tissue Engineering – Part A. 2009;**15**(11):3221-3230

[65] Gu X, Masters KS. Role of the MAPK/ERK pathway in valvular interstitial cell calcification. American Journal of Physiology. 2009;**296**(6):1748-1757

[66] Gu X, Masters KS. Role of the Rho pathway in regulating valvular interstitial cell phenotype and nodule formation. American Journal of Physiology. 2011;**300**(2):448-458

[67] Coombs KE, Leonard AT, Rush MN, Santistevan DA, Hedberg-Dirk EL. Isolated effect of material stiffness on valvular interstitial cell differentiation. Journal of Biomedical Materials Research – Part A. 2017;**105**(1):51-61

[68] Kural MH, Billiar KL. Mechanoregulation of valvular interstitial cell phenotype in the third dimension. Biomaterials [Internet]. 2014;**35**(4):1128-1137. Available from: https://linkinghub.elsevier.com/retrieve/pii/S0142961213012775

[69] Stephens EH, Huynh TN, Cieluch JD, Grande-Allen KJ. Fibronectin-based isolation of valve interstitial cell subpopulations: Relevance to valve disease. Journal of Biomedical Materials Research Part A [Internet]. 2010;**92A**(1):340-349. Available from: https://www.ncbi.nlm.nih.gov/pmc/articles/PMC3624763/pdf/nihms412728.pdf

[70] Quinlan AMT, Billiar KL. Investigating the role of substrate stiffness in the persistence of valvular interstitial cell activation. Journal of Biomedical Materials Research Part A [Internet]. 2012;**23**(1). Available from: https://www.ncbi.nlm.nih.gov/pmc/articles/PMC3624763/pdf/nihms412728.pdf

[71] Gu X, Masters KS. Regulation of valvular interstitial cell calcification by adhesive peptide sequences. Journal of Biomedical Materials Research Part A [Internet]. 2010;**9999**(1). Available from: https://onlinelibrary.wiley.com/doi/10.1002/jbm.a.32660

[72] Porras AM, van Engeland NCA, Marchbanks E, McCormack A, Bouten CVC, Yacoub MH, et al. Robust generation of quiescent porcine valvular interstitial cell cultures. Journal of the American Heart Association. 2017;**6**(3):1-14

[73] Stephens EH, Carroll JL, Grande-Allen KJ. The use of collagenase III for the isolation of porcine aortic valvular interstitial cells: Rationale and optimization. The Journal of Heart Valve Disease. 2007;**16**(2):175-183

[74] Gould RA, Butcher JT. Isolation of valvular endothelial cells. Journal of Visualized Experiments. 2010;**46**:7-12

[75] Cheung WY, Young EWK, Simmons CA. Techniques for isolating and purifying porcine aortic valve endothelial cells. The Journal of Heart Valve Disease. 2008;**17**(6):674-681

[76] Cimini M, Rogers KA, Boughner DR. Aortic valve interstitial cells: An evaluation of cell viability and cell phenotype over time. The Journal of Heart Valve Disease. 2002;**11**(6):881-887

[77] Benton JA, Kern HB, Anseth KS. Substrate properties influence calcification in valvular interstitial cell culture. The Journal of Heart Valve Disease. 2008;**17**(6):689-699

[78] Masjedi S, Amarnath A, Baily KM, Ferdous Z. Comparison of calcification potential of valvular interstitial cells isolated from individual aortic valve cusps. Cardiovascular Pathology [Internet]. 2016;**25**(3):185-194. DOI: 10.1016/j.carpath.2015.12.002

[79] Ma H, Macdougall LJ, GonzalezRodriguez A, Schroeder ME, Batan D, Weiss RM, et al. Calcium signaling regulates valvular interstitial cell alignment and myofibroblast activation in fast-relaxing boronate hydrogels. Macromolecular Bioscience. 2020;**20**(12):1-14

[80] Mabry KM, Lawrence RL, Anseth KS. Dynamic stiffening of poly(ethylene glycol)-based hydrogels to direct valvular interstitial cell phenotype in a three-dimensional environment. Biomaterials [Internet]. 2015;**49**:47-56. DOI: 10.1016/j.biomaterials.2015.01.047

[81] Witt W, Büttner P, Jannasch A, Matschke K, Waldow T. Reversal of myofibroblastic activation by polyunsaturated fatty acids in valvular interstitial cells from aortic valves. Role of RhoA/G-actin/MRTF signalling. Journal of Molecular Cell Cardiology [Internet]. 2014;**74**:127-138. DOI: 10.1016/j.yjmcc.2014.05.008

[82] Kheradvar A, Zareian R, Kawauchi S, Goodwin RL, Rugonyi S. Animal models for heart valve research and development. Drug Discovery Today Disease Model [Internet]. 2017;**24**:55-62. DOI: 10.1016/j.ddmod.2018.04.001

[83] Sider KL, Blaser MC, Simmons CA. Animal models of calcific aortic valve disease. International Journal f Inflammation. 2011;**2011**(Ldl):1-18

[84] Tsang HG, Clark EL, Markby GR, Bush SJ, Hume DA, Corcoran BM, et al. Expression of calcification and extracellular matrix genes in the cardiovascular system of the healthy domestic sheep (*Ovis aries*). Frontiers in Genetics. 2020;**11**:1-22

[85] Merryman WD, Youn I, Lukoff HD, Krueger PM, Guilak F, Hopkins RA, et al. Correlation between heart valve interstitial cell stiffness and transvalvular pressure: Implications for collagen biosynthesis. American Journal of Physiology. 2006;**290**(1):224-231

[86] Jian B, Narula N, Li Q, Mohler ER, Levy RJ. Progression of aortic valve stenosis: TGF-β1 is present in calcified aortic valve cusps and promotes aortic valve interstitial cell calcification via apoptosis. Annals of Thoracic Surgery [Internet]. 2003;**75**(2):457-465. Available from: https://www.sciencedirect.com/science/article/pii/S0003497502043126

[87] Xue Y, Hilaire C, Hortells L, Phillippi JA, Sant V, Sant S. Shape-specific nanoceria mitigate oxidative stress-induced calcification in primary human valvular interstitial cell culture. Cellular and Molecular Bioengineering. 2017;**10**(5):483-500

[88] Fang M, Wang CG, Zheng C, Luo J, Hou S, Liu K, et al. Mir-29b promotes human aortic valve interstitial cell calcification via inhibiting TGF-β3 through activation of wnt3/β-catenin/Smad3 signaling. Journal of Cellular Biochemistry. 2018;**119**(7):5175-5185

[89] Zhang M, Liu X, Zhang X, Song Z, Han L, He Y, et al. MicroRNA-30b is a multifunctional regulator of aortic valve interstitial cells. Journal of Thoracic Cardiovascular Surgery [Internet]. 2014;**147**(3):1073-1080.e2. DOI: 10.1016/j.jtcvs.2013.05.011

[90] Masoumi N, Larson BL, Annabi N, Kharaziha M, Zamanian B, Shapero KS, et al. Electrospun PGS: PCL microfibers align human Valvular interstitial cells and provide tunable scaffold anisotropy. Advanced Healthcare Materials. 2014;**3**(6):929-939

[91] Li XF, Wang Y, Zheng DD, Xu HX, Wang T, Pan M, et al. M1 macrophages promote aortic valve calcification mediated by microRNA-214/TWIST1 pathway in valvular interstitial cells. American Journal of Translational Research. 2016;**8**(12):5773-5783

[92] Venardos N, Nadlonek NA, Zhan Q, Weyant MJ, Reece TB, Meng X, et al. Aortic valve calcification is mediated by a differential response of aortic valve interstitial cells to inflammation. Journal of Surgical Research [Internet]. 2014;**190**(1):1-8. Available from: file:///C:/Users/Carla%0ACarolina/Desktop/Artigos%0Apara%0Aacrescentar%0Ana%0Aqualificação/The%0Aimpact%0Aof%0Abirth%0Aweight%0Aon%0Acardiovascular%0Adisease%0Arisk%0Ain

[93] Sun W, Zhao R, Yang Y, Wang H, Shao Y, Kong X. Comparative study of human aortic and mitral valve interstitial cell gene expression and cellular function. Genomics [Internet]. 2013;**101**(6):326-335. DOI: 10.1016/j.ygeno.2013.03.004

[94] Sainger R, Grau JB, Branchetti E, Poggio P, Seefried WF, Field BC, et al. Human myxomatous mitral valve prolapse: Role of bone morphogenetic protein 4 in valvular interstitial cell activation. Journal of Cell Physiology [Internet]. 2012;**227**(6):2595-2604. Available from: https://onlinelibrary.wiley.com/doi/10.1002/jcp.22999

[95] Artiach G, Carracedo M, Seime T, Plunde O, Laguna-Fernandez A, Matic L, et al. Proteoglycan 4 is increased in human calcified aortic valves and enhances valvular interstitial cell calcification. Cells [Internet]. 2020;**9**(3):684. Available from: https://www.mdpi.com/2073-4409/9/3/684

[96] Albanese I, Yu B, Al-Kindi H, Barratt B, Ott L, Al-Refai M, et al. Role of noncanonical Wnt signaling pathway in human aortic valve calcification. Arteriosclerosis, Thrombosis, and Vascular Biology. 2017;**37**(3):543-552

[97] Latif N, Sarathchandra P, Chester AH, Yacoub MH. Expression of smoothmuscle cellmarkers and co-activators in calcified aortic valves. European Heart Journal. 2015;**36**(21):1335-1345

[98] Galeone A, Brunetti G, Oranger A, Greco G, Di Benedetto A, Mori G, et al. Aortic valvular interstitial cells apoptosis and calcification are mediated by TNF-related apoptosis-inducing ligand. International Journal of Cardiology [Internet]. 2013;**169**(4):296-304. DOI: 10.1016/j.ijcard.2013.09.012

[99] Poggio P, Branchetti E, Grau JB, Lai EK, Gorman RC, Gorman JH, et al. Osteopontin-CD44v6 interaction mediates calcium deposition via phospho-akt in valve interstitial cells from patients with noncalcified aortic valve sclerosis. Arteriosclerosis, Thrombosis, and Vascular Biology. 2014;**34**(9):2086-2094

[100] Vadana M, Cecoltan S, Ciortan L, Macarie RD, Tucureanu MM, Mihaila AC, et al. Molecular mechanisms involved in high glucose-induced valve calcification in a 3D valve model with human valvular cells. Journal of Cellular and Molecular Medicine. 2020;**24**(11):6350-6361

[101] Dittfeld C, Reimann G, Mieting A, Büttner P, Jannasch A, Plötze K, et al. Treatment with XAV-939 prevents in vitro calcification of human valvular interstitial cells. PLoS One. 2018;**13**(12):1-21

[102] Branchetti E, Sainger R, Poggio P, Grau JB, Patterson-Fortin J, Bavaria JE, et al. Antioxidant enzymes reduce DNA damage and early activation of valvular interstitial cells in aortic valve sclerosis. Arteriosclerosis, Thrombosis, and Vascular Biology. 2013;**33**(2):66-74

[103] Lucassen A, Wheeler R. Legal implications of tissue. Annals of the Royal College of Surgeons of England. 2010 Apr;**92**(3):189-192

[104] Ground M, Park YE, Waqanivavalagi S, Callon K, Walker R, Milsom P, et al. Generating robust human valvular interstitial cell cultures: Protocol and considerations. Journal of Molecular Cell Cardiology [Internet]. 2022;**173**(October):118-126. DOI: 10.1016/j.yjmcc.2022.10.008

[105] Cherng S, Young J, Ma H. Alpha-smooth muscle actin (α -SMA). Journal of American Science [Internet]. 2008;**4**(4):7-9. Available from: http://sciencepub.net/american/0404/02_0405_mahongbao-sma_am.pdf

[106] Werner S, Lützkendorf J, Müller T, Müller LP, Posern G. MRTF-A controls

myofibroblastic differentiation of human multipotent stromal cells and their tumour-supporting function in xenograft models. Scientific Reports. 2019;**9**(1):1-12

[107] Chalajour F, Treede H, Ebrahimnejad A, Lauke H, Reichenspurner H, Ergun S. Angiogenic activation of valvular endothelial cells in aortic valve stenosis. Experimental Cell Research. 2004;**298**(2):455-464

[108] Babu AN, Meng X, Zou N, Yang X, Wang M, Song Y, et al. Lipopolysaccharide stimulation of human aortic valve interstitial cells activates inflammation and osteogenesis. The Annals of Thoracic Surgery. 2008;**86**(1):71-76

[109] Poggio P, Sainger R, Branchetti E, Grau JB, Lai EK, Gorman RC, et al. Noggin attenuates the osteogenic activation of human valve interstitial cells in aortic valve sclerosis. Cardiovascular Research. 2013;**98**(3):402-410

[110] Rittié L, Fisher GJ. Isolation and culture of skin fibroblasts. In: Fibrosis Research [Internet]. Totowa, NJ: Humana Press; 2005. pp. 83-98. Available from: http://link.springer.com/10.1385/1-59259-940-0:083

[111] Gao Y, Li N, Xue Q, Fan X, Liu X, Han L. Basic fibroblast growth factor inhibits aortic valvular interstitial cells calcification via Notch1 pathway. Journal of Investigative Medicine. 2022;**2021**:002132

Chapter 2

Perspective Chapter: Two-Dimensional and Three-Dimensional Culture of Human Pluripotent Stem Cells

Qiang Li

Abstract

Human pluripotent stem cells (hPSCs), which include human embryonic stem cells (hESCs) and human induced pluripotent stem cells (hiPSCs), hold immense potential for various biomedical research in both academic and clinical applications. This chapter provides a comprehensive review of culturing techniques for hPSCs, covering two-dimensional (2D) adherent culture, three-dimensional (3D) suspension culture, and the utilization of hydrogel scaffolds in 3D hPSC culture. Furthermore, it explores the application of advanced 3D cell manufacturing techniques to facilitate the production of large quantities of high-quality hPSCs, catering to the needs of advanced biomedical applications. By addressing these topics, this chapter aims to present a comprehensive overview of diverse cultivation methods and their wide-ranging applications in hPSC research, encompassing fundamental studies and advanced biomedical investigations.

Keywords: hPSCs, 2D & 3D cell culture, hydrogel, scaffold, cell manufacturing

1. Introduction

Human pluripotent stem cells (hPSCs) include human embryonic stem cells (hESCs) [1] and human induced pluripotent stem cells (hiPSCs) [2, 3]. These cells are characterized by unique features such as indefinite self-renewal and the remarkable capacity to differentiate into all the cell types in the body. Therefore, the hPSCs represent invaluable cell sources for fundamental stem cell research and a wide range of biomedical applications, including cell therapies, tissue engineering, drug discovery, and disease modeling [4–8]. In addition to hPSCs, tissue-derived adult stem cells can serve as valuable resources for stem cell therapies. These multipotent stem cells, including the stem cells isolated from fetal tissues, mesenchymal stem cells (MSCs), and hematopoietic stem cells (HSCs), exhibit the capacity to differentiate into restricted cell types within their respective lineages [9]. The two-dimensional (2D) adherent culture of hPSCs is a commonly employed in research laboratories for fundamental stem cell studies. The 2D adherent cell culture methods can only produce the cells in limited quantities. However, all biomedical applications require significantly larger quantities of cells [10]. For example, the treatment with one patient of myocardial infarction typically necessitates approximately 10^9 cardiomyocytes. Similarly, the treatment of one diabetic patient requires around 10^9 β cells. In the case of blood transfusions, approximately 10^{12} red blood cells are needed. Moreover, the engineering of a human-size liver and the screening of a library with one million compounds at once typically require approximately 10^{10} cells [11, 12]. Researchers have explored suspension culture methods [13–19] to scale up the production of hPSCs and their derivatives. Furthermore, alternative methods have been developed to enhance the scalability of hPSC culture, including cell microencapsulation within hydrogels such as alginate [20–23] and thermoreversible hydrogels [24, 25]. These methods utilize a supportive matrix that closely mimics the natural extracellular environment, effectively preventing the formation of excessive cell agglomerations and reducing exposure to shear forces generated in the dynamic suspension culture. As a result, they exhibit high efficiency in expanding hPSCs [21, 24]. To facilitate the production of hPSCs on a large scale, implementing three-dimensional (3D) cell manufacturing processes is crucial. Extensive research has been dedicated to implementing 3D cell manufacturing processes for large-scale production [26–30].

2. Two-dimensional (2D) adherent and three-dimensional (3D) suspension culture of hPSCs

The hPSC culture media comprises vital nutrients, growth factors, and supplements that promote cell viability, proliferation, and maintenance of pluripotency. Initially, hPSC culture media contained serum and conditioned media [1, 31]. Currently, xeno-free and fully chemically defined media, such as mTeSR [32, 33] and Essential 8 (E8) [34], are commonly used for the cultivation and propagation of hPSCs. These media formulations eliminate the reliance on animal-derived components, offering a well-defined and controlled environment that supports the optimal growth and maintenance of hPSCs. For example, the E8 medium comprises eight essential components: DMEM/F12, insulin, FGF, L-ascorbic acid, selenium, TGFβ, transferrin, and $NaHCO_3$. DMEM/F12 serves as the base medium, providing a nutrient-rich foundation. Insulin and FGF2 play crucial roles in promoting the survival and proliferation of hPSCs. L-ascorbic acid acts as a stimulant for cell proliferation. Selenium is vital for the sustained expansion of hPSCs. TGFβ is incorporated to enhance pluripotent marker expression and ensure consistent long-term culture stability. When combined with a ROCK inhibitor (RI, Y27632) [35], transferrin improves initial survival and supports high cloning efficiency. $NaHCO_3$ is used for pH adjustment of the medium. These xeno-free and fully defined media efficiently support the expansion of hPSCs, enabling the transition of these cells from basic laboratory research to clinical applications.

The commercial hPSC cell lines, including hESCs and hiPSCs, can be purchased from WiCell Research Institute, the American Type Culture Collection (ATCC), or the European Collection of Authenticated Cell Cultures (ECACC). The equipment and supplies for hPSC maintenance include cell culture incubator, biosafety hood, 4°C refrigerator, −20°C freezer, −80°C freezer, liquid nitrogen dewar, inverted microscope, water bath, centrifuge, cell-freezing container, cryovials, pipette set with various tips, and different sizes of tubes (e.g., 1.5, 15, and 50 mL) along with standard cell culture plates (e.g., 6-well cell culture plates or 24-well cell culture plates) and low-attachment cell culture plates (e.g., low-attachment 6-well cell culture plates or low-attachment V bottom or U bottom 96-well cell culture plates). Additionally, the following materials and reagents are required to culture hPSCs: Essential 8™ medium (Gibco, or STEMCELL Technologies), Matrigel matrix (Thermo Fisher Scientific, Corning Life Sciences, or VWR), Rock inhibitor (Tocris Bioscience, Selleck Chemicals, or MilliporeSigma), Accutase (STEMCELL Technologies, MilliporeSigma, or Invitrogen), EDTA (0.5 M), pH 8.0, RNase-free (Thermo Fisher Scientific, VWR, or Cayman Chemical), DPBS (no calcium, no magnesium, Gibco, MilliporeSigma, or VWR), Dimethyl sulfoxide (Sigma-Aldrich, Thermo Fisher Scientific, or VWR), Penicillin-streptomycin solution (Corning Life Sciences, MilliporeSigma, or VWR), and 70% ethanol.

2.1 2D cell culture methods

The 2D adherent culture of hPSCs in cell culture plates or Petri dishes is a commonly employed method in laboratories for fundamental stem cell studies such as cell expansion and differentiation (**Figure 1**). The culture substrate plays a significant role in supporting hPSC attachment and proliferation, maintaining pluripotency. hPSCs were initially cultured on a layer of feeder cells, which refer to mitotically inactivated mouse embryonic fibroblast (MEF) or human fibroblast feeders that support hPSC growth and maintenance [1, 37]. Due to the significant variability and irreproducibility associated with the feeder-dependent methods, a feeder-independent culture protocol was developed [33], in which the hPSCs were cultured on extracellular matrix (ECM) rather than the feeder cells. Matrigel, an Engelbreth–Holm–Swarm mouse sarcoma tissue extract, has emerged as one of the most extensively utilized extracellular components for feeder-free culture of hPSCs in laboratories [5, 31, 33, 36]. Matrigel is a partially chemically defined and xenogeneic substrate encompassing collagens, laminin, and various other chemical compounds [38]. Therefore,

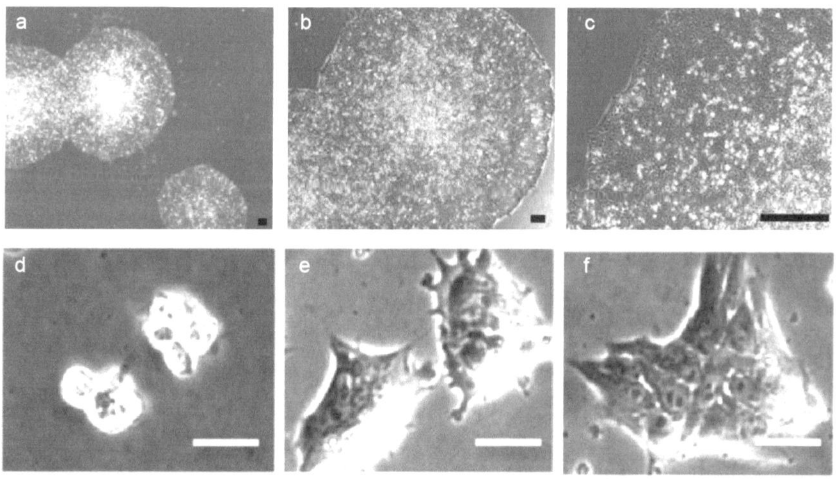

Figure 1.
2D cell culture of human pluripotent stem cells [33, 36]. (a–c) the 2D cell culture is prepared for passaging. The colony and cell morphology when observed through 2× (a), 4× (b), and 10× (c) objectives, respectively. The center of the colonies appears denser and brighter and cells at the center of the colony may appear smaller. Scale bar: 100 μm [33]. (d-f) Following plating, the cells quickly adhere to the plate within a few minutes (d), spread within 2 hours (e), and subsequently grow as colonies within 24 hours (f). Scale bars: 50 μm [36].

xenogenic-free, chemically defined ECM, such as laminin [39, 40], vitronectin [41], and synthetic surfaces [42], was developed to facilitate hPSC expansion and differentiation on the 2D adherent substrates. hPSCs cultured in 2D systems provide valuable insights into their molecular characteristics and differentiation potentials.

The 2D adherent cultures are widely used in laboratories for hPSC expansion and differentiation. The procedures involve several steps to ensure successful cell seeding, cell growth, and maintenance. These steps include (1) thawing frozen cells, (2) seeding cells on ECM-coated culture plates, (3) cell maintenance, (4) cell passaging, and (5) cell freezing for storage. Throughout these processes, it is important to exercise caution to minimize the risk of cell contamination and stress. Adhering to aseptic techniques, maintaining proper culture conditions, and strictly following established protocols will help to preserve the quality of hPSCs, especially for maintaining their pluripotency for long-term culture. Detailed protocols for culturing hPSCs on a 2D surface are provided below, offering step-by-step guidance and best practices.

E8 medium preparation: The E8 basal medium should be stored in a refrigerator at 4°C. Aliquot the E8 supplements into 1-mL tubes and store them in a freezer at – 20°C or – 80°C. To prepare the E8 full medium, mix 48.5 mL of the E8 basal medium with 1 mL of the E8 supplement and 0.5 mL of the penicillin-streptomycin solution.

Matrigel preparation: Aliquot the Matrigel into small volumes (e.g., 200 or 500 μL) and store them in a – 80°C freezer for a long time. The Matrigel was left in a 4°C refrigerator overnight to dissolve, resulting in a viscous liquid. To aliquot the Matrigel, place the Matrigel and all other required tubes in an ice bag in the biosafety hood in case the Matrigel forms the hydrogel during the procedures. To prepare the Matrigel coating medium, dilute the Matrigel with cold DMEM/F12 basal medium to a final concentration of 100 μg/mL. To coat the cell culture plate, add 1 mL of the Matrigel coating solution into one well of the 6-well plate. Place the cell culture plate at room temperature in the biosafety hood for at least 1 hour or place it in the incubator at 37°C for at least 30 minutes.

2.1.1 Thawing frozen cells and seeding on the 2D surface

1. Retrieve one cryovial of hPSCs from the liquid nitrogen dewar. Note: Ensure you have received specific training and wear appropriate personal protective equipment (PPE) before handling liquid nitrogen.

2. Submerge the bottom of the cryovial containing the cells into a 37°C water bath immediately. Gently shake the cells for about 3–5 minutes. Monitor the cells closely, and once the ice has dissolved into small pieces, carefully remove the cryovial from the water bath. Disinfect the exterior of the cryovial by gently pouring 70% ethanol.

3. Place the cells inside a biosafety hood. Utilizing a 1-mL pipette, carefully transfer the cells into a 15-mL tube. Following this, add 9 mL of E8 full medium supplied with 10 μM RI. Gently mix the tube to ensure proper distribution of the cells and medium.

4. Centrifuge the cell suspension at 300 g for 5 minutes.

5. Carefully discard the supernatant, ensuring complete removal. Next, add 2 mL of E8 full medium supplied with 10 μM RI. Resuspend the cells by gently pipetting a couple of times.

6. Remove the Matrigel coating medium from the cell culture plate. Note: The Matrigel coating medium can be recycled and used within 2 weeks.

7. Transfer the 2 mL of cells into one Matrigel-coated well of the 6-well plate. Culture the cells at 37°C and 5% CO_2 in the incubator. Note: To ensure even distribution of the cells, gently shake the cell culture plate back and forth a couple of times.

8. On the second day, replace the medium with 2 mL of fresh E8 full medium. Continue culturing the cells for an additional 3–4 days. Observe the cell morphologies using an inverted microscope and allow the cells to reach approximately 80% confluency before the following passage. Remember to change the medium daily throughout this period.

2.1.2 2D cell passaging (passaging cell clusters using EDTA solution)

1. Before passage, add RI to each well at a final concentration of 10 μM. Then, put the cells back in an incubator for culture for at least 1 hour. Note: RI treatment can significantly improve cell viability after passage.

2. Remove the cell culture medium and add 1 mL of DPBS (without Ca^{2+} or Mg^{2+}) to each well. Gently shake the plate to ensure proper distribution, then carefully remove the DPBS.

3. Add 1 mL of 0.5 mM DETA solution to each well. Allow the cells to remain undisturbed at room temperature for approximately 5–7 minutes.

4. Remove the EDTA solution from each well. Using a 1-mL pipette, gently pipette the cells up and down with 1 mL of E8 full medium supplemented with 10 μM RI.

5. Transfer 200 μL cell solution to a new Matrigel-coated well and add 2 mL fresh E8 full medium supplemented with 10 μM RI.

6. Place the cells in the incubator. To ensure even distribution of the cell clusters within the wells, gently shake the cell culture plate back and forth several times.

7. On the second day, replace the medium with 2 mL of E8 full medium. Continue culturing the cells for an additional 3–4 days. Observe the cell morphologies using an inverted microscope and allow the cells to reach approximately 80% confluency before the following passage. Change the medium daily.

2.1.3 Cell freezing (when ordering new cells, it is necessary to expand them through 2–3 passages and then freeze the cells for banking)

1. Prepare the cell freeze medium by combining 10% DMSO with 90% E8 full medium supplemented with 10 μM RI. Note: Add the DMSO into the E8 medium slowly, drop by drop.

2. The EDTA-based procedure described above was used to dissociate the cells into cell clusters. Note: Freezing cell clusters instead of single cells can enhance the efficiency of cell recovery.

3. Collect the cell clusters into one 15-mL tube. Centrifuge the cells at 300 g for 5 minutes.

4. Remove the supernatant from the tube.

5. Add 1 mL of cell freeze medium to the tube. Next, resuspend the cell clusters by gently pipette a couple of times. Finally, transfer the resulting cell solution into a cryogenic vial.

6. Place the cryogenic vial into the cell-freezing container.

7. Transfer the cell freezing container to a − 80°C freezer and leave it overnight.

8. On the second day, transfer the frozen cells to a liquid nitrogen dewar for long-term storage.

In summary, 2D cell culture methods are widely used in laboratories for basic stem cell research. For hPSC maintenance, it is essential to exercise caution to minimize the risk of cell contamination and cell stress. Preserving the high quality and pluripotency of hPSCs requires strict adherence to aseptic techniques, meticulous maintenance of culture conditions, and rigorous adherence to established protocols. Despite their extensive use, 2D culture approaches have inherent limitations. These systems fail to fully replicate the intricate 3D microenvironment *in vivo*, which hampers accurate modeling of cell-cell interactions, morphogenesis, and organ development. As a result, researchers have developed 3D suspension culture methods to address these questions.

2.2 3D suspension culture methods

In laboratories, 3D suspension culture methods have also emerged as valuable tools for studying hPSCs' expansion and differentiation in a more physiologically relevant environment (**Figure 2**). This approach allows cells to interact with neighboring cells and the extracellular matrix, promoting cell-cell interaction and tissue-like organization. The 3D culture enables the formation of more complex structures, such as embryoid bodies (EBs) or organoids, which are self-organized 3D tissues with multiple cell types. Organoids derived from hPSCs have demonstrated remarkable cellular organization and functionality, offering great potential for cell therapies, disease modeling, and drug screening. The 3D suspension culture is widely used for hPSC and hPSC-derived organoid cultures [8, 43–48]. In this system, the cells can be cultured in low-attachment cell culture plates or spinner flasks. The detailed protocols for culturing hPSCs on 3D suspension are provided below, offering step-by-step guidance and best practices.

2.2.1 2D to 3D suspension passage cells

1. Before passage, add RI to each well of the cells at a final concentration of 10 μM. Subsequently, return the cells to an incubator at 37°C and allow them to incubate for at least 1 hour.

2. Discard the cell culture medium from each well.

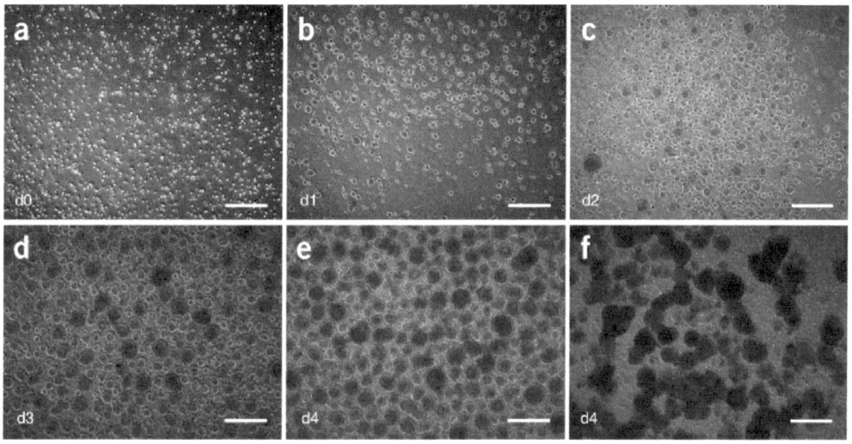

Figure 2.
3D suspension culture of human pluripotent stem cells [19]. (a–e) Single cells were seeded in the 3D suspension culture on day 0 (a); the formation and growth of cell spheroids from day 1 to day 4 (b–e). (f) Undesirable cell spheroid agglomeration formed in the center of the dish. Scale bars: 500 μm.

3. Add 1 mL of DPBS to each well, gently shaking the plate, and then remove the DPBS from each well. Note: It is necessary to perform a single wash of the cells with DPBS. The resident cell culture medium will neutralize the Accutase activity.

4. Add 1 mL of Accutase solution to each well. Place the cells at 37°C in the incubator for 3–5 minutes.

5. Observe the cell statue under the miscopy. When you see many cells detached from the bottom of the plate, add 2 mL of E8 full medium supplemented with 10 μM RI. Gently pipette the cells into single cells using a 1-mL pipette.

6. Perform cell counting by combining 10 μL of the cell solution with 10 μL of trypan blue. Transfer 10 μL of this mixture to a cell counting plate. Utilize the automated cell counter to measure cell viability and concentration.

7. Centrifuge the single-cell solution at 300 g for 5 minutes.

8. Carefully discard the supernatant from the tube. Add 1 mL of E8 full medium supplemented with 10 μM RI. Resuspend the cells by pipetting a couple of times.

9. Calculate the cell number needed for the following experiments.

2.2.2 For 3D suspension culture in low-attachment 96-well plate (many organoid culture protocols begin by generating hPSC spheroids using a low-attachment V bottom or U bottom 96-well plate)

1. Calculate the cell concentration and dilute the cells in E8 full medium supplemented with RI. Seed around 3000–6000 cells per 200 μl medium per well in the low-attachment 96-well plate.

2. Put the cells at 37°C in an incubator for culture. The cells will go down to the bottom of the well and form one cell cluster within 1 hour.

3. On the second day, the cells should form a uniform spheroid in each well. Carefully remove 150 μL medium from each well and add 150 μL E8 full medium.

4. Change the medium daily and process to organoid differentiation on day 3–5 by replacing the medium with an organoid-specific differentiation medium.

2.2.3 For 3D suspension culture in low-attachment 6-well plate

1. Transfer $1{\times}10^5$ cells into a single well of the low-attachment 6-well plate. Add 2 mL of fresh E8 full medium supplemented with 10 μM RI.

2. Place the cell plate onto a shaker inside the incubator for suspension culture. The cells will form small cell spheroids within 24 hours.

3. Passage the cells on days 5–7. The final concentration of the cells in the 3D suspension culture is about 1–2 million cells per mL.

2.2.4 3D to 3D suspension passage cells

1. Add RI to the cells at a final concentration of 10 μM. Subsequently, return the cells to the incubator and culture them for at least **1** hour.

2. Collect the cells and transfer them to 15-mL tubes.

3. Centrifuge the cells at 200 g for 3 minutes.

4. Remove the supernatant, wash the cells once using DPBS, and add 5–10 Accutase solution to digest the cell spheroids.

5. Put the sample at 37°C for about 10–15 minutes and dissociate them into single cells using a 1-mL pipette by carefully pipetting up and down several times.

6. Add equal E8 full medium supplemented with RI to neutralize the Accutase activity.

7. Perform the cell counter and seed the cells at a concentration of $1{\times}10^5$ cells.

Biomedical applications require large quantities of cells of high quality. To scale up the production of hPSCs, researchers have explored suspension culture methods using trimethyl ammonium-coated polystyrene microcarriers [14] or Matrigel-coated microcarriers [13]. However, it has been observed that the cell expansion rate tends to decrease gradually over long-term culture on microcarriers, and there can be challenges associated with detaching the cells from these microcarriers [13, 14]. To address these challenges, researchers have developed 3D suspension culture techniques for hPSCs without using microcarriers [15–19]. hPSCs have been successfully cultured as floating aggregates in static conditions using low-attachment cell culture plates or in suspension culture with regular agitation of the culture vessel, such as spinner flasks. Currently, the 3D suspension culture of hPSCs and hPSC-derived organoids in spinner flasks is also commonly applied in laboratories.

2.3 3D culture with hydrogel scaffolds

Hydrogel scaffolds like alginate [20–23], or thermoreversible hydrogels [24, 25] have been developed to culture hPSCs in a more physiologically relevant microenvironment. These hydrogel scaffolds can be used as physical barriers to isolate cells from agglomeration and hydrodynamic stresses generated from the dynamic suspension. These methods enhance the scalability of hPSC culture and exhibit high efficiency in expanding hPSCs [21, 24]. For example, a thermoreversible hydrogel-based cell culture system has been developed to expand and differentiate multiple cell lines of hPSCs [24, 25]. In this system, single cells were seeded within this thermoreversible PNIPAAm-PEG hydrogel matrix (**Figure 3**). As the hydrogel is very soft, the single cells can clonally grow into uniform spheroids within 3–5 days by deforming the scaffolds to create spaces. The hydrogel eliminating the cell agglomeration and hydrodynamic stresses could significantly improve the culture efficiency. The hydrogel scaffold enabled serial expansion of hPSCs with good cell viability, growth rate, yield, and purity. Moreover, after expansion, the hydrogel can be easily dissolved into liquid by adding cold DPBS to the well. Then, the cell spheroids can be easily harvested and centrifuged for the following passage procedures. This method provides an efficient, defined, scalable, and good manufacturing practice-compatible approach for hPSC production and differentiation [24].

Additionally, alginate hydrogels offer tunable mechanical properties and can encapsulate hPSCs to form 3D structures. Alginate hydrogel tube cell culture system

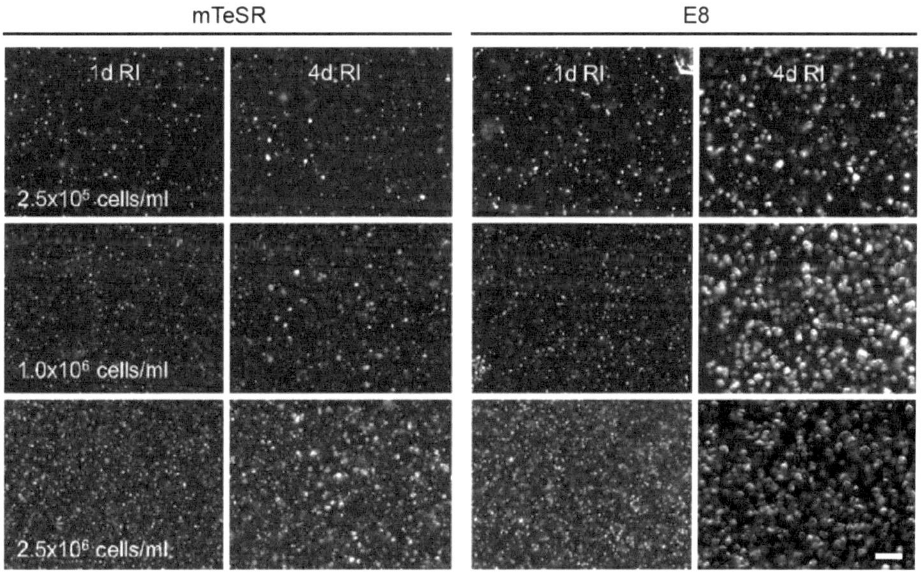

Figure 3.
3D culture of hPSCs in thermoreversible PNIPAAm-PEG hydrogels [24]. Phase images depict cell morphologies following culture in PNIPAAm-PEG hydrogel. Single-cell seeding densities were 2.5×10^5, 1.0×10^6, or 2.5×10^6 cells per mL, respectively, and cultures were maintained for 4 days in either mTeSR or E8 medium with 1-day or 4-day RI. Scale bar: 250 μm.

(AlgTubes) [21–23] provides cells with a culture microenvironment that aligns with their physiological needs. In this system, hPSCs are seeded in microscale alginate hydrogel tubes (**Figure 4**). Within 24 hours, these hPSCs can form small cell clusters, enhancing cell viability post-seeding. The diameter of the tubes can be adjusted to approximately 200–400 μm, allowing nutrients and metabolic waste to diffuse freely through the hydrogel tube walls. The design of AlgTubes represents a paradigm shift in cell culture efficiency, as evidenced by improvements in various aspects such as growth rate, viability, yield, and genetic and phenotype stability. Furthermore, AlgTubes can be easily adapted for scalable expansion in bioreactors, showcasing the high potential for producing hPSCs and their derivatives with exceptional quality and quantity, making them ideal for various biomedical applications.

2.4 3D cell manufacturing for advanced biomedical applications

HPSCs hold great promise for a range of cutting-edge biomedical applications, all of which necessitate substantial quantities of high-quality cells [11, 12]. (1) hPSCs can differentiate into diverse cell types, such as cardiomyocytes and neurons, with applications in regenerative medicine and organ transplantation. For instance, treatment of myocardial infarction in a single patient often demands roughly 10^9 cardiomyocytes. (2) hPSCs can be derived from individual patients, paving the way

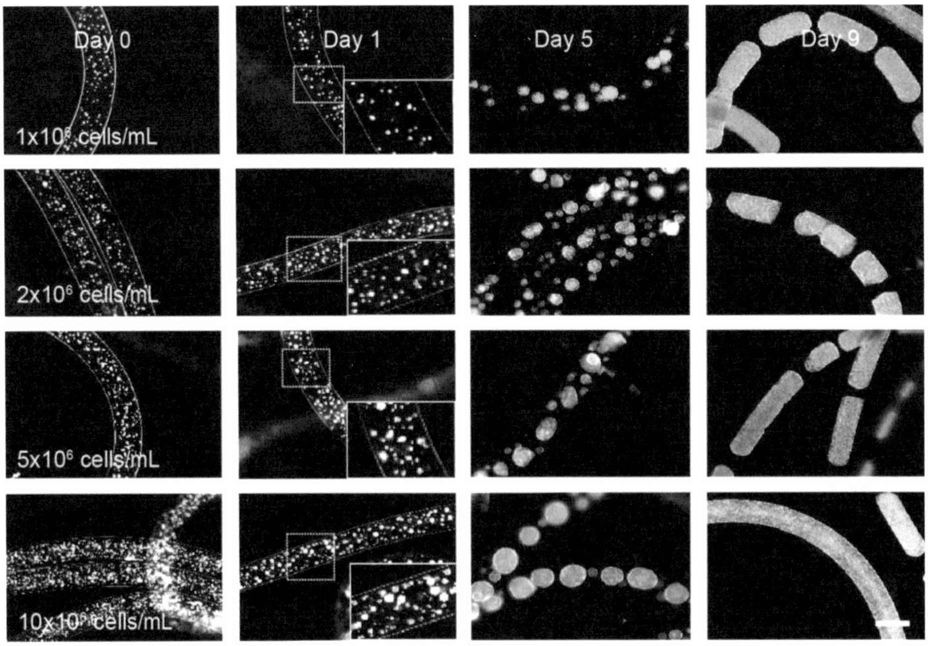

Figure 4.
3D culture of hPSCs in Algtubes [21]. Phase images illustrate the cell morphologies during culture in Algtubes. Cells were seeded at densities of 1×10^6, 2×10^6, 5×10^6, or 10×10^6 cells per mL and were cultured for 9 days. Within 24 hours, individual cells aggregated to form small clusters. These clusters subsequently grew into spheroids by day 5, which further evolved into fibrous cell masses by day 9.

for personalized disease modeling and high-throughput drug screening. Notably, screening a library containing one million compounds at once typically demands around 10^{10} cells. (3) hPSCs offer the potential to produce substantial quantities of blood cells, including platelets. For example, approximately 10^{12} red blood cells are required for blood transfusions.

Advancing the mass production of hPSC-based therapeutics toward a scalable and cost-efficient process will facilitate widespread adoption in all biomedical applications. The field of 3D cell manufacturing has opened exciting possibilities for large-scale production of hPSCs and their derivatives. Developing the 3D cell manufacturing processes requires incorporating approaches that allow for the precise control of culture conditions, such as nutrient and oxygen supply, pH, and mechanical forces, as well as precise monitor the cell conditions, including cell viability, cell states, and cell function [26–29]. Extensive research has been dedicated to implementing 3D cell manufacturing processes using bioreactors for large-scale production of hPSCs [27, 30, 49, 50]. For instance, bioreactors such as the Cellspin Integra Biosciences spinner flask [51], the NDS Technologies Bioreactor [52], and the Able Bioreactor [53] have successfully generated clinically relevant cell numbers. While the agitation of culture in bioreactors can improve mass transport and reduce cell agglomeration, it can also introduce hydrodynamic stresses, such as shear stresses, which may negatively impact hPSC expansion and differentiation. The variations in the stirred-tank bioreactor are influenced by numerous factors, including the bioreactor design, such as impeller geometry and size, as well as vessel geometry and size. Therefore, there is a critical need for automated 3D cell manufacturing in advanced biomedical applications, enabling the large-scale production of hPSCs in accordance with current good manufacturing practices (cGMP) and strict safety assessment criteria. In summary, advancements in this field continuously broaden our comprehension of hPSCs and their wide-ranging potential in addressing diverse human diseases and industrial demands.

3. Conclusion

This chapter has provided a comprehensive overview of the different culturing methods used in stem cell research and their biomedical applications. The 2D cell culture methods have been extensively employed in laboratories for fundamental stem cell research. Detail protocols for hPSC culture in 2D adherent substrates were summarized. Furthermore, 3D suspension culture methods have demonstrated their value in basic stem cell research and advanced biomedical studies. The utilization of hydrogel scaffolds, including alginate and thermoreversible hydrogels, has enabled the development of 3D culture systems for hPSCs expansion and differentiation. Additionally, the application of 3D cell manufacturing techniques has shown promise in producing large quantities of high-quality hPSCs for advanced biomedical applications. This chapter has offered readers a comprehensive understanding of the diverse culturing methods and their roles in stem cell research, ranging from fundamental studies to cutting-edge biomedical advancements. This knowledge serves as a foundation for readers interested in delving deeper into stem cell biology and biomedical applications, empowering them to make valuable contributions to the field.

4. Troubleshooting

Problem 1: Matrigel forms hydrogel during handling.

Solution 1: Thaw the frozen Matrigel at 4°C overnight. Ensure all handling steps, including aliquoting and preparing the Matrigel coating solution, are performed while keeping the Matrigel on ice.

Problem 2: In 2D culture, cell colonies cannot be detached from the plate.

Solution 2: Appropriately extend EDTA treatment time.

Problem 3: In 2D culture, there is a low cell survival rate after cell plating.

Solution 3: Ensure RI is added before cell passage and continue including RI in the cell culture medium within 24 hours after passage.

Problem 4: In 3D suspension culture, lots of large cell aggregations form.

Solution 4: Appropriately increase agitation rate.

Appendices and nomenclature

hPSCs	human pluripotent stem cells
hESCs	human embryonic stem cells
hiPSCs	human-induced pluripotent stem cells
MSCs	mesenchymal stem cells
HSCs	hematopoietic stem cells
2D	two-dimensional
3D	three-dimensional
E8	essential 8
RI	ROCK inhibitor
MEF	mouse embryonic fibroblast
ECM	extracellular matrix
DMSO	dimethyl sulfoxide
PPE	personal protective equipment
EBs	embryoid bodies
AlgTubes	alginate hydrogel tube cell culture system
cGMP	current good manufacturing practices

Author details

Qiang Li
John A. Paulson School of Engineering and Applied Sciences, Harvard University, Boston, MA, USA

*Address all correspondence to: qiangli@seas.harvard.edu

References

[1] Thomson JA et al. Embryonic stem cell lines derived from human blastocysts. Science. 1998;**282**(5391):1145-1147

[2] Takahashi K et al. Induction of pluripotent stem cells from adult human fibroblasts by defined factors. Cell. 2007;**131**(5):861-872

[3] Yu J et al. Induced pluripotent stem cell lines derived from human somatic cells. Science. 2007;**318**(5858):1917-1920

[4] Avior Y, Sagi I, Benvenisty N. Pluripotent stem cells in disease modelling and drug discovery. Nature Reviews. Molecular Cell Biology. 2016;**17**(3):170-182

[5] Chen KG et al. Human pluripotent stem cell culture: Considerations for maintenance, expansion, and therapeutics. Cell Stem Cell. 2014;**14**(1):13-26

[6] Tabar V, Studer L. Pluripotent stem cells in regenerative medicine: Challenges and recent progress. Nature Reviews. Genetics. 2014;**15**(2):82-92

[7] Engle SJ, Puppala D. Integrating human pluripotent stem cells into drug development. Cell Stem Cell. 2013;**12**(6):669-677

[8] Lancaster MA, Knoblich JA. Organogenesis in a dish: Modeling development and disease using organoid technologies. Science. 2014;**345**(6194)

[9] Kimbrel EA, Lanza R. Next-generation stem cells-ushering in a new era of cell-based therapies. Nature Reviews. Drug Discovery. 2020;**19**(7):463-479

[10] Kropp C, Massai D, Zweigerdt R. Progress and challenges in large-scale expansion of human pluripotent stem cells. Process Biochemistry. 2017;**59**:244-254

[11] Badylak SF, Taylor D, Uygun K. Whole-organ tissue engineering: Decellularization and recellularization of three-dimensional matrix scaffolds. Annual Review of Biomedical Engineering. 2011;**13**:27-53

[12] Desbordes SC, Studer L. Adapting human pluripotent stem cells to high-throughput and high-content screening. Nature Protocols. 2013;**8**(1):111-130

[13] Oh SKW et al. Long-term microcarrier suspension cultures of human embryonic stem cells. Stem Cell Research. 2009;**2**(3):219-230

[14] Phillips BW et al. Attachment and growth of human embryonic stem cells on microcarriers. Journal of Biotechnology. 2008;**138**(1-2):24-32

[15] Amit M et al. Suspension culture of undifferentiated human embryonic and induced pluripotent stem cells. Stem Cell Reviews. 2010;**6**(2):248-259

[16] Steiner D et al. Derivation, propagation and controlled differentiation of human embryonic stem cells in suspension. Nature Biotechnology. 2010;**28**(4):361-364

[17] Olmer R et al. Long term expansion of undifferentiated human iPS and ES cells in suspension culture using a defined medium. Stem Cell Research. 2010;**5**(1):51-64

[18] Singh H et al. Up-scaling single cell-inoculated suspension culture of human embryonic stem cells. Stem Cell Research. 2010;**4**(3):165-179

[19] Zweigerdt R et al. Scalable expansion of human pluripotent stem cells in suspension culture. Nature Protocols. 2011;**6**(5):689-700

[20] Chan C et al. Microencapsulation technology: A powerful tool for integrating expansion and cryopreservation of human embryonic stem cells. PLoS One. 2011;**6**(8)

[21] Li Q et al. Scalable and physiologically relevant microenvironments for human pluripotent stem cell expansion and differentiation. Biofabrication. 2018;**10**(2)

[22] Lin H et al. Hydrogel-based bioprocess for scalable manufacturing of human pluripotent stem cell-derived neural stem cells. ACS Applied Materials & Interfaces. 2018;**10**(35):29238-29250

[23] Lin H et al. Engineered microenvironment for manufacturing human pluripotent stem cell-derived vascular smooth muscle cells. Stem Cell Reports. 2019;**12**(1):84-97

[24] Lei Y, Schaffer DV. A fully defined and scalable 3D culture system for human pluripotent stem cell expansion and differentiation. Proceedings. National Academy of Sciences. United States of America. 2013;**110**(52)

[25] Lin H, Li Q, Lei Y. Three-dimensional tissues using human pluripotent stem cell spheroids as biofabrication building blocks. Biofabrication. 2017;**9**(2)

[26] Silva M et al. Generating iPSCs: Translating cell reprogramming science into scalable and robust biomanufacturing strategies. Cell Stem Cell. 2015;**16**(1):13-17

[27] Polanco A, Kuang B, Yoon S. Bioprocess technologies that preserve the quality of iPSCs. Trends in Biotechnology. 2020;**38**(10):1128-1140

[28] Aijaz A et al. Biomanufacturing for clinically advanced cell therapies. Nature Biomedical Engineering. 2018;**2**(6):362-376

[29] Fernandes TG et al. Stem cell bioprocessing for regenerative medicine. Journal of Chemical Technology and Biotechnology. 2014;**89**(1):34-47

[30] Tristan CA et al. Robotic high-throughput biomanufacturing and functional differentiation of human pluripotent stem cells. Stem Cell Reports. 2021;**16**(12):3076-3092

[31] Xu C et al. Feeder-free growth of undifferentiated human embryonic stem cells. Nature Biotechnology. 2001;**19**(10):971-974

[32] Ludwig TE et al. Derivation of human embryonic stem cells in defined conditions. Nature Biotechnology. 2006;**24**(2):185-187

[33] Ludwig TE et al. Feeder-independent culture of human embryonic stem cells. Nature Methods. 2006;**3**(8):637-646

[34] Chen G et al. Chemically defined conditions for human iPSC derivation and culture. Nature Methods. 2011;**8**(5):424-429

[35] Watanabe K et al. A ROCK inhibitor permits survival of dissociated human embryonic stem cells. Nature Biotechnology. 2007;**25**(6):681-686

[36] Beers J et al. Passaging and colony expansion of human pluripotent stem cells by enzyme-free dissociation in chemically defined culture conditions. Nature Protocols. 2012;**7**(11):2029-2040

[37] Richards M et al. Human feeders support prolonged undifferentiated

growth of human inner cell masses and embryonic stem cells. Nature Biotechnology. 2002;**20**(9):933-936

[38] Aisenbrey EA, Murphy WL. Synthetic alternatives to Matrigel. Nature Reviews Materials. 2020;**5**(7):539-551

[39] Rodin S et al. Clonal culturing of human embryonic stem cells on laminin-521/E-cadherin matrix in defined and xeno-free environment. Nature Communications. 2014;**5**(1)

[40] Rodin S et al. Monolayer culturing and cloning of human pluripotent stem cells on laminin-521–based matrices under xeno-free and chemically defined conditions. Nature Protocols. 2014;**9**(10):2354-2368

[41] Braam SR et al. Recombinant vitronectin is a functionally defined substrate that supports human embryonic stem cell self-renewal via αVβ5 integrin. Stem Cells. 2008;**26**(9):2257-2265

[42] Saha K et al. Surface-engineered substrates for improved human pluripotent stem cell culture under fully defined conditions. Proceedings. National Academy of Sciences. United States of America. 2011;**108**(46):18714-18719

[43] Eiraku M et al. Self-organizing optic-cup morphogenesis in three-dimensional culture. Nature. 2011;**472**(7341):51-56

[44] Drakhlis L, Devadas SB, Zweigerdt R. Generation of heart-forming organoids from human pluripotent stem cells. Nature Protocols. 2021;**16**(12):5652-5672

[45] Lancaster MA et al. Cerebral organoids model human brain development and microcephaly. Nature. 2013;**501**(7467):373-379

[46] Morizane R et al. Nephron organoids derived from human pluripotent stem cells model kidney development and injury. Nature Biotechnology. 2015;**33**(11):1193-1200

[47] Wimmer RA et al. Human blood vessel organoids as a model of diabetic vasculopathy. Nature. 2019;**565**(7740):505-510

[48] Velasco S et al. Individual brain organoids reproducibly form cell diversity of the human cerebral cortex. Nature. 2019;**570**(7762):523-527

[49] Goldring CEP et al. Assessing the safety of stem cell therapeutics. Cell Stem Cell. 2011;**9**(2)

[50] Nath SC, Harper L, Rancourt DE. Cell-based therapy manufacturing in stirred suspension bioreactor: Thoughts for cGMP compliance. Frontiers in Bioengineering and Biotechnology. 2020;**8**

[51] Haraguchi Y, Matsuura K, Shimizu T, Yamato M, Okano T. Simple suspension culture system of human iPS cells maintaining their pluripotency for cardiac cell sheet engineering. Journal of Tissue Engineering and Regenerative Medicine. 2015;**9**:1363-1375

[52] Nath SC, Horie M, Nagamori E, Kino-Oka M. Size- and time-dependent growth properties of human induced pluripotent stem cells in the culture of single aggregate. Journal of Bioscience and Bioengineering. 2017;**124**:469-475

[53] Nath SC, Tokura T, Kim MH, Kino-Oka M. Botulinum hemagglutinin-mediated in situ break-up of human induced pluripotent stem cell aggregates for high-density suspension culture. Biotechnology and Bioengineering. 2018;**115**:910-920

Chapter 3

Perspective Chapter: Investigating Cancer Tumor Microenvironment *In Vitro* – Co-Culture Studies on Adipocytes and Cancer Cells

Ozge Rencuzogullari, Pelin Ozfiliz-Kilbas, Enes Bal and Burcu Ayhan-Sahin

Abstract

The tumor microenvironment increases the growth and invasion of cancer cells, makes classical chemotherapy applications inadequate, and is associated with a poor cancer prognosis. Recent studies reveal that cancer stroma supports tumor growth and metastasis and develops resistance to chemotherapy. *In vitro* co-culture techniques are widely used to study cross-talk between tumor microenvironment cells such as adipocytes, endothelial cells, fibroblasts, macrophages, and cancer cells. Co-culture techniques are classified into two main categories: indirect and direct methods. Transwell (indirect) co-culture of mature adipocytes with cancer cells has shown cancer cell viability, growth, proliferation, invasion, and metastases. This chapter covers the general methods of co-culture studies and will emphasize the results obtained on the co-culture of adipocytes and cancer cells.

Keywords: tumor microenvironment (TME), cell culture, co-culture, adipocytes, organoids

1. Introduction

Cancer is a multidisciplinary disease that occurs when the balance between cell proliferation and death is disrupted in the direction of proliferation. After excessive proliferation, tumor mass invades vessels and metastasizes to adjacent tissues [1]. In other words, a tumor colony is a group of cells that have gained a higher proliferation rate and can invade other tissues and metastasis [2]. Due to its high metastasis capacity, although it changes according to the type, the mortality rate is high in many cases [3]. Cancer is named according to the tissue or organ it originates from. Indeed, this terminology is insufficient most of the time, and each cancer type is also subdivided in compliance with molecular marker, histological staining status, and cell type. About 200 cancer subtypes have been identified by the National Cancer Institute (NCI), but the exact number of subtypes is still a mystery. Subtyping in cancer cases is essential because the subtype affects the therapy model [4]. The accumulation

IntechOpen

of genetic and environmental risk factors triggers the initiation of carcinogenesis. Although there are cancer-type-specific markers, KRAS, BRAF, PIK3CA, epithelial growth factor receptor (EGFR), p53, and c-myc are the most common mutated genes in various cancers [5].

Surgery supported by chemotherapy is still the primary treatment method for solid tumors. Still, the adverse effects of chemotherapy reduce the life quality of patients because of the toxicity of drugs to healthy cells and the tumor cells. Radiation therapy and immunotherapy can also be used for treatment. The main drawback of developing new therapies with minimal adverse effects is the intra- and peri-tumor heterogeneity [3]. The tumor contains highly proliferating cancerous cells and is surrounded by different types of cells and tissues, defined as a tumor microenvironment (TME). Recent studies concerning tumor biology showed that TME has a significant role in drug resistance, tumor development, and malignancy [6]. Intra-tumor heterogeneity also affects the classification and subtyping of tumors. In some cases, intra-tumor solid heterogeneity complicates the type and prevents identifying the case, whether a known subtype or a new one [4]. Therefore, it is essential to define the tumor microenvironment to assess the subtype, malignancy, and metastasis profile of a tumor and study in cultures for new therapeutic approaches.

This chapter summarizes the important concepts of TME and potential applications in cell culture. As TME is quite complex, this chapter focuses on the studies of TME in adipocytes and cancer cells. This chapter summarizes the components of TME and the developing cell culture applications that show great potential to model various cancers. We also summarize the co-culture applications in cancer research. This chapter also provides further insights into novel cell-culture-based TME studies.

2. Cellular and non-cellular components of tumor microenvironment (TME)

TME, defined as the environment of a tumor that tumor cells interact with, is composed of cellular and noncellular compartments [7]. The cellular compartment contains mainly fibroblasts, endothelial cells, adipocytes, stem cells, stellate cells, and the innate and adaptive immune cells such as B lymphocytes, T lymphocytes, macrophages, dendritic cells, natural killer (NK) cells, neutrophils, and myeloid-derived suppressor cells (MDSC) (**Figure 1**) [8]. These cells are laid on a scaffold formed by an extracellular matrix (ECM), which includes proteins, proteoglycans, glycoproteins, and chemicals secreted by cellular compartments such as growth factors, cytokines, and chemokines [9]. Tumor cells interact with other cellular components, and the extracellular matrix provides a structure to hold all these elements together. Interactions between cellular immune cells and non-cellular components account for the tumor heterogeneity and complexity of TME and limitations for targeted therapies in patients [10]. As a result of this interaction, the cellular environment also shows some genetic and morphological changes. It causes conditions supporting the tumor's development, contributing to metastasis and drug resistance [11]. Due to its contribution to tumor development, metastasis, and drug resistance, the tumor microenvironment is an important topic that has been popular recently by researchers and offers a new perspective on cancer treatment [8].

It was first discovered in the 1890s when tumor and tumor microenvironment interacted, emerging with the theory "the seeds and the soil" of English surgeon Stephen Paget [12]. This theory that metastasis depends on communication between

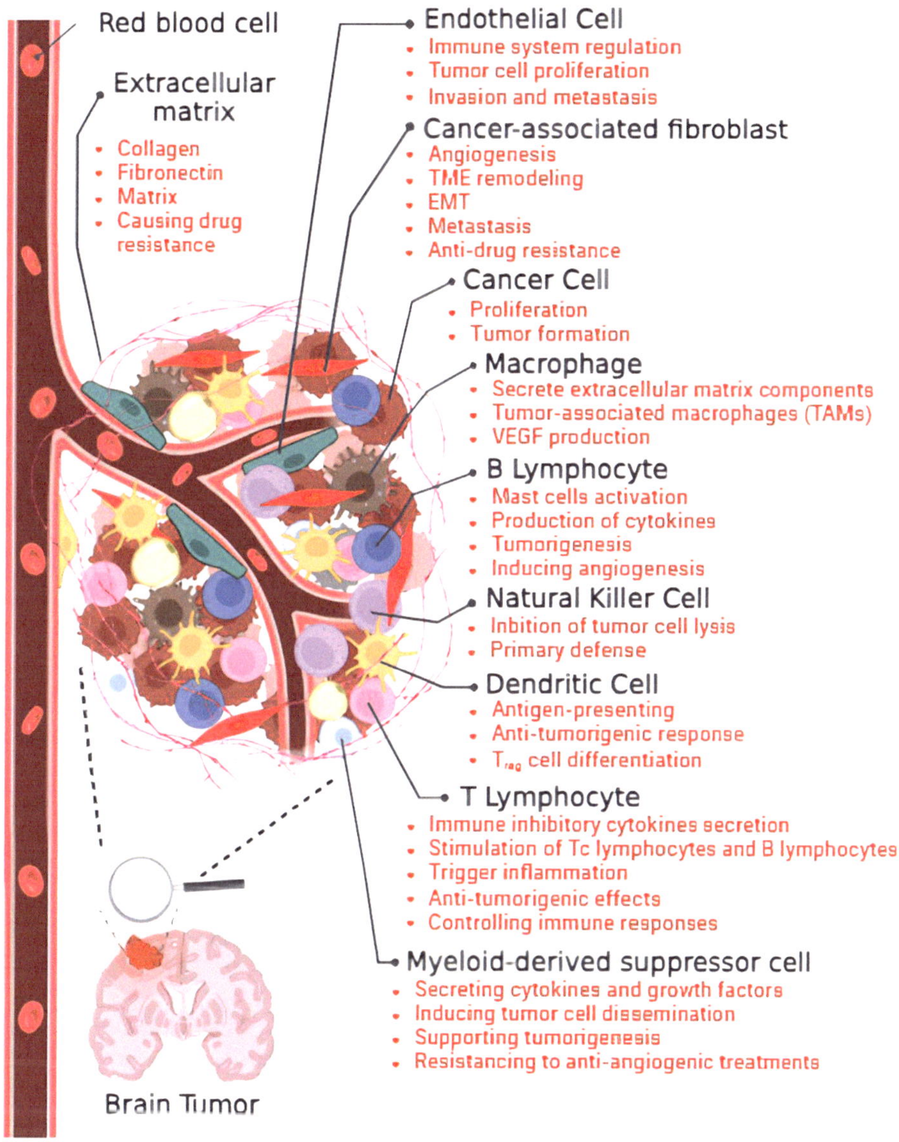

Figure 1.
Illustration of cellular and non-cellular of the tumor microenvironment.

certain cancer cells as the "seeds" and specific tissue microenvironments as the "soil" still retains its thought [13]. In further studies, it has become important to investigate the role of TME-associated cellular responses in carcinogenesis. Dysregulated immune responses are the main characteristics of inflammatory tissue microenvironment. TME exhibits an increase in inflammatory regulators such as tumor-infiltrating cells, T cells, macrophages, dendritic cells, and natural killer cells that are the cause of chronic inflammation to promote tumor progression, carcinogenesis, and metastasis [14]. A suppressed immune system, defined as immunosuppression, is one of the potential adverse effects of several cancers.

2.1 Cellular components of tumor microenvironment

2.1.1 Fibroblasts

Cells of connective tissue that are neither epithelial nor endothelial nor immune cells are referred to as fibroblasts [15]. Fibroblasts are substantial cells involved in extracellular matrix formation by synthesizing collagen, laminin, and fibronectin and thus contribute significantly to wound healing [16]. Resident fibroblasts, resident endothelial cells, and epithelial cells can be activated in tumorigenesis and differentiate into cancer-associated fibroblasts (CAFs) in response to growth factors and cytokines. Cancer-associated fibroblasts are the primary cell type in the stromal part of the tumor microenvironment [17]. These differentiated fibroblasts contribute to inflammation, angiogenesis, TME remodeling, epithelial-mesenchymal transition (EMT), metastasis and anti-drug resistance *via* producing matrix metalloproteinases (MMPs), growth factors (TGF-β, IGF, HGF, VEGF), cytokines and chemokines (IL-6, CCL7, CXCL2) and inhibiting natural killer cells, and cytotoxic T lymphocytes [9, 16, 18, 19]. Therefore, they promote tumor development and progression and are shown to be related to poor prognosis in pancreatic and breast cancers [17].

2.1.2 Endothelial cells

Endothelial cells are the chief cells that line the blood vessels, forming a cellular layer between the blood and the tissue, which organizes the interchange of materials from each side. Endothelial cells differentiate into tumor endothelial cells (TECs) in tumors that organize and play critical roles in immune system regulation, tumor cell proliferation, invasion, and metastasis by expression of inhibitory receptors for immune cells, vascular cell adhesion molecule 1 (VCAM1), and Notch1 signaling [10, 17].

2.1.3 Adipocytes

Adipocytes convert glucose and fatty acid to triglycerides in a mechanism called lipogenesis and, in this way, provide the energy demand of other cells [20]. In the tumor microenvironment, cancer-associated adipocytes (CAAs) are likewise energy storage cells [21] by supplying lipids to tumorigenic cells [10]. Apart from lipid and energy supply, they release various metabolites (fatty acids), hormones (leptin, adiponectin), proteins (collagen, VEGF, and matrix metalloproteinases), inflammatory factors (CCL2), and cytokines (IL-8), which are referred to as adipokines [9, 22, 23]. Through these adipokines, they regulate macrophage differentiation [9], immune cell homeostasis [23], immune evasion and tumor progression [9, 22], ECM remodeling [22], metastasis [24], and angiogenesis *via* increasing vascularization [25]. Many cancer types, such as breast, ovarian, colon, and prostate cancer, are primarily known for high adipocyte concentration; therefore, it is important to understand the interactions of the cells in TME with adipocytes to improve the therapeutics [26].

2.1.4 B lymphocytes

Although B lymphocytes' primary function is antibody secretion upon antigen recognition, in the tumor microenvironment, they activate mast cells and promote the production of cytokines when they interact with the antigens of tumor cells [27]. B lymphocytes have mainly antitumor functions, whereas when specific chemokines

(CXCL13) and cytokines (IL-10) are present, they promote tumorigenesis and induce angiogenesis [9]. Thus, the role and function of B lymphocytes in the tumor microenvironment differ in accordance with the chemical secretions in the environment and tumor type, which affects the tumor's prognosis status [28].

2.1.5 T lymphocytes

T lymphocytes constitute one of the immune system's main groups of adaptive immune response. T lymphocytes have many subgroups, of which $CD8^+$ cytotoxic T lymphocytes (T_c) are responsible for the detection of tumor antigens and for destroying the tumor cells [18]. In the tumor microenvironment, as well as T_c lymphocytes, $CD4^+$ T helper lymphocytes (T_h) also play essential promoting roles as both stimulation of T_c lymphocytes and B lymphocytes *via* secreting various cytokines [9]. On the contrary, another subgroup of T helper lymphocytes, T_h17 lymphocytes, exerts double-sided effects. T_h17 lymphocytes trigger inflammation and angiogenesis on the one hand and have anti-tumorigenic effects on the other [29].

The common immunosuppressor T cells, regulatory T cells (T_{regs}), are typically identified near tumors and suppress anticancer immune responses [30]. In normal cells, T_{regs} sustain self-tolerance mechanisms by preventing the proliferation of T lymphocytes and induction of immunity against self-antigens [31]. Studies indicated that the $FOXP3^+CD4^+CD25^+$ subtype of T_{reg} cells is instrumental in controlling immune responses in the tumor microenvironment. Tumor cells and macrophages in the tumor microenvironment secrete CCL22 chemokine, whose receptor, CCR4, was expressed by T_{reg} cells and causes the accumulation of T_{reg} cells in TME [30]. T_{regs} contribute to immunosuppression by secreting immune inhibitory cytokines such as transforming growth factor (TGF-β) and interleukin 10 (IL-10), which inhibit the anti-tumor activity of killer cells [27].

2.1.6 Macrophages

Macrophages are the phagocytic cells in the innate immune system [32]. However, when recruited to the tumor microenvironment with certain cytokines or differentiated from circulating monocytes, they can acquire different phenotypes and are referred to as tumor-associated macrophages (TAMs) after that [17]. These different phenotypes are known as M1 anti-tumor macrophages and M2 pro-tumor macrophages, which are determined by the polarization status of the TAMs [9]. TAMs, particularly the M2 phenotype, are the cells that are the most infiltrated immune system cells in TME [33]. Besides, since they secrete extracellular matrix components, they participate in ECM remodeling [19].

2.1.7 Dendritic cells

Dendritic cells are the antigen-presenting cells in the immune system that detect foreign antigens, present them to T lymphocytes and induce adaptive immune response [34]. In the tumor microenvironment, different subtypes of dendritic cells can be infiltrated. Classical dendritic cells and plasmacytoid dendritic cells are responsible for anti-tumorigenic response. On the other hand, tumor cells secrete CCL-2, CXCL1, and CXCL5, thus preventing dendritic cell maturation [9]. Further, dendritic cells promote T_{reg} cell differentiation by secreting cytokines [10]. In this manner, dendritic cells' function depends on the chemicals secreted in TME [22].

2.1.8 Natural killer (NK) cells

Natural killer (NK) cells are the primary defense cells of the immune system against tumor cells [35]. When they encounter tumor cells in the bloodstream, natural killer cells essentially destroy them, whereas they cannot function effectively in the tumor microenvironment [22]. Although the dendritic cells recruit natural killer cells to the tumor microenvironment through various chemokines such as IFN-γ, they cannot correctly mature [9].

2.1.9 Neutrophils

Neutrophils, widespread in circulation [35], are the cells of the innate immune system responsible for the phagocytosis of dead cells or pathogens and for presenting antigens to adaptive immune system cells [9, 36]. In the case of tumorigenesis, neutrophils are recruited to the tumor site. In the early stages of tumorigenesis, these neutrophils are differentiated into the anti-tumor N1 subtype under the influence of IFN-γ [37, 38]. N1 subtype neutrophils release cytokines and reactive oxygen species, thus inducing apoptotic death of tumor cells [22]. In later stages of tumorigenesis, cancer-associated fibroblasts (CAFs) secrete TGF-β that promotes the differentiation of N2 subtypes and suppression of N1 subtype [28]. These N2 neutrophils secrete cytokines as IL-1β, growth factors as vascular endothelial growth factor (VEGF), and matrix metalloproteinases as MMP-9, in this way, stimulate inflammation, angiogenesis, matrix remodeling, and invasion [22, 28, 37].

2.1.10 Myeloid-derived suppressor cells (MDSCs)

MDSCs secrete cytokines and growth factors; in this way, they suppress immune cells and induce tumor cell dissemination and renewal of tumor cells [27]. They originate from myeloid progenitor cells in the bone marrow and then migrate to the tumor microenvironment to support tumorigenesis, angiogenesis, and metastasis [9]. In TME, they also differentiate into tumor-associated macrophages (TAMs), which act together responsible for resistance to anti-angiogenic treatments [17].

2.2 Non-cellular components of tumor microenvironment

Non-cellular components of the tumor microenvironment involve mainly the extracellular matrix (ECM), which is different from the matrix of normal tissue in that it enables the neoplastic cells to metastasize. Collagen and fibronectin are the main components of ECM in solid tumors and ensure the endurance of the matrix [17]. Factors secreted from cellular components of the tumor microenvironment, such as cytokines, chemokines, proteases, growth factors, polysaccharides, and integrins, are also significant members, and these components provide the communication and interaction of cells with each other in TME [39, 40]. The importance of non-cellular components of the tumor microenvironment is that they organize not only the scaffold where the tumor mass resides but also regulate the polarization and the fate of the cellular components [41]. Non-cellular components of ECM also participate in drug resistance by providing a dense environment and disallowing the drugs to penetrate [17].

To summarize, the cancer tumor microenvironment is characterized by abundant stroma, extracellular matrix, stellate cells, cancer-associated fibroblasts, various immune cells, and cytokines are associated with a hypoxic environment, high vascularity and intense immunosuppression, proliferation, metastasis, and drug resistance of several cancers [36]. Various studies explained the direct interaction between tumor cells and immune cells, but the relationship between immune cells and TME needs to be clarified in the cell fate decision.

3. Cell culture applications studying TME

To improve personalized therapy and the prognosis of cancer, extensive and varied investigations are performed from experimental to clinical applications. However, the complexity and heterogeneity of TME contribute to cancer development, progression, and drug resistance. Thus, developing cell culture applications is required to enlighten the interaction network of TME and cancer cells.

Since it was first introduced in 1907 for nerve fibers [42], cell culture techniques have been eased in cancer research to mimic the *in vivo* evolvement and reaction of tumors *in vitro* [43]. Classical two-dimensional (2D) cell cultures allow the cell to grow only in two dimensions; in other words, cells can only grow along the surface of the culture flask (**Figure 2**) [44]. These cultures may be conducted in adherent cultures and suspension cultures, of which the latter imitates the non-adhesive cells, such as hematologic malignancies [45], to resemble the natural environment in blood or lymph [43]. In adherent cultures, cells are attached to the plastic surface of flasks or dishes as a single layer and grow, interacting with the surface's neighboring cells and protein coating. These cultures provide a cost-effective, high-performance, and reproducible way to study *in vitro* [46]. For many years, these cultures have been used to investigate the cytotoxicity of drugs in

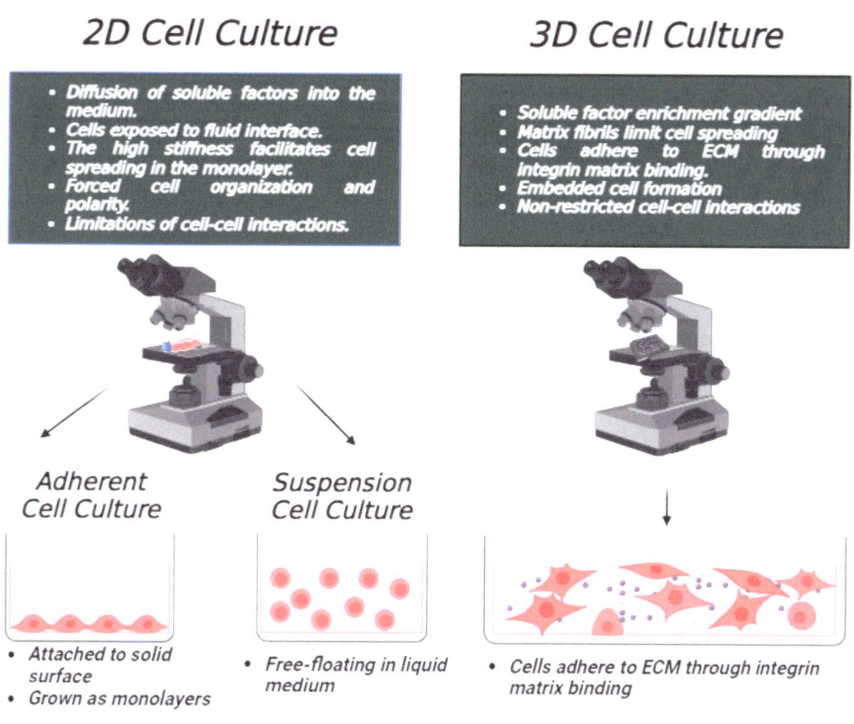

Figure 2.
Comparison of 2D vs. 3D cell culture.

pharmacology studies, the migration and invasion characteristics of the cells in understanding the underlying mechanisms, the genome modifications in enlightening the genetic basis of cancer cells, and so on [47]. However, studying just one cell type in a 2D cell culture is nowhere near the interactions between cellular and non-cellular components of the tumor microenvironment [48]. Therefore, new techniques and applications must be developed to mimic the *in vivo* tumor microenvironment better.

3.1 *In vitro* co-culture applications for TME and cancer cell interactions

The co-culture method is the direct or indirect interaction of multiple cell types in an identical culture environment [49]. Through the interaction of various cell types, co-culture systems regulate the properties of individual cells. Cellular mechanisms, including paracrine signaling, cell-to-cell communication, and modeling epithelial-mesenchymal transition (EMT), are better estimated by co-culturing numerous cell lines into a single three-dimensional (3D) model [50].

Two types of cells are defined as "target cells" and "assisting cells" responsible for the co-culture environment. Target cells act to build up the engineered tissue and are responsible for the mechanical and metabolical functions of the tissue. Target cells are guided by assisting cells to exhibit various desired characteristics such as proliferation, differentiation, matrix production, or secretion [49].

Co-culture aims to form tissue-like characters of cells *in vitro* through cell–cell communication of multiple cell types in direct or indirect interaction. Co-cultures are used to explore three cell-to-cell communication patterns: cell–cell adhesion, including adherence, gap and tight junctions, cell-ECM adhesion, and paracrine signaling *via* soluble factors.

Co-culture techniques have been essential for researching the interaction between different cell types under diverse conditions, such as cancer cell growth, differentiation, proliferation, and metastasis. The complexity of the *in vivo* environment, where several cells interact and signal concurrently, is not fully mimicked by one cell type-based *in vitro* cell cultures due to inadequate prediction of signaling cascades. Therefore, several co-culture models have been improved to mimic TME, which are less expensive than *in vivo* animal models.

3.1.1 Direct co-cultures

Direct co-cultures allow two or more different cells to make physical connections with each other while mixing in the 2D or 3D cell culture environment. Mixed populations of monolayer cells on slides or in dishes are standard 2D cell culture methods [51].

Due to their simplicity and ease of control, 2D culture systems are acceptable as the easiest way to investigate cellular interactions, such as examining adhesion molecules, cytokine production, and juxtracrine signals. However, the composition of native tissues can be modeled using 3D culture systems, which is accomplished by culturing multiple cells in materials such as fibrin, agarose, collagen, or alginate [49].

3.1.2 Indirect co-cultures

Additionally, conditioned medium is widely employed in indirect co-cultures. Multiple cell types are separated through good inserts like Transwell or Boyden chamber permeable membranes or pre-conditioned media. Utilizing permeable membrane inserts allows secreted soluble factors to pass through the membrane, only enabling signaling *via* the cell secretome. Alternatively, utilizing previously created conditioned media from one of the co-culture cell types are used [52].

Indirect 2D co-culture systems allow a better explanation of specific cellular functions and features that occur specifically to cell types. However, in indirect 3D co-culture systems, multiple cell types are separated using hydrogel or collagen matrix, which allows better mimicking of native tissue structures and eliminates the requirement for direct cell-to-cell contact between cell types [49].

Transwell plates [53, 54], microfluidic platforms [55–58], or solid supports like Petri dishes [59, 60], 3D scaffolds [61], hydrogels [62], or microarrays [63] are often utilized choices for co-culturing populations that are somewhat separated from one another (**Figure 3**). For co-cultures, many custom microfluidic devices have been

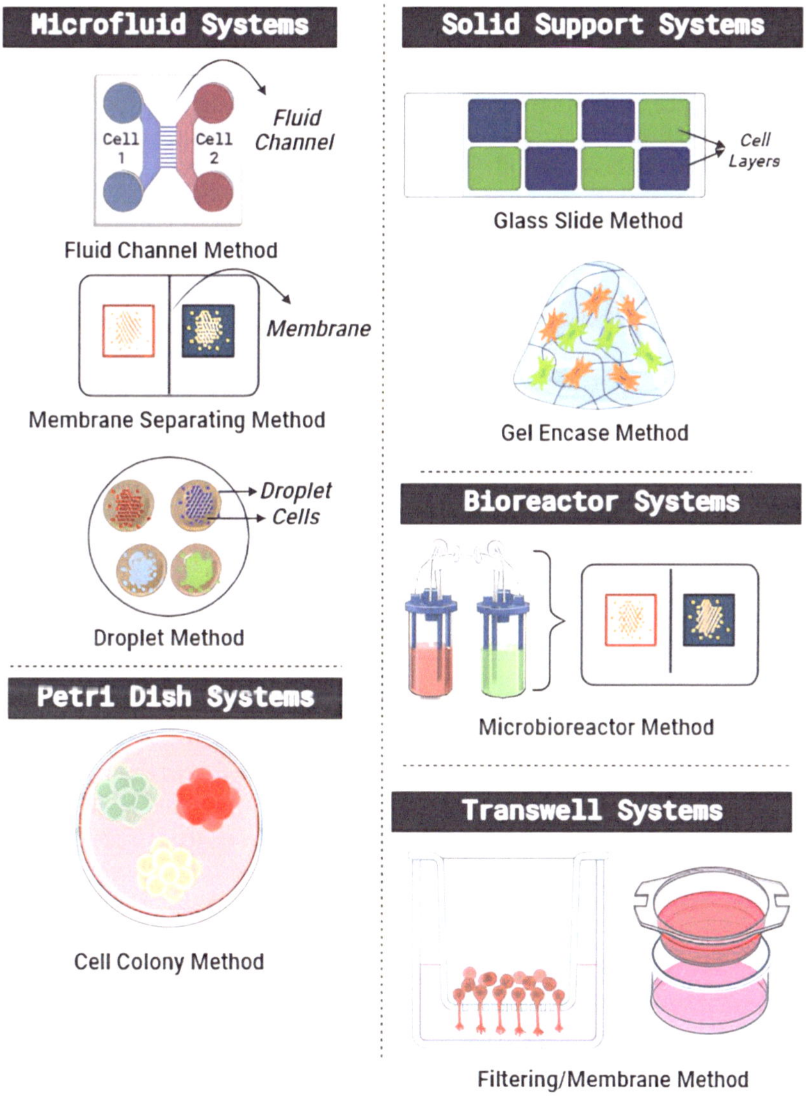

Figure 3.
Available technologies for co-culture [64].

developed [65–69]. The possibility for high throughput in microfluidics is one of its key benefits. Microfluidic setups might not represent larger volumes as complexity rises [56], and this benefit may be undermined. Techniques used in monoculture modified for co-culture include cell movement experiments. Most traditional laboratory cultivation focuses on maintaining pure cultures of certain species [65, 70]. Although there have been recent developments that could be further utilized, such as microcarrier beads, the micro-Petri dish [71], diffusion chambers, dialysis reactors, and other techniques [70, 72], the cultivation of monocultures on agar plates and in liquid medium has always followed the same paradigm [71, 72].

Cell-to-cell direct co-cultures have some limitations and need more optimization steps. Therefore, indirect co-cultures are preferred as they provide reproducible results. However, indirect co-cultures cannot capture cell–cell adhesion, integrin, and notch-type signaling.

3.2 Adipocytes-related cancer studies

Since TME has become widely recognized as a significant contributor to tumor progression and cancer, it has gained importance to determine the roles of individual TME components in cancer. Adipocytes, endothelial cells, fibroblasts, macrophages, and muscle cells are cell types commonly studied in co-culture systems [73]. Adipocytes were subsequently recognized as a significant TME component in tumor development, angiogenesis, and metastasis through the secretion of soluble factors [74]. Adipocytes were once considered the body's energy reserves, but today, it is recognized that they have a role in various pathological processes, such as inflammation. In particular, they impact gene expression and cell motility in TME [75].

Mammalian epithelial cells are surrounded by an extracellular environment comprising adipocytes and other stromal cells, including fibroblasts, and endothelial and inflammatory cells. Adipocytes and adipocyte precursor cells are a substantial percentage of breast TME, which may play a significant role in the interaction between stromal and ductal epithelial cells and function in the secretion of growth factors, hormones, and adipokines [76].

According to the studies in cancer metastasis, it was determined that free fatty acids (FFAs) released from adipocytes through lipolysis may have been an energy source for cancer cells through mitochondrial fatty acid oxidation (FAO) [77]. In culture conditions, FFAs from adipocytes to cancer cells enhance FAO and support tumor cell proliferation, migration, and invasion. Elevating lipid load in adipocytes increases FFA transfer and is stored in tumor cells with lipid droplets, which is also associated with enhancing CAA-mediated proliferative effects in the TME [78].

In the first studies investigating the role of TME in breast cancer, the role of fibroblasts, macrophages, and other inflammatory cells was investigated. Still, the role of adipocytes was not recognized. Further studies in breast cancer have shown that the adipocyte-rich environment can promote cancer proliferation. It was shown that murine mammary carcinoma SP-1 cells injected subcutaneously into adipose-rich tissues in mice promoted tumor growth and metastasis. Still, neither growth nor metastasis is observed in tissues with low levels of adipose. These findings were reinforced by the discovery that estrogen regulates the adipocyte-associated growth and proliferation of SP-1 cells [79, 80].

It has been determined by DNA microarray studies that adipokines secreted from adipocytes regulate cell survival, proliferation, invasion, and angiogenesis by inducing IGF2, FOS, JUN, cyclin D1, MMP1, ATF3, and NFkβ transcriptional target molecules. In addition, the induction of these genes was specific to factors released by adipocytes and cannot be observed in other breast stromal cells [74].

Previous studies showed that co-culture of differentiated murine preadipocytes with mammary tumor cells caused the delipidation of adipocytes with the increase of proinflammatory cytokines and matrix remodeling proteins while decreasing adipocyte differentiation markers, including resistin, adiponectin, and hormone-sensitive lipase (HSL). These adipocyte cells, triggered by tumor cells and changed phenotypically and functionally, are considered cancer-associated adipocytes (CAA) [81]. In addition, bidirectional contact between adipocytes and tumor cells changes the shape of adipocytes into adipocyte-derived fibroblasts, which are the parts of the CAAs and induce invasion, migration, and metastasis [82]. Studies have shown that interleukins produced by CAAs such as IL6, IL1β, TNFα, and MMP-11 are functionally responsible for supporting the invasion capacity of tumor cells and augmenting the proinflammatory TME properties. It was shown that pancreatic cancer cells linked to obesity release cytokines such as IL1β, which promotes infiltration of immune cells and leads to cancer progression and drug resistance [83]. Numerous pro-inflammatory cytokines, such as tumor necrosis factor alfa (TNF-α) and plasminogen activator inhibitor-1 (PAI-1), are produced by adipocytes and overexpressed in obese patients [84].

The methods by which stromal cells in TME alter tumor properties in different microenvironments are initiated to be clarified by several direct or indirect studies. Stromal cell-secreted factors such as matrix metalloproteases (MMPs) and growth factors regulate the cancer cell decision. As mentioned in previous studies, the interaction between adipocytes and cancer cells causes the change in the phenotype of adipocytes into CAA and delipidation with the overexpression of inflammatory cytokines but increases in the adipocyte differentiation markers [81]. Secretion of inflammatory cytokines such as IL-6, IL1β, or leptin promotes cancer development and metastasis in several cancer cells.

Even if the interaction between adipocytes and cancer cells is known, these cell–cell interactions are complicated, and it is also known that they cannot occur by the effect of a single secreted molecule or a signal. Therefore, molecular mechanisms of heterotypic signaling in bidirectional crosstalk between adipocytes and cancer cells are frequently investigated with *in vitro* and *in vivo* studies to enlighten carcinogenesis and metastasis [76]. Adipocytes and breast cancer cells are immediately near one another due to invasive breast tumors that penetrate the basement membrane and infiltrate fibrous tissue barriers, enabling paracrine interaction between two cell types [85].

To understand how adipocytes affected the motility of epithelial and cancer cells and the role of adipocytes on the cancer-related gene expressions on co-culture systems in TME, adipocyte-conditioned media treated MCF-10A mammary epithelial cells and invasive MCF-10CA1 carcinoma cells were used. The first step of the study was the differentiation of primary human breast preadipocytes into lipid-accumulating mature adipocytes. Preadipocytes were cultured in a differentiation medium for seven days to do this experiment. After differentiation, adipocytes were maintained in DMEM-F12 containing 1% (penicillin–streptomycin fungizone) PSF, 10% (fetal bovine serum) FBS, and five ng/ml insulin for a week. To compare the role of TME in normal and cancer cell motility, cells were treated with adipocyte-conditioned media. For the formation of conditioned medium (CM), preadipocytes were serum-starved in DMEM: F12 containing 1%PSF for 24 h. Meanwhile, MCF-10A and MCF-10CA1 cells were serum-starved overnight and treated with CM from preadipocytes and mature adipocytes for 24 h. It was found that mature adipocyte-conditioned media increases cancer cell motility and migration. In addition, mature adipocyte-CM increases phosphorylated-Akt levels in MCF-10CA cells. However, angiogenesis (fibroblast growth factor receptor-2, interferon-a, and platelet-derived growth factor β-peptide) and matrix metalloproteinases (MMP-1, −2, and − 9) genes were downregulated in adipocyte CM-treated MCF-10A cells, in adipocyte-CM treated MCF-10ACA1 cancer cells, PAI-1, TGF-B-1 expression upregulated but MMP1 and angiopoietin-1 were downregulated. These data confirmed the positive effect of adipocytes on cell survival, proliferation, migration, and tumorigenesis in breast cancer (**Figure 4**) [86].

The dysregulation of adipokine synthesis and secretion is a significant risk factor for obesity, and Type 2 diabetes mellitus (T2DM) results from obesity. Given that T2DM is caused by insulin resistance, insulin promotes adipokine production, leading to cancer development [87]. Considering T2DM is related to cancer, metformin is used as an oral antidiabetic drug explaining the role of adipocytes in inflammation-associated cancer development. To find out how adipokine production is associated with obesity and T2DM, the molecular mechanism in insulin resistance and breast cancer development needs to be clarified. Similar to other studies, preadipocytes were differentiated for 14 days, and after differentiation, adipocyte cells were propagated in low glucose (5 mM) conditions for 48 h. Then, MCF-7 cells were treated with

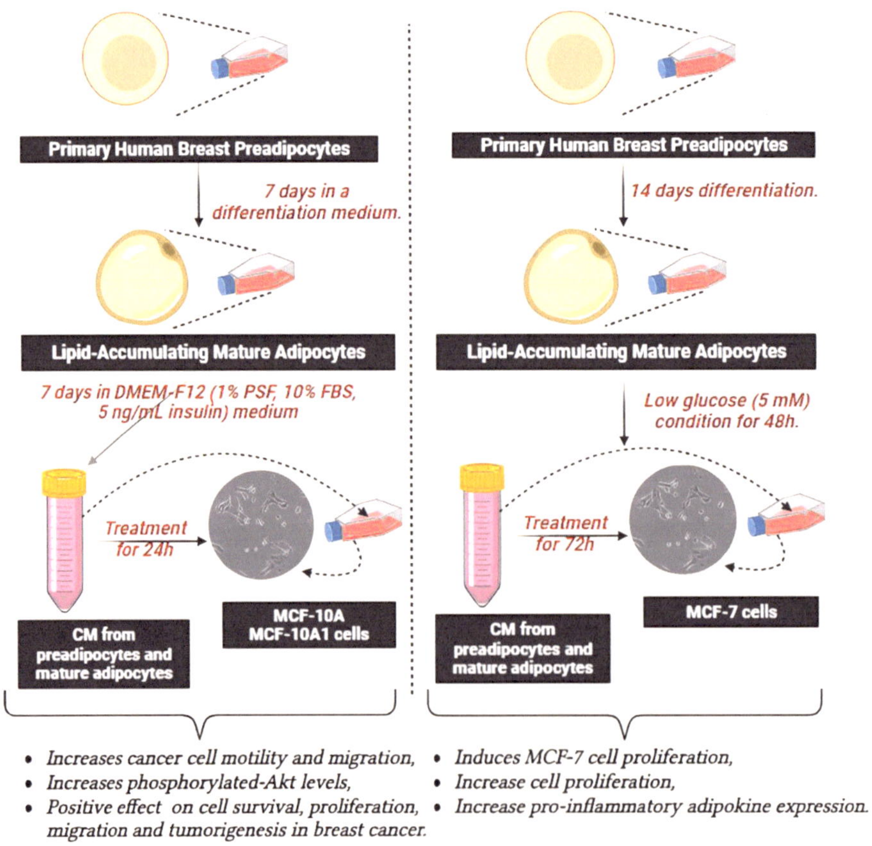

Figure 4.
A workflow example in adipocytes-related cancer studies.

preadipocyte and adipocyte-CM for 72 h. It was found that adipocyte-derived CM significantly induced MCF-7 cell proliferation that preadipocyte-derived CM did not. In addition, IL-6 expression was increased in differentiated mature adipocytes, which are downregulated by metformin. Findings suggested that conditioned media from adipocytes increase cell proliferation and pro-inflammatory adipokine expression, which are decreased by metformin treatment (**Figure 4**) [88].

Utilizing direct or indirect 2D co-culture, in which tumor and stromal cells are cultured together on a flat, has begun to be inadequate for explaining the complex intercellular interactions between heterogeneous cancer, immune and stromal cells, and between cells and ECM [89]. However, compared to 2D culture, 3D culture is the most commonly effective method for constructing tissue architecture lacking monolayer cultures and better recapitulating the effect of TME. The 3D cell culture enables the investigation of cell function, gene expression, and paracrine signaling effects.

Organoids are one of the improved *in vitro* tissue engineering and 3D-cell culture applications that represent the structure and function of the organ *in vivo*, such as a population of self-organizing stem cells developing into specific complex organs [90]. Organoids can be produced through cell signaling pathways regulating self-organization and differentiation of embryonic stem cells (ESCs), induced pluripotent stem cells (iPSCs), and adult stem cells (ASCs) for the establishment of complex organs. For successful organoid formation and growth, improved cell culture conditions activate cellular signaling pathways triggered by intrinsic factors and extrinsic mediators, such as extracellular matrix (ECM), to maintain stem cell function [91]. Since the cancer organoids have the same characteristics and genetic background as the original tumors, co-culture cancer organoid models have been sufficiently developed to mimic the heterogeneous environment *in vitro* to investigate the complex interaction network of cellular and non-cellular components in TME. However, maintaining different cancer organoids requires cancer-specific procedures and methods.

In recent years, many studies have been conducted on organoids for cancer research. In such a study, biopsies of metastatic lesions of colorectal cancer patients were collected, and patient-derived organoids (PDOs) were generated in cultures. These PDOs were tested to consider the sensitivity of patients to chemotherapy and observed that PDOs can be used to foresee which patients bearing metastatic colorectal cancer are insensitive to irinotecan-based chemotherapy [92]. In another study, organoid models were formed from human breast epithelial samples and used to understand the molecular mechanisms underlying breast cancer oncogenesis. Four breast cancer-associated tumor suppressor genes were knocked out by CRISPR-Cas9 in mammary progenitor cells, and observed that at least three tumor suppressor genes must be blocked for breast tumorigenesis [93]. Wang et al. carried out a study with PDOs from hepatocellular carcinoma (HCC) patients to demonstrate the relationship between the presence of CD44 surface marker and resistance to sorafenib treatment in HCC patients. They also indicated that blocking Hedgehog signaling can reverse this sorafenib resistance [94]. Another study used PDOs generated from colorectal cancer patients for genomic and transcriptomic analysis to assess multi-drug response. It is indicated that these PDOs can be used to predict the response of the patient to drug treatment and develop personalized therapies [95].

The method is frequently initiated by mechanical or enzymatic digestion of tumor tissues obtained from the minimum necrotic margins into pieces about 1 mm in diameter. Following the breaking down of tumor samples, it is required to seed the cells with supplemented media onto a 3D Matrigel scaffold matrix, which consists of ECM proteins such as laminin, entactin, proteoglycan, and collagen IV and can be improved by tumor-specific growth factors such as Wnt3A, transforming growth factor beta (TGF-B) receptor inhibitor, and Noggin of epidermal growth factor (EGF) [96]. Organoid cells are co-cultured with TME components through this process.

4. Conclusion

Identifying the complex network between stromal cells and cancer cells in the TME is a potential hallmark to enlighten cancer progression and metastasis. This chapter summarizes the role of TME in various cancers by describing current co-culture cell culture applications.

The critical ideas of TME and its possible uses in cell culture are outlined in this chapter. This chapter concentrates on the research on TME in adipocytes and cancer cells since TME is highly complicated. The components of TME and emerging cell culture applications that have a great deal of promise to represent different malignancies are outlined in this chapter. We also provide a summary of co-culture's uses in cancer research. Additionally, this chapter offers more information on cutting-edge TME investigations based on cell culture.

Acknowledgements

We acknowledge İrem Nur Ateş and Deniz Turan for their valuable contribution to constructing this chapter.

Conflict of interest

The authors declare no conflict of interest.

Author details

Ozge Rencuzogullari, Pelin Ozfiliz-Kilbas*, Enes Bal and Burcu Ayhan-Sahin
Department of Molecular Biology and Genetics, Istanbul Kultur University, Istanbul, Turkey

*Address all correspondence to: p.ozfiliz@iku.edu.tr

References

[1] Kontomanolis EN, Koutras A, Syllaios A, et al. Basic principles of molecular biology of cancer cell-molecular cancer indicators. Journal of Balkan Union of Oncology. 2021;**26**:1723-1734

[2] Vaghari-Tabari M, Ferns GA, Qujeq D, et al. Signaling, metabolism, and cancer: An important relationship for therapeutic intervention. Journal of Cellular Physiology. 2021;**236**:5512-5532

[3] Zaimy MA, Saffarzadeh N, Mohammadi A, et al. New methods in cancer diagnosis and gene therapy of cancer-based on nanoparticles. Cancer Gene Therapy. 2017;**24**:233-243

[4] Song Q, Merajver SD, Li JZ. Cancer classification in the genomic era: Five contemporary problems. Human Genomics. 2015;**9**:27

[5] Mendiratta G, Ke E, Aziz M, et al. Cancer gene mutation frequencies for the U.S. population. Nature Communications. 2021;**12**:5961

[6] Hanahan D. Hallmarks of cancer: New dimensions. Cancer Discovery. 2022;**12**:31-46

[7] Ren B, Cui M, Yang G, et al. The tumor microenvironment participates in the metastasis of pancreatic cancer. Molecular Cancer. 2018;**17**:108

[8] Wang S, Li Y, Xing C, et al. Tumor microenvironment in chemoresistance, metastasis and immunotherapy of pancreatic cancer. American Journal of Cancer Research. 2020;**10**:1937-1953

[9] Bożyk A, Wojas-Krawczyk K, Krawczyk P, et al. Tumor microenvironment—A short review of cellular and interaction diversity. Biology (Basel). 2022;**11**:929

[10] Mun J-Y, Leem S-H, Lee JH, et al. Dual relationship between stromal cells and immune cells in the tumor microenvironment. Frontiers in Immunology. 2022;**13**:864739

[11] Xu Z, Vonlaufen A, Phillips PA, et al. Role of pancreatic stellate cells in pancreatic cancer metastasis. The American Journal of Pathology. 2010;**177**:2585-2596

[12] Paget S. The distribution of secondary growths in cancer of the breast. Lancet. 1889;**133**:571-573

[13] Fidler IJ. The pathogenesis of cancer metastasis: The 'seed and soil' hypothesis revisited. Nature Reviews. Cancer. 2003;**3**:453-458

[14] Colotta F, Allavena P, Sica A, et al. Cancer-related inflammation, the seventh hallmark of cancer: Links to genetic instability. Carcinogenesis. 2009;**30**:1073-1081

[15] Chen Y, McAndrews KM, Kalluri R. Clinical and therapeutic relevance of cancer-associated fibroblasts. Nature Reviews. Clinical Oncology. 2021;**18**:792-804

[16] Shiga K, Hara M, Nagasaki T, et al. Cancer-associated fibroblasts: Their characteristics and their roles in tumor growth. Cancers (Basel). 2015;**7**:2443-2458

[17] Wu T, Dai Y. Tumor microenvironment and therapeutic response. Cancer Letters. 2017;**387**:61-68

[18] Arneth B. Tumor microenvironment. Medicina (B Aires). 2019;**56**:15

[19] Guo S, Deng C-X. Effect of stromal cells in tumor microenvironment on metastasis initiation. International Journal of Biological Sciences. 2018;**14**:2083-2093

[20] Church C, Horowitz M, Rodeheffer M. WAT is a functional adipocyte? Adipocytes. 2012;**1**:38-45

[21] Pallegar NK, Christian SL. Adipocytes in the tumor microenvironment. Advances in Experimental Medicine and Biology (Cham: Springer). 2020:1-13

[22] Anderson NM, Simon MC. The tumor microenvironment. Current Biology. 2020;**30**:R921-R925

[23] Kedia-Mehta N, Finlay DK. Competition for nutrients and its role in controlling immune responses. Nature Communications. 2019;**10**:2123

[24] Wu Q, Li B, Li Z, et al. Cancer-associated adipocytes: Critical players in breast cancer progression. Journal of Hematology & Oncology. 2019;**12**:95

[25] Wagner M, Steinskog ES, Wiig H. Blockade of lymphangiogenesis shapes tumor-promoting adipose tissue inflammation. The American Journal of Pathology. 2019;**189**:2102-2114

[26] Mukherjee A, Bilecz AJ, Lengyel E. The adipocyte microenvironment and cancer. Cancer Metastasis Reviews. 2022;**41**:575-587

[27] Liu Q, Liao Q, Zhao Y. Chemotherapy and tumor microenvironment of pancreatic cancer. Cancer Cell International. 2017;**17**:68

[28] Galli F, Aguilera JV, Palermo B, et al. Relevance of immune cell and tumor microenvironment imaging in the new era of immunotherapy. Journal of Experimental & Clinical Cancer Research. 2020;**39**:89

[29] Pan Y, Yang W, Tang B, et al. The protective and pathogenic role of Th17 cell plasticity and function in the tumor microenvironment. Frontiers in Immunology. 2023;**14**:1192303

[30] Nishikawa H, Sakaguchi S. Regulatory T cells in tumor immunity. International Journal of Cancer. 2010;**127**:759-767

[31] Kondělková K, Vokurková D, Krejsek J, et al. Regulatory T cells (Treg) and their roles in immune system with respect to immunopathological disorders. Acta Medica (Hradec Kral Czech Republic). 2010;**53**:73-77

[32] Hirayama D, Iida T, Nakase H. The phagocytic function of macrophage-enforcing innate immunity and tissue homeostasis. International Journal of Molecular Sciences. 2017;**19**:92

[33] Chen D, Zhang X, Li Z, et al. Metabolic regulatory crosstalk between tumor microenvironment and tumor-associated macrophages. Theranostics. 2021;**11**:1016-1030

[34] Kvedaraite E, Ginhoux F. Human dendritic cells in cancer. Science Immunology. 2022;7:70

[35] Corn KC, Windham MA, Rafat M. Lipids in the tumor microenvironment: From cancer progression to treatment. Progress in Lipid Research. 2020;**80**:101055

[36] Hinshaw DC, Shevde LA. The tumor microenvironment innately modulates cancer progression. Cancer Research. 2019;**79**:4557-4567

[37] Xiao Y, Yu D. Tumor microenvironment as a therapeutic target

in cancer. Pharmacology & Therapeutics. 2021;**221**:107753

[38] Coffelt SB, Wellenstein MD, de Visser KE. Neutrophils in cancer: Neutral no more. Nature Reviews. Cancer. 2016;**16**:431-446

[39] Mbeunkui F, Johann DJ. Cancer and the tumor microenvironment: A review of an essential relationship. Cancer Chemotherapy and Pharmacology. 2009;**63**:571-582

[40] Mongiat M, Andreuzzi E, Tarticchio G, et al. Extracellular matrix, a hard player in angiogenesis. International Journal of Molecular Sciences. 2016;**17**:1822

[41] Zamarron BF, Chen W. Dual roles of immune cells and their factors in cancer development and progression. International Journal of Biological Sciences. 2011;**7**:651-658

[42] Harrison RG, Greenman MJ, Mall FP, et al. Observations of the living developing nerve fiber. The Anatomical Record. 1907;**1**:116-128

[43] Kapałczyńska M, Kolenda T, Przybyła W, et al. 2D and 3D cell cultures – A comparison of different types of cancer cell cultures. Archives of Medical Science. 2018;**14**:910

[44] Charwat V, Egger D. The Third Dimension in Cell Culture: From 2D to 3D Culture Formats. Springer Nature, Springer International Publishing AG; 2018. pp. 75-90

[45] Flobak Å, Skånland SS, Hovig E, et al. Functional precision cancer medicine: Drug sensitivity screening enabled by cell culture models. Trends in Pharmacological Sciences. 2022;**43**:973-985

[46] Langhans SA. Three-dimensional in vitro cell culture models in drug discovery and drug repositioning. Frontiers in Pharmacology. 2018;**9**:6

[47] Ballav S, Deshmukh AJ, Siddiqui S, et al. Two-dimensional and three-dimensional cell culture and their applications. In: Cell Culture - Advanced Technology and Applications in Medical and Life Sciences. IntechOpen; 2022. DOI: 10.5772/intechopen.100382

[48] Ajjarapu SM, Tiwari A, Kumar S. Applications and utility of three-dimensional In vitro cell culture for therapeutics. Future Pharmacology. 2023;**3**:213-228

[49] Paschos NK, Brown WE, Eswaramoorthy R, et al. Advances in tissue engineering through stem cell-based co-culture. Journal of Tissue Engineering and Regenerative Medicine. 2015;**9**:488-503

[50] Bin LY, Kim EM, Byun H, et al. Engineering spheroids potentiating cell-cell and cell-ECM interactions by self-assembly of stem cell microlayer. Biomaterials. 2018;**165**:105-120

[51] Nishiofuku M, Yoshikawa M, Ouji Y, et al. Modulated differentiation of embryonic stem cells into hepatocyte-like cells by coculture with hepatic stellate cells. Journal of Bioscience and Bioengineering. 2011;**111**:71-77

[52] Marino S, Bishop RT, de Ridder D, et al. 2D and 3D in vitro co-culture for cancer and bone cell interaction studies. Methods in Molecular Biology. 2019;**1914**:71-98

[53] Hatherell K, Couraud PO, Romero IA, et al. Development of a three-dimensional, all-human in vitro model of the blood-brain barrier using mono-, co-, and tri-cultivation transwell models.

Journal of Neuroscience Methods. 2011;**199**:223-229

[54] Miki Y, Ono K, Hata S, et al. The advantages of co-culture over mono cell culture in simulating in vivo environment. The Journal of Steroid Biochemistry and Molecular Biology. 2012;**131**:68-75

[55] Moraes C, Mehta G, Lesher-Perez SC, et al. Organs-on-a-chip: A focus on compartmentalized microdevices. Annals of Biomedical Engineering. 2012;**40**:1211-1227

[56] Wu MH, Bin HS, Bin LG. Microfluidic cell culture systems for drug research. Lab on a Chip. 2010;**10**:939-956

[57] Lin B, Levchenko A. Microfluidic technologies for studying synthetic circuits. Current Opinion in Chemical Biology. 2012;**16**:307-317

[58] El-Ali J, Sorger PK, Jensen KF. Cells on chips. Nature. 2006;**442**:403-411

[59] Harcombe W. Novel cooperation experimentally evolved between species. Evolution. 2010;**64**:2166-2172

[60] Kerr B, Riley MA, Feldman MW, et al. Local dispersal promotes biodiversity in a real-life game of rock-paper-scissors. Nature. 2002;**418**:171-174

[61] Campbell JJ, Davidenko N, Caffarel MM, et al. A multifunctional 3D co-culture system for studies of mammary tissue morphogenesis and stem cell biology. PLoS One. 2011;**6**:9

[62] Albrecht DR, Underhill GH, Wassermann TB, et al. Probing the role of multicellular organization in three-dimensional microenvironments. Nature Methods. 2006;**3**:369-375

[63] Felton EJ, Copeland CR, Chen CS, et al. Heterotypic cell pair co-culturing on patterned microarrays. Lab on a Chip. 2012;**12**:3117-3126

[64] Goers L, Freemont P, Polizzi KM. Co-culture systems and technologies: Taking synthetic biology to the next level. Journal of the Royal Society, Interface. 2014;**11**:20140065

[65] Park J, Kerner A, Burns MA, et al. Microdroplet-enabled highly parallel co-cultivation of microbial communities. PLoS One. 2011;**6**:2

[66] Hesselman MC, Odoni DI, Ryback BM, et al. A multi-platform flow device for microbial (co-) cultivation and microscopic analysis. PLoS One. 2012;7:5

[67] Hyun JK, Boedicker JQ, Jang WC, et al. Defined spatial structure stabilizes a synthetic multispecies bacterial community. Proceedings of the National Academy of Sciences of the United States of America. 2008;**105**:18188-18193

[68] Jeddi I, Saiz L. Three-dimensional modeling of single stranded DNA hairpins for aptamer-based biosensors. Scientific Reports. 2017;7:1-13

[69] Dalmau N, Jaumot J, Tauler R, et al. Epithelial-to-mesenchymal transition involves triacylglycerol accumulation in DU145 prostate cancer cells. Molecular BioSystems. 2015;**11**:3397-3406

[70] Sabra W, Dietz D, Tjahjasari D, et al. Biosystems analysis and engineering of microbial consortia for industrial biotechnology. Engineering in Life Sciences. 2010;**10**:407-421

[71] Ingham CJ, Sprenkels A, Bomer J, et al. The micro-petri dish, a million-well growth chip for the culture and high-throughput screening of microorganisms. Proceedings of the National Academy of Sciences of the United States of America. 2007;**104**:18217-18222

[72] Alain K, Querellou J. Cultivating the uncultured: Limits, advances and future challenges. Extremophiles. 2009;**13**:583-594

[73] Dar SA, Srivastava PP, Rather MA, et al. Molecular and computational analysis of ghrelin, growth hormone Secretagogues receptor and mRNA expression of growth-related genes after exogenous administered ghrelin peptide in Labeo rohita. International Journal of Biological Macromolecules. 2020;**142**:756-768

[74] Iyengar P, Combs TP, Shah SJ, et al. Adipocyte-secreted factors synergistically promote mammary tumorigenesis through induction of anti-apoptotic transcriptional programs and proto-oncogene stabilization. Oncogene. 2003;**22**:6408-6423

[75] Church FC, Carter JC. Obesity and breast cancer: The roles of peroxisome proliferator-activated receptor-γ and plasminogen activator inhibitor-1. PPAR Research. 2009;**2009**:345320

[76] Rajala MW, Scherer PE. Minireview: The adipocyte–at the crossroads of energy homeostasis, inflammation, and atherosclerosis. Endocrinology. 2003;**144**:3765-3773

[77] Nieman KM, Kenny HA, Penicka CV, et al. Adipocytes promote ovarian cancer metastasis and provide energy for rapid tumor growth. Nature Medicine. 2011;**17**:1498-1503

[78] Balaban S, Shearer RF, Lee LS, et al. Adipocyte lipolysis links obesity to breast cancer growth: Adipocyte-derived fatty acids drive breast cancer cell proliferation and migration. Cancer & Metabolism. 2017;**5**:1

[79] Elliott BE, Tam S-P, Dexter D, et al. Capacity of adipose tissue to promote growth and metastasis of a murine mammary carcinoma: Effect of estrogen and progesterone. International Journal of Cancer. 1992;**51**:416-424

[80] Schmeichel KL, Weaver VM, Bissell MJ. Structural cues from the tissue microenvironment are essential determinants of the human mammary epithelial cell phenotype. Journal of Mammary Gland Biology and Neoplasia. 1998;**3**:201-213

[81] Dirat B, Bochet L, Dabek M, et al. Cancer-associated adipocytes exhibit an activated phenotype and contribute to breast cancer invasion. Cancer Research. 2011;**71**:2455-2465

[82] Bochet L, Lehuédé C, Dauvillier S, et al. Adipocyte-derived fibroblasts promote tumor progression and contribute to the desmoplastic reaction in breast cancer. Cancer Research. 2013;**73**:5657-5668

[83] Incio J, Liu H, Suboj P, et al. Obesity-induced inflammation and desmoplasia promote pancreatic cancer progression and resistance to chemotherapy. Cancer Discovery. 2016;**6**:852-869

[84] Alessi MC, Bastelica D, Morange P, et al. Plasminogen activator inhibitor 1, transforming growth factor-beta1, and BMI are closely associated in human adipose tissue during morbid obesity. Diabetes. 2000;**49**:1374-1380

[85] Lorincz AM, Sukumar S. Molecular links between obesity and breast cancer. Endocrine-Related Cancer. 2006;**13**:279-292

[86] Carter JC, Church FC. Mature breast adipocytes promote breast cancer cell motility. Experimental and Molecular Pathology. 2012;**92**:312-317

[87] Wellen KE, Hotamisligil GS. Inflammation, stress, and diabetes.

The Journal of Clinical Investigation. 2005;**115**:1111-1119

[88] Grisouard J, Dembinski K, Mayer D, et al. Targeting AMP-activated protein kinase in adipocytes to modulate obesity-related adipokine production associated with insulin resistance and breast cancer cell proliferation. Diabetology and Metabolic Syndrome. 2011;**3**:16

[89] Mcmillin DW, Negri JM, Mitsiades CS. The role of tumour–stromal interactions in modifying drug response: Challenges and opportunities. Nature Reviews. Drug Discovery. 2013;**12**:217-228

[90] Dutta D, Heo I, Clevers H. Disease modeling in stem cell-derived 3D organoid systems. Trends in Molecular Medicine. 2017;**23**:393-410

[91] Corrò C, Novellasdemunt L, Li VSW. A brief history of organoids. American Journal of Physiology. Cell Physiology. 2020;**319**:C151-C165

[92] Ooft SN, Weeber F, Dijkstra KK, et al. Patient-derived organoids can predict response to chemotherapy in metastatic colorectal cancer patients. Science Translational Medicine. 2019;**11**:513

[93] Dekkers JF, Whittle JR, Vaillant F, et al. Modeling breast cancer using CRISPR-Cas9-mediated engineering of human breast organoids. Journal of the National Cancer Institute. 2020;**112**:540-544

[94] Wang S, Wang Y, Xun X, et al. Hedgehog signaling promotes sorafenib resistance in hepatocellular carcinoma patient-derived organoids. Journal of Experimental & Clinical Cancer Research. 2020;**39**:22

[95] Papaccio F, García-Mico B, Gimeno-Valiente F, et al. Proteotranscriptomic analysis of advanced colorectal cancer patient derived organoids for drug sensitivity prediction. Journal of Experimental & Clinical Cancer Research. 2023;**42**:8

[96] Aboulkheyr Es H, Montazeri L, Aref AR, et al. Personalized cancer medicine: An organoid approach. Trends in Biotechnology. 2018;**36**:358-371

Chapter 4

Perspective Chapter: *In Vitro* Contracting Cardiomyogenic Models from Whole Fish Embryos and Larvae – Method, Properties, and Applications

Bianka Grunow and Valeria Di Leonardo

Abstract

Heart diseases remain a leading cause of mortality worldwide. The development of effective treatments and interventions relies on a deep understanding of cardiac biology. Traditional two-dimensional (2D) cell cultures and animal models fall short in replicating crucial physiological and pathological features of cardiac tissue. In response, 3D cardiac models have emerged, offering a more faithful replication of the native heart tissue's architecture and functionality in a controlled environment. Although technical hurdles limit the widespread adoption of *in vitro* 3D models, they hold promise for advancing cardiovascular research. This chapter provides a description of the development of 3D spontaneously contracting cardiac primary cultures derived from fish embryos and larvae, presenting an easily accessible model for diverse applications, including the investigation of viral heart infections, as well as biomedical, pharmacological, and cardiology research. In this chapter, we will highlight the importance of *in vitro* model systems for modern cardiac research. Additionally, we will provide an overview of the protocol and results concerning the creation of *in vitro* 3D heart-like cell aggregates using enzymatically digested whole fish embryos/larvae. These aggregates exhibit long-term stability and spontaneous contractions, making them promising candidates for high-throughput screening.

Keywords: cardiac model, 3D culture, heart, fish virology, pharmacology

1. Introduction

Cell cultures represent a fundamental *in vitro* technology in several research fields [1–3]. Since these experimental systems have become part of everyday research, they have contributed to important discoveries, including key findings in cardiomyocyte mechanisms concerning the cardiovascular field [3]. Cardiomyocytes are the main components of heart tissue and have their origin in mesodermal stem cells [4]. A lot of work in cardiovascular research has been focused on the development of cardiomyogenic models to study various conditions, e.g., heart diseases and healing capacity [5].

In vivo models of animal species like rats and pigs are currently used as the gold standard to understand pathology mechanisms and explore novel treatments within the context of a whole organism [3, 6], as the cardiovascular apparatus is characterized by a high degree of complexity, which means that a number of cardiovascular diseases are usually multifactorial [7, 8]. However, these kinds of experimental *in vivo* models have some important limitations that make them alone insufficient in certain circumstances such as the discovery and development of new therapies [3, 6]. The main disadvantages of *in vivo* models are in fact the ethical concerns, the high costs, and the low throughput. As an alternative, *in vitro* systems are good platforms for high-throughput drug efficiency and toxicity tests and are widely used to study biological mechanisms in a faster, low-cost, and reproducible way [6]. Taken together, *in vivo* and *in vitro* systems allowed the study of complex biological mechanisms and, during recent years, new emerging vertebrate models enriched the tools of researchers engaged in cardiovascular studies for human and veterinary medicine [9].

Since the 1990s, teleost fishes, in particular, have grown in popularity because of the introduction of zebrafish (*Danio rerio*) as a commonly used animal model in several research fields due to biological similarities with humans and generally good conservation and expression of important target genes. Specifically, zebrafish is a small freshwater cyprinid that is cheap and easy to grow and produces a high amount of external transparent embryos, making it an efficient model for multi-disciplinary studies in cardiac research, covering a range of topics including, but not limited to, development, genetic biology, and regeneration [10]. In addition to zebrafish research, in general, fish cell cultures of the mainly remunerative fish species are increasingly established due to their relevance in the important economic sector of aquaculture and fisheries, and the necessity to obtain more cost-effective model systems for research in this field as well [11].

For these reasons, multiple *in vitro* cardiomyogenic models from different mammalian and teleost species were developed for basic and applied research during the last decades [3, 9].

The technology used to produce cardiomyogenic models can vary according to the cell source and the use of 2D or 3D culture conditions. Among the variety of experimental techniques, this book chapter will focus on the cultivation methods for 3D autonomously beating cardiomyocytes (defined as "SCC—Spontaneously Contracting Cell aggregate", ZebraFish Heart AggregateZFHA, and SCPC—Salmon Cardiac Primary Cultures") using whole fish larvae from the pre- and post-hatch stage as the main cell source, including derivation outcomes from different species. In addition, the following sections explore the potential applications of 3D spontaneously beating cardiomyogenic fish cell cultures, with emphasis on basic research regarding their use to study resilience to climate change, and in applied research in pharmacology and fish virology.

1.1 Overview of cardiomyogenic *in vitro* models

The main approaches employed to generate *in vitro* heart models can be categorized based on the type of biological sample utilized. These include the utilization of whole isolated heart organs [12], sections of heart tissue [13], explant cultures [14], cultivation of induced pluripotent stem cells (IPS) [15], as well as the culture of embryonic stem cells, and embryoid bodies [16]. However, if these approaches are carried out using adult heart tissue from mammals, the regenerative capacity is not sufficient. In addition, obtaining biological samples from humans is challenging due to costs and ethical considerations.

Furthermore, the general limited availability of differentiated cardiomyocytes in 3D models poses a significant constraint in cardiovascular research. A promising tool for research involves the use of induced pluripotent stem cells to generate fully developed cardiomyocytes. Nonetheless, this technology is still in its early stages and faces limitations in terms of differentiation efficiency, thereby reducing the model's reliability [17]. Over the past few years, a growing body of research has focused on utilizing cardiac scaffold-free spheroids for drug testing and toxicology purposes. Numerous studies have been published during this time, often employing a combination or co-culture of various cell types including rodent or human primary cardiomyocytes, human induced pluripotent stem cell-derived cardiomyocytes (hiPSC-CMs), fibroblasts, stem cells, and endothelial cells [17–23]. An inherent problem with previous *in vitro* cardiac test systems derived from various organisms such as mice, birds, amphibians, and fish is their limited activity. These test systems show either no contraction or only in response to electrical stimulation or only brief contractions, lasting from a few minutes to a few hours or days, which limits their use in the study of contractile activity of cardiac tissue [24]. Another disadvantage of conventional cell cultures is their lack of resemblance to organotypic structures because, in most cases, *in vitro* models are available as individual cells or as adherent cells organized in two-dimensional layers.

It is widely recognized that the physiological environment of cells is characterized by the presence of an extracellular matrix and complex interactions based on biochemical and mechanical stimuli [1, 3, 17]. Specifically, cardiomyocytes are naturally and constantly subjected to shear stress from blood flow, mechanical deformations, electrical impulses, and calcium transients, which are mostly absent in the artificial environment of classical two-dimensional (2D) cell cultures. For these reasons, there is a need to achieve experimental three-dimensional (3D) models with the ability to mimic the complex and dynamic microenvironment found in living tissues [17].

The advantages of 3D cell cultures over conventional 2D cell cultures are manifold. In 3D cultures, cells can interact with each other and their environment, allowing for the formation of cell-cell contacts, cell-matrix interactions, and the development of tissue-like structures. In particular, cells in 3D cultures can communicate more effectively through direct cell-cell contacts such as gap junctions. This enables the exchange of important molecular signals, growth factors, and cytokines, leading to improved cell differentiation, functionality, and responsiveness to external stimuli. This better mimics the conditions and cellular behavior *in vivo*, leading to more accurate and reliable experimental results.

Consequently, 3D cultures provide a valuable platform for the study of cell–cell interactions, tissue development, disease progression, and drug treatments [17, 22].

Since 2006, there has been a growing interest in utilizing fish cells for a more detailed study of cardiac physiology and human cardiac research [24–26]. Researchers have highlighted significant differences between murine and human cardiac electrophysiology, while noting the remarkable similarity between fish and human cardiac electrophysiology in terms of action potential properties and thus ion channels [24, 25]. Compared to the human models, 3D spontaneously beating cardiomyogenic cultures from fishes are not dependent on clinical samples, and several experimental analyses confirmed the presence of fully developed cardiomyocytes organized in adult tissue phenotype and owning many biological characteristics in common with human cardiomyocytes [27–34]. The establishment of *in vitro* model systems with fish cells is therefore important not only for basic research in fish biology and virology but also for research in human medicine and pharmacology.

2. Establishment and cultivation of 3D spontaneously beating heart aggregates from fish

It is possible to obtain autonomously contracting 3D cardiomyogenic aggregates from both pre- and post-hatching whole fish larvae in a simple and efficient way (**Figure 1**). The main lab equipment and materials needed are in line with common primary cell culture isolation protocols and are summarized in **Tables 1** and **2**. At this point, it is significant to use larvae that are at least at eye point stage and still within the yolk sac phase so that they are not actively feeding. Larvae of several salmonid species like rainbow trout (*Oncorhynchus mykiss*), brown trout (*Salmo trutta*),

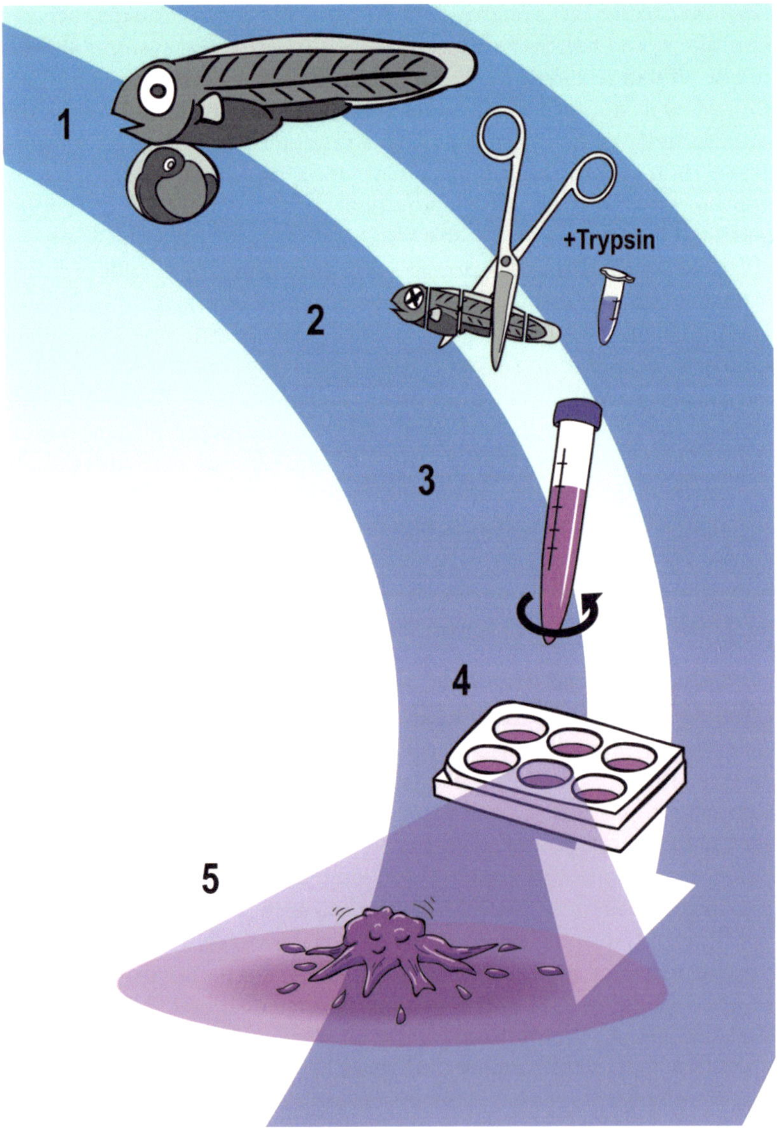

Figure 1.
Illustration of the establishment of spontaneously contracting cell aggregates (SCCs)/ZebraFish heart aggregates (ZFHAs) and Salmon cardiac primary cultures (SCPCs).

Category	Equipment
Lab equipment	Laminar Flow Sterile workbench Autoclave Sterilizator Incubator (for temperatures between 15 and 30°C) Centrifuge Microscope
Instruments	Forceps Scissors Pipettes Pipetboy
Solutions/reagents	Fetal calf serum Cell culture medium (Leibovitz 15 or DMEM) Phosphate-Buffered Saline Solution (PBS) Amphotericin Gentamycin Kanamycin Penicillin/Streptomycin Trypsin–EDTA Solution
Cell culture plastic	Pipette tips (from 10 µl up to 5 ml) Cell culture plates (6-well up to 24-well plates) Falcons (50 ml and 15 ml) 1.5 ml vials

Table 1.
Overview of the cell lab equipment required for the establishment of SCCs, ZFHA, and SCPCs.

Protocol step	Material	Name
Washing	Washing solution	Dulbecco's phosphate-buffered saline (DPBS)
Dissociation	Enzymatic solution	Trypsin solution 0,1%
		Trypsin solution 2,5%
Centrifugation	Neutralizing solution (culture medium + FBS)	Dulbecco modified eagle medium (DMEM)
		L-15 medium (Leibovitz's)
		Fetal Bovine Serum (FBS)
Cell seeding	Multi-well plate	6-well culture plate
		12-well culture plate
		24-well culture plate
	Culture medium	Dulbecco modified eagle medium (DMEM)
		L-15 medium (Leibovitz's)
	Culture supplement	Fetal Bovine Serum (FBS) 20%
	Antibiotics	Penicillin (100 U/ml)
		Streptomycin (0.1 mg/ml)

Table 2.
Summary of the main solutions/reagents used for the establishment and cultivation of SCCs, ZFHA, and SCPCs divided per protocol step.

maraena whitefish (*Coregonus maraena*), Atlantic salmon (*Salmo salar*), but also zebrafish (*Danio rerio*) or sturgeon species like *Acipenser oxyrinchus* were used to produce these cardiomyogenic *in vitro* systems [27–34].

1. Depending on the pre- or post-hatched larvae, the specimen needs to be isolated out of the egg. The eggshell is gently scored and pulled apart, ensuring the well-being of the animals. Subsequently, in accordance with ethical guidelines and animal welfare regulations, the animals are euthanized. After euthanasia, the larvae are washed three times in PBS (phosphate-buffered saline) to reduce contamination.

2. The most common enzymatic dissociation protocol involves the use of 0,1% trypsin solution for 1 to 3 minutes at room temperature, together with the induced mechanical dissociation using scissors. Depending on the fish's size, the size of the scissors needs to be adapted; for ZF larvae, micro-scissors were used.

3. The digestion is then stopped with the addition of the double or triple amount of culture medium with 20% fetal bovine serum (FBS). The immediately following centrifugation for 5 minutes at around 130 g allows the collection of isolated cells and small explants from the processed larvae. The growth medium used is alternatively Dulbecco modified eagle medium (DMEM) with 20% FBS or L-15 (Leibovitz's) medium with 20% FBS, both added with 100 U/ml penicillin and 0.1 mg/ml streptomycin.

4. Depending on the size of the fish larvae and their growth potential, primary cell cultures are then cultivated in 24/12/6-well plates. Cells from salmonids and sturgeon can be cultivated in a six-well plate, and from smaller fish species like the zebrafish, the use of a 12 or 24-well plate is preferable. Additionally, the cultivation temperatures depend on the species: cells of cold-water fish species, like Atlantic salmon, rainbow trout, maraena whitefish, or Atlantic surgeon, should be cultivated at 15–22°C, and warm-water fish species like the zebrafish at 28°C.

5. After isolation, explants and cells are generally attached after 24–48 h. Using the described protocol, the generation of spontaneously and autonomously beating cell aggregates, experimentally identified as fully developed cardiomyocytes, was possible for all the cited species (Video 1, https://bit.ly/3SRb9Wb).

2.1 Troubleshooting—contamination

Contamination in primary fish cell cultures poses a significant challenge to maintaining the integrity of research experiments and ensuring the reliability of results/outcomes. Here, we outline a systematic and scientific approach to mitigating and preventing contamination, both bacterial and fungal. Before initiating cell cultures, clean and sterilize all equipment and glassware. Just work under the sterile bench and use gloves that were previously cleaned with a 70% ethanol solution. Additionally, the contamination can be avoided/prevented or reduced through careful washing steps before preparation.

Upon identification of contamination, take the following steps:

1. Avoid cross-contamination by using separate sterile pipettes and tools for each well.

2. Wash the affected culture wells twice with sterile PBS. If fungal contamination is localized in single spot, use a sterile pipette to carefully aspirate fungal material from the contaminated culture well.

3. Initially, supplement the culture medium with double amount of the standard antibiotic concentration for the first few days.

4. If contamination persists or worsens, consider discarding individual contaminated wells while preserving uncontaminated cultures.

5. Implement a rigorous regimen of washing every 2 days and maintaining the elevated antibiotic concentration for at least 1 week.

6. After this period, gradually reduce the antibiotic concentration while closely monitoring cultures.

7. Continue monitoring for a minimum of 3 weeks to ensure complete eradication of contaminants.

2.2 Troubleshooting—absence of cardiomyogenic aggregates

The isolation protocol described in Section 2 is generally easy to perform compared to common cell isolation practices, but good technical execution is nevertheless essential to ensure the presence of functionally active cardiomyogenic aggregates. The following are some tips to consider in order to maximize the yield of heart aggregates during the isolation process.

1. Dissociation technique: When performing dissociation, ensure that you target the specific area of the original heart in small pieces. Cardiomyogenic aggregates are believed to originate from the original heart progenitor cells. Therefore, the likelihood of achieving a high yield may depend on the preservation of the portions of the original embryonic heart (refer to Chapter 3.1 for more details).

2. Overzealous microscopy examination: It is crucial to limit the frequency of microscopy inspections. Frequent disturbances by removing cultures from the incubator can decrease the occurrence of contracting aggregates. Cells require a period of rest to develop and maintain these structures. Therefore, just check maximum once per day.

3. The wells must be carefully examined at an appropriate speed to discern contractions.

3. Characteristics

3.1 General properties

A first work in 2010 proved the possibility of obtaining SCCs from rainbow trout by applying the previously described protocol [27]. In rainbow trout, it was shown that a short time before hatching, the highest ratio of SCCs can be gained with up to five SCCs from one fish larva [29]. In zebrafish, here called zebrafish heart aggregate (ZFHA), larvae should be used 3-4 days post-fertilization (dpf) [31].

The 3D SCCs are able to mimic myocardial tissue organization after a few days in culture. After 7 days in culture, most of the SCCs are developed [29]. Analysis of

SCCs, larval heart, and adult fish heart suggested that the embryonic progenitors of heart tissue are able to differentiate *in vitro* in fully developed adult-like cardiomyocytes while the heart from larval stage revealed immature and unorganized sarcomere structures within the cells, so that heart cells of larvae are not fully differentiated. These kinds of observations were verified by electron microscopic analysis for SCCs generated from salmonids [29, 32]. Additional confirmation of the theory was performed with zebrafish Tg (cmlc2:eGFP) larvae whose fluorescence indicates the expression of eGFP attached to cmcl2 in the heart [34]. This fluorescence signal was also present in the *in vitro* ZFHA indicating the further differentiation and propagation of heart progenitor cells [34].

This work, in particular, demonstrated the possibility of cultivating active cardiomyocytes from rainbow trout or other fish species in the long term over several months of constant beating frequency, **Figure 2** [29]. These important results show the possibility of obtaining highly specialized adult-like cardiac structures from whole larval cell cultures and the possibility of using fish heart aggregates as *in vitro* specification models in developmental biology.

3.2 Cell anatomy

The cell anatomy of SCC, which includes various cell types of the heart, has been extensively characterized using advanced imaging techniques such as confocal microscopy and electron microscopy, providing detailed information about the ultrastructure and spatial organization of cells [27–34].

Transmission electron microscopy confirmed the presence of well-developed sarcomeres within the beating cell aggregates with a length ranging from 1.4 to 1.6 μm consistent with adult cardiac tissue structure [27, 29, 31]. Moreover, the reported presence of T-tubuli structures in the z-bands and near the sarcoplasmic reticulum indicates the importance of ion flux and action potential [29].

Moreover, immunohistochemical analysis within SCCs revealed the presence of well-organized sarcomeres and heart-specific proteins [27–31]. LC-MS mass spectrometry analysis confirmed the previously reported observations by showing that 80% of the isolated proteins from 3D SCCs/ZFHA were also expressed in the atrium

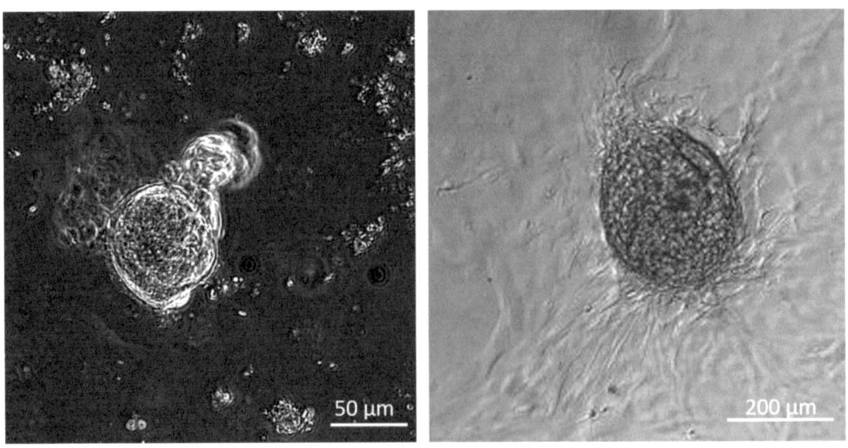

Figure 2.
SCC established from zebrafish (Danio rerio) and rainbow trout (Oncorhynchus mykiss).

and/or ventricular tissue of adult animals of the same species [31]. Studies on SCCs generated from larvae of zebrafish, rainbow trout, and Atlantic salmon have demonstrated important aspects of cell morphology within SCCs [27–34]. In fact, it has been shown that *in vitro* heart aggregates exhibit a syncytium-like structure with a high number of mitochondria, cell contact, and myofilaments, reflecting their contractile capacity, active metabolism, and long-term stability [27–34].

Pacemaker cells are the primary contractile cells responsible for generating the long-lasting rhythmic contractions observed in SCCs. Two important proteins associated with pacemaker cell function are HCN4 (Hyperpolarization-activated cyclic nucleotide-gated channel 4) and Connexin45 (Cx45). HCN4 is a key ion channel protein, which is responsible for the "funny" current (If), which plays a critical role in the generation of pacemaker potentials and the spontaneous depolarization phase of the action potential. Cx45 is a gap junction protein that facilitates electrical coupling between pacemaker cells and helps synchronize their activity. Both have been found in the SCCs [28].

Taken together, these results suggest the high reliability of SCCs in the *in vitro* reproduction of heart tissue development and organization, highlighting their potential use as 3D cardiac model systems.

3.3 Cell physiology

In vitro systems featuring functional ion channels serve as invaluable resources for scientific research. Ion channels, which are transmembrane proteins forming pores, facilitate the movement of electrically charged particles (ions) across cell membranes. They play a crucial role in regulating electrical signaling within cells. By employing *in vitro* systems that replicate ion channels, researchers gain the ability to investigate their functionality and regulation under carefully controlled conditions [35].

Electrophysiological techniques enable the recording and analysis of electrical signals generated by cells. Intracellular action potentials refer to the electrical impulses that occur within individual cells, while extracellular field potentials represent the summation of electrical activity from multiple nearby cells. These measurements provide valuable insights into cellular communication, ion channel function, and tissue-level electrical properties. Electrophysiological measurements of intracellular action potentials and extracellular field potentials thus provide valuable information about cellular electrophysiology. These techniques have multiple applications and contribute to our understanding of cardiac function in general, and more specifically arrhythmias, ion channelopathies, and drug effects on cardiac cells [36].

In the case of SCCs, several studies have demonstrated their electrophysiological properties. These studies have shown that SCCs can beat autonomously for up to 6 months without electrical stimulation. Electrophysiological analysis using sharp electrodes revealed that the cardiomyocyte characteristics of the SCC are very similar to human cardiomyocytes *in vivo* [28, 30]. The action potential exhibits a distinctive plateau phase and a rapid final repolarization phase. Initial tests have shown the reaction on several drugs (please see Section 4.2.1.) as well as on temperature. The contraction frequency of the SCCs correlated highly with temperature [29]. A steady and significant increase in their frequency with increasing temperature and vice versa was observed [29], as found in the *in vivo* condition in fish and mammals [37–39].

Overall, *in vitro* systems and electrophysiological measurements in fish models offer valuable insights into ion channel function, cardiac electrophysiology, and drug effects. These models contribute to our understanding of cardiac disorders and aid in the development of therapeutic interventions in human and veterinary medicine.

4. Applications

In line with the described properties of Section 3, SCCs represent a potential investigative tool in numerous research fields and allow the study of different aspects of cardiac tissue, which are consistent with its complexity. The following is a nonexhaustive description of some of the main potential areas of application according to the experimentally proven properties of SCCs to date.

4.1 Basic science: fish heart research

4.1.1 Climate model

In vitro cardiac models are valuable tools for studying the impact of climate change on cardiac health and function [40–43]. These models enable researchers to investigate the effects of various environmental factors associated with climate change on the heart in a controlled laboratory setting. This includes exploring the effects of temperature fluctuations and changes in oxygen levels on cardiac cells, allowing researchers to manipulate temperature and oxygen concentrations to simulate hypoxic or hyperoxic conditions to overall mimic different climate scenarios *in vitro*. In this way, the effects of temperature and oxygen changes on cardiomyocyte viability, contractility, electrical activity, and overall cardiac function can be assessed, allowing further investigation of the biological mechanisms underlying cardiomyocyte behavior under adverse environmental conditions [37–44]. Moreover, extensive research has demonstrated the influence of temperature on cardiac contractility across vertebrates [24, 37, 41, 42]. Therefore, the need to use *in vitro* heart model in the field of climate change is also of great importance to researchers working on fish health, aquaculture, and biodiversity conservation. Indeed, ectothermic animals experience temperature changes due to their environment that result in a significant remodeling of cardiac tissue [40, 42]. The changes in the heart usually primarily affect contractile activity and metabolism. Fish living in temperate climates are particularly susceptible to significant and long-term seasonal temperature changes [45–47]. Among these fish species, salmonids like rainbow trout can remain active between 4 and 24°C [42, 48]. Also, minnow fishes like zebrafish can experience a broad change in temperature of about 10°C considering fluctuations between winter and summer [49, 50]. For carp SCCs, an increase in culture temperature from 21 to 31°C caused a significant decrease in the contraction frequencies [51]. Due to their biology, fishes are provided with mechanisms to protect heart function from temperature changes. In the past, the comprehension of biological responses to lower temperatures has been the main target of experimental research. However, higher temperatures are becoming one of the main global concerns, making the biological responses to high temperatures another important field to investigate [42]. To date, several biological mechanisms in response to sharp temperature changes are already known and, in some cases, they are similar to human responses. For example, hypothermia leads to bradycardia and higher blood viscosity in both humans and fishes [41, 42]. However, some important differences between mammals and fishes can help to reach a deeper understanding of temperature change resilience strategies. For example, it has been shown that trout cardiac actin-myosin ATPase activity was more Ca^{2+} sensitive than that of rats within their respective physiological temperature [42, 52]. In addition, researchers reported that the Ca^{2+} quantity required by trout cardiac muscle to reach half-maximal tension was about one-tenth of that of rat cardiac tissue [42]. To conclude, fish *in vitro* heart models, which can be generated from several fish species, could provide key findings into the cellular and molecular mechanisms underlying temperature-induced cardiac stress. A deeper understanding of these processes can contribute to improve health resilience in case of unfavorable environmental conditions in response to increasing global temperatures and extreme climate events.

4.2 Applied science

4.2.1 Fish cell culture in human research/pharmacology

There is a growing interest in the use of fish cells as *in vitro* model for human cardiac research because of their physiological similarity [24–26]. The electrocardiogram (ECG) of the most widely used murine cardiac model exhibits some important electrophysiological similarities to humans, including a P wave (atrial depolarization), PQ interval (conduction of impulses from the atria through AV nodes, and the His-Purkinje system to the ventricles), and the QRS complex (ventricular depolarization). However, the action potential of mouse cardiomyocytes differs significantly from that of humans [26, 53, 54], with notable differences in cardiac ion channels [24–26, 55]. One notable drawback of murine cardiac models is the absence of an ERG channel homolog (Either-a-gogo-Related-Gene channel), which limits the modeling of repolarization disturbances and the applicability of pharmacological testing. Additionally, myosin light chains (MLC2a and MLC2v) exhibit distinct development patterns in mouse cardiomyocytes. The absence of surface marker protein SIRPA in mice leads to a 10-fold higher tolerance to drugs that may be harmful to humans [34, 56].

Despite these disadvantages, fish models, such as rainbow trout and zebrafish, have gained popularity in ion channel research, as the basic properties of many ion channels are similar in fish and humans. The ERG channel present in fish models is important in drug research, especially for studying drug-induced torsades tachycardia, a specific type of ventricular arrhythmia associated with the inhibition of the rapid delayed rectifier current (I_{Kr}) and the prolongation of the QT interval on the electrocardiogram [57].

To further advance pharmacological research, the developed *in vitro* cardiac model SCC offer significant advantages, including the potential for high-throughput screening of various factors like small molecules, nucleic acids, proteins, and lipids [34]. Zebrafish research, in particular, benefits from having a fully sequenced genome, enabling the study of cardiovascular mutants to better understand the progression of cardiovascular diseases. In addition, SCCs derived from trout as well as zebrafish, hold promise for studying ion channel disruptions and identifying suitable drugs for safety screening [28, 30, 31, 34]. SCCs can be maintained in cell culture for extended periods while retaining their functionality and exhibiting contraction rates similar to the human heart [28–30]. Moreover, the pharmacological responses of L-type calcium channels using Isoproterenol and K_{ATP}-channels using Rilmakalim have been successfully demonstrated [28].

The existence of a hERG channel in these 3D SCCs further enhances their significance. In [30], extracellular field potential recordings using a multielectrode array (MEA) showed a significant prolongation of field potential duration after administration of common hERG potassium channel blockers. The addition of terfenadine at 10 μm and dofetilide at 1 μm resulted in a 2-fold and 10-fold increase in field potential durations, respectively. This response of SCCs to these drugs highlights their suitability for *in vitro* I_{Kr} assays, as required by Guideline S7B [58].

4.2.2 Fish virology

Another application of cell cultures can be found in fish virology. Fish virology is a significant research field due to the substantial economic losses caused by fish viruses in aquaculture [59]. Outbreaks of fish diseases can have severe economic and social impacts. For instance, in Chile, outbreaks of infectious salmon anemia virus (ISAV) between 2007 and 2009 resulted in losses of approximately 2 billion US dollars and 15,000 jobs in the aquaculture industry [45, 60].

The Atlantic salmon is a species known to be vulnerable to several cardiac diseases, including cardiomyopathy syndrome (CMS) caused by piscine myocarditis virus, heart and skeletal muscle inflammation (HSMI) induced by piscine reovirus, infectious salmon anemia (ISA) caused by infectious salmon anemia virus (ISAV), and pancreas disease (PD) induced by salmonid alphavirus (also known as salmon pancreas disease virus or SPDV) [33, 61–63].

SPDV has been detected in salmonids such as Atlantic salmon (*S. salar*), rainbow trout (*O. mykiss*), and brown trout (*S. trutta*) for over three decades [61, 64], and more recently, it has also been found in some non-salmonid marine species like the dab (*Limanda limanda*) [65–67]. The relatively limited host range of SPDV compared to terrestrial alphaviruses suggests that temperature sensitivity likely plays a role in virus replication [68]. SPDV is associated with two diseases impacting the salmonid aquaculture industry [69, 70]. Pancreas disease (PD) in Atlantic salmon was first recognized in 1976 and described in 1984 [71]. A similar disease in rainbow trout, initially reported in France in 1994, is called as Sleeping Disease (SD) [72]. The mortality rates associated with these diseases vary, but even the survivors' experience significant growth reduction, leading to economic losses.

The impact of fish viruses can vary depending on the specific viral strain and the type of host cells involved. Different fish species and cell types may exhibit varying susceptibility and responses to viral infections, leading to diverse outcomes in terms of disease severity and transmission dynamics. Understanding these host-virus interactions at the cellular level is crucial for effective fish virology research and the development of targeted control measures. In this research area as well, there is an effort to transition to *in vitro* models in order to reduce the use of a high number of experimental animals [45].

The SCC *in vitro* model system has therefore already been tested for use in fish virology [32, 33, 73]. Since the main focus of research here was on Atlantic salmon, the model system was renamed SCPCs (Salmon Cardiac Primary Cultures) [32]. SCPCs have proven to be valuable tools in fish virology, enabling the study of pathogens such as Salmon Pancreas Disease Virus (SPDV) and Infectious Salmon Anemia virus (ISA virus) and their effects on cardiac cells. SPDV and ISA viruses are significant pathogens that can cause severe diseases and economic losses in salmonid aquaculture. SCPCs provide a controlled and specific environment to investigate the interactions between these viruses and cardiac cells. Researchers have utilized SCPCs to examine various aspects of viral infection, including viral replication dynamics, viral entry, cellular tropism, and the subsequent molecular and cellular responses of cardiac cells. In the case of SPDV, SCPCs have shed light on the pathogenesis of the virus in cardiac tissues, including its effects on cell viability, morphology, and function [32, 33].

Moreover, SCPCs offer the advantage of representing a more biologically relevant model compared to traditional cell lines, as they maintain the characteristics and functionality of primary cardiac cells [32]. This allows for a better understanding of the virus-host interactions within the natural cardiac environment. Additionally, SCPCs may serve as a platform for testing potential antiviral interventions, such as antiviral compounds or vaccines, to assess their efficacy in controlling viral replication and minimizing virus-induced cardiac damage in the future [33].

In conclusion, SCPCs have emerged as valuable *in vitro* models for studying SPDV, ISA virus, and other fish viral diseases. They provide insights into the molecular and cellular events underlying viral pathogenesis in cardiac tissues, aiding in the development of effective control strategies and therapeutic interventions. The utilization of SCPCs could contribute to mitigating the impact of these viral pathogens on salmonid aquaculture and improving the management of related diseases [32, 33, 73].

Conclusion

In vitro cardiomyogenic models using whole fish larvae and embryos provide a versatile platform for investigating cardiac biology and pathology. Their unique properties and applications make them valuable tools in cardiac research for both fish and humans. With further advances and refinements, these models hold great promise for unraveling the complexities of cardiac diseases and advancing therapeutic strategies, as well as examining adaptations in heart cell anatomy and physiology due to climate changes.

Acknowledgements

We would like to thank Dr. George Franz for illustrating the cartoon.

Conflict of interest

The authors declare no conflict of interest.

Funding

The publication of this article was funded by the Open Access Fund of the FBN.

Notes/thanks/other declarations

Special thanks go to Prof. Charli Kruse, who gave me the possibility to invent this model system. Further, I would like to thank Prof. C. Kruse and Dr. Daniel Rapoport from the University of Luebeck in Germany for their support in testing the SCC-model system with regard to pharmacological application. Further, I would like to thank Prof. Holly Shiels from the University of Manchester, UK, for the possibility to adapt the SCCs establishment also on zebrafish larvae/embryos and establish the ZFHA. Moreover, I would like to thank Dr. Patricia Noguera from Marine Scotland in Aberdeen, who tested this model system for application in fish virology.

This model system has been patented and is registered under the following number: WO2011029584A1.

Acronyms and abbreviations

2D	two-dimensional
3D	three-dimensional
AP	action potential
Cx	connexin
CMS	cardiomyopathy syndrome
DMEM	Dulbecco modified eagle medium
ECG	electrocardiogram
ERG channel	either-a-gogo-related-gene channel
FP	field potential
FBS	fetal bovine serum
L-15	leibovitz's medium
HCN4	hyperpolarization-activated cyclic nucleotide-gated channel 4
HSMI	heart and skeletal muscle inflammation
Ikr	rapid delayed rectifier current
IPS	induced pluripotent stem cells
ISA	infectious salmon anemia
MLC2a and MLC2v	myosin light chains
PD	pancreas disease
SD	sleeping diseaseSD
SCC	spontaneously contracting cell aggregate
SCPC	salmon cardiac primary culturesSCPC
SIRPA	surface marker protein
SPDV	salmon pancreas disease virus
ZFHA	zebrafish heart aggregate

Author details

Bianka Grunow* and Valeria Di Leonardo
Research Institute for Farm Animal Biology, Fish Growth Physiology, Dummerstorf, Germany

*Address all correspondence to: grunow@fbn-dummerstorf.de

References

[1] Shigehisa A, Toshiaki T, Hajime S, Shuji T. Progress in cell culture systems for pathological research. Pathology International. 2016;**10**:554-562. DOI: 10.1111/pin.12443

[2] Abbas M, Moradi F, Hu W, Regudo KL, Osborne M, Pettipas J, et al. Vertebrate cell culture as an experimental approach – Limitations and solutions. Comparative Biochemistry and Physiology, Part B. 2021;**254**:110570. DOI: 10.1016/j.cbpb.2021.110570

[3] Inbody SC, Sinquefield BE, Lewis JP, Horton RE. Biomimetic microsystems for cardiovascular studies. American Journal of Physiology. Cell Physiology. 2021;**320**(5):C850-C872. DOI: 10.1152/ajpcell.00026.2020

[4] Parisi C, Vashisht S, Winata CL. Fish-Ing for enhancers in the heart. International Journal of Molecular Sciences. 2021;**22**(8):3914. DOI: 10.3390/ijms22083914

[5] Lien CL, Harrison MR, Tuan TL, Starnes VA. Heart repair and regeneration: Recent insights from zebrafish studies. Wound Repair and Regeneration. 2012;**20**(5):638-646. DOI: 10.1111/j.1524-475X.2012.00814

[6] Chen J, Zhang X, Millican R, Lynd T, Gangasani M, Malhotra S, et al. Recent progress in *in vitro* models for atherosclerosis studies. Frontiers in Cardiovascular Medicine. 2022;**8**:790529. DOI: 10.3389/fcvm.2021.790529

[7] Kramer F, Just S, Zeller T. New perspectives: Systems medicine in cardiovascular disease. BMC Systems Biology. 2018;**12**(1):57. DOI: 10.1186/s12918-018-0579-5

[8] Pasipoularides A. Linking genes to cardiovascular diseases: Gene action and gene–environment interactions. Journal of Cardiovascular Translational Research. 2015;**8**(9):506-527. DOI: 10.1007/s12265-015-9658-9

[9] Stoyek MR, Quinn TA. One fish, two fish, red fish, blue fish*: Zebrafish as a model for cardiac research. Progress in Biophysics and Molecular Biology. 2018;**138**:1-2. DOI: 10.1016/j.pbiomolbio.2018.11.003

[10] González-Rosa JM. Zebrafish models of cardiac disease: From fortuitous mutants to precision medicine. Circulation Research. 2022;**130**:1803-1826. DOI: 10.1161/CIRCRESAHA.122.320396

[11] Goswami M, Yashwanth BS, Trudeau V, Lakra WS. Role and relevance of fish cell lines in advanced *in vitro* research. Molecular Biology Reports. 2022;**49**:2393-2411. DOI: 10.1007/s11033-021-06997-4

[12] Yeo JM, Tse V, Kung J, Lin HY, Lee YT, Kwan J, et al. Isolated heart models for studying cardiac electrophysiology: A historical perspective and recent advances. Journal of Basic and Clinical Physiology and Pharmacology. 2017;**28**(3):191-200. DOI: 10.1515/jbcpp-2016-0110

[13] Ou Q, Jacobson Z, Abouleisa RRE, Tang XL, Hindi SM, Kumar A, et al. A physiological biomimetic culture system for pig and human heart slices. Circulation Research. 2019;**125**(6):628-642. DOI: 10.1161/CIRCRESAHA.119.314996

[14] Cao J, Poss KD. Explant culture of adult zebrafish hearts for epicardial

regeneration studies. Nature Protocols. 2016;**11**(5):872-881. DOI: 10.1038/nprot.2016.049

[15] Burnett SD, Blanchette AD, Chiu WA, Rusyn I. Human induced pluripotent stem cell (Ipsc)-derived cardiomyocytes as an *in vitro* model in toxicology: Strengths and weaknesses for hazard identification and risk characterization. Expert Opinion on Drug Metabolism & Toxicology. 2021;**17**(8):887-902. DOI: 10.1080/17425255.2021.1894122

[16] Mummery CL, Zhang J, Ng S, E, Elliott DA, Elefanty AG, Kamp TJ. Differentiation of human embryonic stem cells and induced pluripotent stem cells to cardiomyocytes: A methods overview. Circulation Research. 2012;**111**(3):344-358. DOI: 10.1161/CIRCRESAHA.110.227512

[17] Zuppinger C. 3D cardiac cell culture: A critical review of current technologies and applications. Frontiers in Cardiovascular Medicine. 2019;**26**(6):87. DOI: 10.3389/fcvm.2019.00087

[18] Polonchuk L, Chabria M, Badi L, Hoflack J-C, Figtree G, Davies MJ, et al. Cardiac spheroids as promising *in vitro* models to study the human heart microenvironment. Scientific Reports. 2017;7:7005. DOI: 10.1038/s41598-017-06385-8

[19] Archer CR, Sargeant R, Basak J, Pilling J, Barnes JR, Pointon A. Characterization and validation of a human 3D cardiac microtissue for the assessment of changes in cardiac pathology. Scientific Reports. 2018;**8**:10160. DOI: 10.1038/s41598-018-28393-y

[20] Ravenscroft SM, Pointon A, Williams AW, Cross MJ, Sidaway JE. Cardiac non-myocyte cells show enhanced pharmacological function suggestive of contractile maturity in stem cell derived cardiomyocyte microtissues. Toxicological Sciences. 2016;**152**:99-112. DOI: 10.1093/toxsci/kfw069

[21] Verheijen M, Schrooders Y, Gmuender H, Nudischer R, Clayton O, Hynes J, et al. Bringing *in vitro* analysis closer to *in vivo*: Studying doxorubicin toxicity and associated mechanisms in 3D human microtissues with PBPK-based dose modelling. Toxicology Letters. 2018;**294**:184-192. DOI: 10.1016/j.toxlet.2018.05.029

[22] Richards DJ, Coyle RC, Tan Y, Jia J, Wong K, Toomer K, et al. Inspiration from heart development: Biomimetic development of functional human cardiac organoids. Biomaterials. 2017;**142**:112-123. DOI: 10.1016/j.biomaterials.2017.07.021

[23] Kehat I, Gepstein A, Spira A, Itskovitz-Eldor J, Gepstein L. High-resolution electrophysiological assessment of human embryonic stem cell-derived cardiomyocytes: A novel in vitro model for the study of conduction. Circulation Research. 2002;**91**(8):659-661. DOI: 10.1161/01.res.0000039084.30342.9b

[24] Shiels HA, Calaghan SC, White E. The cellular basis for enhanced volume-modulated cardiac output in fish hearts. The Journal of General Physiology. 2006;**128**(1):37-44. DOI: 10.1085/jgp.200609543

[25] Brette F, Luxan G, Cros C, Dixey H, Wilson C, Shiels HA. Characterization of isolated ventricular myocytes from adult zebrafish (Danio rerio). Biochemical and Biophysical Research Communications. 2008;**374**(1):143-146. DOI: 10.1016/j.bbrc.2008.06.109

[26] Nemtsas P, Wettwer E, Christ T, Weidinger G, Ravens U. Adult zebrafish heart as a model for human heart? An electrophysiological study. Journal of Molecular and Cellular Cardiology. 2010;**48**(1):161-171. DOI: 10.1016/j.yjmcc.2009.08.034

[27] Grunow B, Ciba P, Rakers S, Klinger M, Anders E, Kruse C. In vitro expansion of autonomously contracting, cardiomyogenic structures from rainbow trout Oncorhynchus mykiss. Journal of Fish Biology. 2010;**76**:427-434. DOI: 10.1111/j.1095-8649.2009.02535

[28] Grunow B, Wenzel J, Terlau H, Langner S, Gebert M, Kruse C. In vitro developed spontaneously contracting cardiomyocytes from rainbow trout as a model system for human heart research. Cellular Physiology and Biochemistry. 2011;**27**(1):1-12. DOI: 10.1159/000325212

[29] Grunow B, Schmidt M, Klinger M, Kruse C. Development of an In vitro cultivated, spontaneously and long-term contracting 3D heart model as a robust test system. Cell Science & Therapy. 2012;**3**:1. DOI: 10.4172/2157-7013.1000116

[30] Mehnert JM, Brandenburger M, Grunow B. Electrophysiological characterization of spontaneously contracting cell aggregates obtained from rainbow trout larvae with multielectrode arrays. Cellular Physiology and Biochemistry. 2013;**32**:1374-1385. DOI: 10.1159/000356576

[31] Grunow B, Mohamet L, Shiels HA. Generating an *in vitro* 3D cell culture model from zebrafish larvae for heart research. The Journal of Experimental Biology. 2015;**218**(Pt 8):1116-1121. DOI: 10.1242/jeb.118224. Epub 2015 Feb 24

[32] Noguera P, Klinger M, Örün H, Grunow B, del-Pozo J. Ultrastructural insights into the replication cycle of salmon pancreas disease virus (SPDV) using salmon cardiac primary cultures (SCPCs). Journal of Fish Diseases. 2021;**44**:1-11. DOI: 10.1111/jfd.13518

[33] Noguera PA, Grunow B, Klinger M, Lester K, Collet B, del-Pozo J. Atlantic salmon cardiac primary cultures: An in vitro model to study viral host pathogen interactions and pathogenesis. PLoS One. 2017;**12**(7):e0181058. DOI: 10.1371/journal.pone.0181058

[34] Hodgson P, Ireland J, Grunow B. Fish, the better model in human heart research? Zebrafish heart aggregates as a 3D spontaneously cardiomyogenic in vitro model system. Progress in Biophysics and Molecular Biology. 2018;**138**:132-141. DOI: 10.1016/j.pbiomolbio.2018.04.009

[35] Restrepo-Angulo I, De Vizcaya-Ruiz A, Camacho J. Ion channels in toxicology. Journal of Applied Toxicology. 2010;**30**(6):497-512. DOI: 10.1002/jat.1556

[36] Liu J, Laksman Z, Backx PH. The electrophysiological development of cardiomyocytes. Advanced Drug Delivery Reviews. 2016;**96**:253-273. DOI: 10.1016/j.addr.2015.12.023

[37] Gamperl AK, Thomas AL, Syme DA. Can temperature-dependent changes in myocardial contractility explain why fish only increase heart rate when exposed to acute warming? The Journal of Experimental Biology. 2022;**225**(4):jeb243152. DOI: 10.1242/jeb.243152

[38] Riyanto M, Mawardi W, Khoerunnisa FI. Temperature effect on heart rate performance of eel (Anguila bicolor). IOP Conference Series: Earth and Environmental Science. 2022;**1033**:012029. DOI: 10.1088/1755-1315/1033/1/012029

[39] Anttila K, Couturier CS, Øverli Ø, Johnsen A, Marthinsen G, Nilsson GE, et al. Atlantic salmon show capability for cardiac acclimation to warm temperatures. Nature Communications. 2014;**5**:4252. DOI: 10.1038/ncomms5252

[40] Liu C, Yavar Z, Sun Q. Cardiovascular response to thermoregulatory challenges. American Journal of Physiology. Heart and Circulatory Physiology. 2015;**309**:H1793-H1812. DOI: 10.1152/ajpheart.00199.2015

[41] Khraishah H, Alahmad B, Ostergard RL, AlAshqar A, Albaghdadi M, Vellanki N, et al. Climate change and cardiovascular disease: Implications for global health. Nature reviews. Cardiology. 2022;**19**(12):798-812. DOI: 10.1038/s41569-022-00720-x

[42] Keen AN, Klaiman JM, Shiels HA, Gillis TE. Temperature-induced cardiac remodelling in fish. The Journal of Experimental Biology. 2017;**220**(Pt 2):147-160. DOI: 10.1242/jeb.128496

[43] Pelster B. Environmental influences on the development of the cardiac system in fish and amphibians. Comparative Biochemistry and Physiology. Part A, Molecular & Integrative Physiology. 1999;**124**(4):407-412. DOI: 10.1016/s1095-6433(99)00132-4

[44] Sakaguchi K, Hinata Y, Kagawa Y, Iwasaki K, Tsuneda S, Shimizu T, et al. Low-temperature culturing improves survival rate of tissue-engineered cardiac cell sheets. Biochemistry and Biophysics Reports. 2018;**14**:89-97. DOI: 10.1016/j.bbrep.2018.04.001

[45] Grunow B, Strauch SM. Status assessment and opportunities for improving fish welfare in animal experimental research according to the 3R-guidelines. Reviews in Fish Biology and Fisheries. 2023;**33**:1075-1093. DOI: 10.1007/s11160-023-09781-8

[46] Asch RG, Stock CA, Sarmiento JL. Climate change impacts on mismatches between phytoplankton blooms and fish spawning phenology. Global Change Biology. 2019;**25**(8):2544-2559. DOI: 10.1111/gcb.14650. Epub 2019 May 31

[47] Beamish RJ. Climate Change and Northern Fish Populations. Canadian Special Publication of Fisheries and Aquatic Sciences 121. Ottawa, Ontario, Canada: National Research Council of Canada; 1995. 739 p

[48] Michaelsen J, Fago A, Bundgaard A. High temperature impairs mitochondrial function in rainbow trout cardiac mitochondria. The Journal of Experimental Biology. 2021;**224**(9):jeb242382. DOI: 10.1242/jeb.242382

[49] Schaefer J, Ryan A. Developmental plasticity in the thermal tolerance of zebrafish Danio rerio. Journal of Fish Biology. 2006;**69**:722-734. DOI: 10.1111/j.1095-8649.2006.01145.x

[50] Angiulli E, Pagliara V, Cioni C, Frabetti F, Pizzetti F, Alleva E, et al. Increase in environmental temperature affects exploratory behaviour, anxiety and social preference in Danio rerio. Scientific Reports. 2020;**10**:5385. DOI: 10.1038/s41598-020-62331-1

[51] Rastgar S, Ardeshir RA, Movahedinia AM, Zabihi E, Salati AP, Salamat N. Spontaneously contracting cell aggregates derived from grass carp heart as a promising tool in *in vitro* heart research. Cytotechnology. 2019;**71**(1):261-266. DOI: 10.1007/s10616-018-0281-x

[52] Churcott CS, Moyes CD, Bressler BH, Baldwin KM, Tibbits GF.

Temperature and pH effects on Ca^{2+} sensitivity of cardiac myofibrils: A comparison of trout with mammals. The American Journal of Physiology. 1994;**267**(1 Pt 2):R62-R70. DOI: 10.1152/ajpregu.1994.267.1.R62

[53] Sabir IN, Killeen MJ, Grace AA, Huang CLH. Ventricular arrhythmogenesis: Insights from murine models. Progress in Biophysics and Molecular Biology. 2008;**98**(2-3):208-218. DOI: 10.1016/j.pbiomolbio.2008.10.011

[54] Schrickel JW, Bielik H, Yang A, Schimpf R, Shlevkov N, Burkhardt D, et al. Induction of atrial fibrillation in mice by rapid transesophageal atrial pacing. Basic Research in Cardiology. 2002;**97**:452-460. DOI: 10.1107/s00395-002-0374-7

[55] Milan DJ, Jones IL, Ellinor PT, CA MR. *In vivo* recording of adult zebrafish electrocardiogram and assessment of drug-induced QT prolongation. American Journal of Physiology. Heart and Circulatory Physiology. 2006;**291**(1):H269-H273. DOI: 10.1152/ajpheart.00960.2005

[56] Price PS, Keenan RE, Swartout JC. Characterizing interspecies uncertainty using data from studies of anti-neoplastic agents in animals and humans. Toxicology and Applied Pharmacology. 2008;**233**(1):64-70. DOI: 10.1016/j.taap.2008.03.026

[57] Yap YG, Camm AJ. Drug induced qt prolongation and torsades de pointes. Heart. 2003;**89**:1363-1372. DOI: 10.1113/JP270526

[58] EMEA. ICH Topic S 7 B the Nonclinical Evaluation of the Potential for Delayed Ventricular Repolarization (QT Interval Prolongation) by Human Pharmaceuticals. EMEA; 2005. Available from: https://www.ema.europa.eu/en/ich-s7b-non-clinical-evaluation-potential-delayed-ventricular-repolarization-qt-intervalCPMP/ICH/423/02

[59] Jones AE, Munro LA, Green DM, Morgan KL, Murray AG, Norman R, et al. The contact structure of great Britain's salmon and trout aquaculture industry. Epidemics. 2019;**28**:100342. DOI: 10.1016/j.epidem.2019.05.001

[60] Mardones FO, Perez AM, Valdes-Donoso P, Carpenter TE. Farm-level reproduction number during an epidemic of infectious salmon anemia virus in southern Chile in 2007-2009. Preventive Veterinary Medicine. 2011;**102**(3):175-184. DOI: 10.1016/j.prevetmed.2011.07.005

[61] McLoughlin MF, Graham DA. Alphavirus infections in salmonids – A review. Journal of Fish Diseases. 2007;**30**(9):511-531. DOI: 10.1111/j.1365-2761.2007.00848.x

[62] Haugland O, Mikalsen AB, Nilsen P, Lindmo K, Thu BJ, Eliassen TM, et al. Cardiomyopathy syndrome of Atlantic salmon (*Salmo salar* L.) is caused by a double-stranded RNA virus of the Totiviridae family. Journal of Virology. 2011;**85**(11):5275-5286. DOI: 10.1128/JVI.02154-10

[63] Finstad ØW, Falk K, Løvoll M, Evensen Ø, Rimstad E. Immunohistochemical detection of piscine reovirus (PRV) in hearts of Atlantic salmon coincide with the course of heart and skeletal muscle inflammation (HSMI). Veterinary Research. 2012;**43**(1):27. DOI: 10.1186/1297-9716-43-27

[64] Villoing S, Béarzotti M, Chilmonczyk S, Castric J, Brémont M. Rainbow trout sleeping disease virus is an

atypical alphavirus. Journal of Virology. 2000;**74**(1):173-183. DOI: 10.1128/jvi.74.1.173-183.2000

[65] Snow M, Black J, Matejusova I, McIntosh R, Baretto E, Wallace IS, et al. Detection of salmonid alphavirus RNA in wild marine fish: Implications for the origins of salmon pancreas disease in aquaculture. Diseases of Aquatic Organisms. 2010;**91**(3):177-188. DOI: 10.3354/dao02265

[66] Bruno DW, Noguera PA, Black J, Murray W, Macqueen DJ, Matejusova I. Identification of a wild reservoir of salmonid alphavirus in common dab Limanda limanda, with emphasis on virus culture and sequencing. Aquaculture Environment Interactions. 2014;**5**:89-98. DOI: 10.3354/aei00097

[67] Simons J, Bruno DW, Ho YM, Murray W, Matejusova I. Common dab, Limanda limanda (L.), as a natural carrier of salmonid alphavirus (SAV) from waters off North-West Ireland. Journal of Fish Diseases 2016;39(4): 507-510. DOI: 10.1111/jfd.12376

[68] Graham DA, Wilson C, Jewhurst H, Rowley H. Cultural characteristics of salmonid alphaviruses—Influence of cell line and temperature. Journal of Fish Diseases. 2008;**31**(11):859-868. DOI: 10.1111/j.1365-2761.2008.00946.x

[69] Maclachlan NJ, Dubovi EJ. Fenner's Veterinary Virology. 4th ed. Elsevier; 2010. 534 p. ISBN: 9780123751584

[70] Weston JH, Welsh MD, McLoughlin MF, Todd D. Salmon pancreas disease virus, an alphavirus infecting farmed Atlantic salmon, *Salmo salar* L. Virology. 1999;**256**(2):188-195. DOI: 10.1006/viro.1999.9654

[71] Munro ALS, Ellis AE, McVicar AH, McLay AH, Nedham EA. An exocrine pancreas disease in farmed Atlantic salmon in Scotland. Helgoländer Meeresuntersuchungen. 1984:571-586

[72] Boucher P, Castric J, Baudin-Laurencin F. Observation of virus like particles in rainbow trout Oncorhynchus mykiss infected with sleeping disease virulent material. Bulletin of the European Association of Fish Pathologists. 1994;**14**:215-216

[73] Noguera P, Collet B, Klinger M, Örün H, Del Pozo J. Use of salmon cardiac primary cultures (SCPCs) of different genotypes for comparative kinetics of mx expression, viral load and ultrastructure pathology, after infection with salmon pancreas disease virus (SPDV). Fish & Shellfish Immunology. 2018;**72**:181-186. DOI: 10.1016/j.fsi.2017.10.059

Section 2

Assessing Cytotoxicity of Herbal Medicines

Chapter 5

In Vitro Assessment of Antioxidant Capacity, Potential to Induce Autophagy/Mitophagy, and Mitochondrial Turnover of Natural Compounds for Cytotoxicity Studies

Madhulika Tripathi, Manish Kumar Verma and Brijesh Kumar Singh

Abstract

The dynamic process of intracellular reduction-oxidation status (redox homeostasis) is influenced by various factors, with mitochondria being one of the most significant contributors. Mitochondria play a crucial role in the bioenergetic pathway, fulfilling the metabolic energy demands of cells. To maintain increased energy requirements, mitochondrial biogenesis and fusion are employed, while decreased energy demands or damaged mitochondria are addressed through fission and autophagic removal, known as mitophagy. Any disruption in these adaptive responses can compromise redox homeostasis and cellular function, and make cells more vulnerable to oxidative stress, resulting in oxidative DNA damage, inflammatory responses, and apoptotic/anti-apoptotic reactions. Such dysregulation contributes to the development of "free radical diseases" like metabolic disorders and cancer. Traditional medicines and herbs (possessing antioxidant and autophagic properties) have been utilized for centuries in the treatment of various diseases; however, it is only recently that researchers have begun to investigate their molecular, cellular, and tissue-level modes of action. Nevertheless, concerns about their cytotoxicity have also arisen. This manuscript focuses on the current technological advancements in assessing the properties of plant-based natural compounds. Both cell-free and cell-based methods are employed to evaluate the therapeutic potential of these compounds, allowing for their scientific evaluation and validation.

Keywords: antioxidant potential, autophagy, mitophagy, cytotoxicity, natural compound

1. Introduction

Past decades have perceived growing interest of studies investigating the role of Reactive Oxygen Species (ROS) and its amelioration with natural products [1–4]. There are increasing evidences that suggest increased oxidative stress from ROS plays critical role in several pathophysiologies and causes Endoplasmic Reticulum stress and mitochondrial dysfunction. The intracellular redox balance (reduction–oxidation status) is a dynamic system which maintains the cellular homeostasis and may change via many factors [5–7]. Mitochondria, being one of the most important organelles, play a crucial role in maintaining intracellular source of energy. Any alteration in mitochondrial dynamics can cause an overproduction of ROS, which urges the induction of oxidative DNA damage and up- or down-regulation of phosphatases, proliferative/anti-proliferative factors, apoptotic/anti-apoptotic factors, etc. [5]. However, physiological levels of ROS are crucial for a proper cell function (i.e. intracellular signaling, inflammation, and immune function) [8]. Moreover, mitochondrial dysfunction and redox imbalance can contribute to a wide range of pathologies under a wide group termed as "free radical diseases" (e.g. cancer, neurodegeneration, atherosclerosis, inflammation, etc.) [5]. Altered cellular homeostasis by oncogenes, mitochondrial dysfunction, and/or oxidative stress leads to an increased genomic instability and the target cells can adapt to the changing environment [9].

Oxygen is vital for most unicellular and multicellular organisms, paradoxically it is also a key in generating Reactive Oxygen and Nitrogen Species (ROS/RNS). Endogenously, ROS are normal by-products of mitochondrial metabolism and energy production [5]. Exogenous sources include smoking, pollution, physical exercise etc. In metabolic disorders or cancer, ROS/RNS homeostasis compromises leading to oxidative stress [10, 11]. To maintain the fine balance and to counteract the harmful effects taking place in the cell, biological system has evolved itself with some strategies like prevention of oxidative damage, repair mechanism, finally the most important is the antioxidant defense mechanisms. Cells are protected from oxidative stress by several antioxidant mechanisms both inside and outside the mitochondria. These include manganese superoxide dismutase (Mn-SOD) in the mitochondria, copper/zinc-SOD in the cytosol, and extracellular-SOD in the interstitium [12]. Classically antioxidants were defined as molecules that inhibit or prevent oxidation of a substrate and are required to control any excess in these ROS/RNS and to maintain Redox homeostasis.

Plant-based natural products have been basis of treatment of many human diseases. The extracts of several plants have been used as therapeutic agents. Spices and herbs that are sources of natural antioxidants, can protect against oxidative stress and thus play preventive role in many diseases [13]. The medicinal properties of folk plants are mainly attributed to the presence of flavonoids, other organic, and inorganic compounds such as coumarins, phenolic acids, and antioxidant micronutrients and are able to counter various metabolic disorders, cancers, and COVID-19 [14–17].

1.1 Role of mitochondrial oxidative stress in metabolic disorders and cancer

Metabolism refers to breaking down of food, in a highly-regulated manner, into simpler units of proteins, carbohydrates, and fats in order to generate energy for the organism. Metabolic syndrome is considered to be a progressive pathophysiological state which is clinically manifested by a cluster of interrelated risk factors such as abdominal obesity, atherogenic dyslipidemia, increased blood pressure, insulin resistance, and pro-inflammatory response [18, 19]. "Cancer metabolism" refers to the alterations in cellular metabolic pathways that are evident in cancer cells compared with most normal tissue cells. Major metabolic changes in cancer cells are increased glycolysis and reduced oxidative phosphorylation coupled with increased biosynthetic intermediates needed for cell growth and proliferation [20]. Likewise, the most common cellular patho-physiologies of metabolic disorders are increased oxidative stress, impaired mitochondrial functions, and cellular repair mechanism (autophagy and apoptosis), etc. [21–23].

Mitochondria are importantly placed in the history of cancer and have been attributed to play a crucial role in metabolic disorders [24, 25]. Although, mitochondria are the center for bioenergetics and biosynthesis pathways, they have also become a critical source of oxidative stress by generating ampule amount of superoxide in ETC [5]. Hence, mitochondrial redox control is utmost important for oxidative phosphorylation, ATP synthesis, and ROS production. The production of ROS in mitochondria is tightly regulated by the mitochondrial superoxide dismutase (SOD2) and glutathione-peroxidase (GPx), as well as by catalase and non-enzymatic antioxidants [5]. Excessive mitochondrial ROS and mitochondrial dysfunctions can promote metabolic−/age-related disorders (NAFLD, diabetes, atherosclerosis, neurodegeneration, etc.) and cancer through suppression of complex I, induction of oxidative mitochondrial DNA damage, increased excessive calcium ion influx, alterations in autophagy [5, 26–28]. Therefore, Phyto molecules/plant-based compounds having antioxidant potential and/or autophagic induction may provide beneficial effects in above-mentioned diseases that are associated with ROS and oxidative stress.

1.2 Therapeutic potential of natural compounds in metabolic disorders and cancer

Natural products have always been of immense importance in health and disease treatment throughout the human evolution [1–4]. Traditionally several natural products have been used to treat different types of human illnesses including general injuries, wound healing, and pain [29, 30]. Natural products have been shown to have pharmacological properties to regulate cell signaling pathways that cause mitogenic and cytotoxic reactions leading to disease pathologies. Usage of modern technological tools such as genomics, proteomics, and metabolomics paved wider use of these natural products. Nowadays, mostly natural products are processed and have been developed as potential pharmacological agents with effective anti-oxidative, anti-mitotic, anti-microbial, anti-inflammatory, anti-angiogenic, and anti-carcinogenic properties [13, 31]. Different civilizations have been relying on plants or plant-based products to improve health. This also holds remarkable therapeutic alternatives to synthetic drugs. Millions of Europeans, Africans, Asians, Australians, and North Americans use plant-based medicine system as health supplements for their general health wellbeing. However, in developing countries, ayurveda and TCM, the traditional Indian and Chinese medicine is often the primary source of health care [32].

Many medicinal plants and natural products are considered by the public as a safe, natural, and cost-effective alternative to synthetic drugs without explicit proof by randomized controlled clinical trials [32]. Recently, there is an added attention in the validation of the efficacy and safety of natural products [32].

Several preclinical studies and non-scientific data accumulated over decades of research confirm the use of medicinal plants and plant-based products in the treatment or prevention of metabolic disorders. These products have gained popularity in western societies; however, their expected health benefits are often not backed by rigorous clinical trials. Therefore, it is essential to provide robust scientific evidence on their clinical efficacy and safety.

2. *In vitro* tools and techniques for the assessments of natural compounds

2.1 Assessment of antioxidant potential of plant extract/natural compound in a cell free system

2.1.1 Total antioxidant capacity by ABTS decolorization assay

Principle: The assay is based on the ability of antioxidant molecules to quench the stable ABTS (2,2′-azino-bis (3-ethylbenzthiazoline-6-sulfonic acid)) radical [33]. A blue-green solution of ABTS cation radical ($ABTS^{•+}$) is formed by the loss of an electron from stable ABTS radical and absorbs at 743 nm. In the assay, addition of Trolox decolorizes the solution resulted by the quenching the hydrogen atom by nitrogen atom.

Methodology (for 96-well plate format):

i. $ABTS^{•+}$ radical cation is produced by reacting $ABTS^{•+}$ stock solution (7 mM) with 2.45 mM potassium persulfate in the ratio of 1:0.5.

ii. Allow the above-prepared mixture to stand in the dark at room temperature for 12–16 h.

iii. Take 80 μl of ABTS solution prepared in step I and add 2.91 ml of ethanol to it.

iv. Add 10 μl of plant extract (usually 1 mg/ml) or pure plant-based chemical and read it on a spectrophotometer immediately at 734 nm for 4–5 min.

v. 10 μl trolox was used as a standard. The decolorization of chromophore was measured on spectrophotometer.

vi. Data should be presented as trolox equivalent antioxidant capacity (TEAC)/mg extract.

2.1.2 Nitric oxide (NO) quenching capacity

Principle: Nitric oxide (NO) interacted with oxygen to produce stable products, nitrite and nitrate that are key to generate RNS. Scavenger of NO competed with oxygen, leading to a reduced production of nitrite. The concentration of nitrite assays by the method of Feelisch and Noack [34] by reacting nitrite with griess reagent with which it forms a stable product which is measured at 542 nm on a spectrophotometer.

Methodology (for 96-well plate format):

i. Add 40 μl of phosphate buffer (20 mM, pH 7.4) to 10 μl of extract (1 mg/ml).

ii. To the solution prepared in step I, add 50 μl of sodium nitroprusside (10 mM).

iii. Incubate in dark at 37°C for 2.5 h.

iv. Subsequently add 100 μl of griess reagent to it and after 30 min of incubation.

v. Read the product microplate reader at 542 nm.

vi. Capacity of extract to inhibit this stable product is taken as NO quenching capacity/mg extract.

Application: Oxidative stress plays a key role on pathophysiology and aging [35–37]. Many phytochemicals/active compounds (polyphenols, flavonoids, terpenes, etc.) have antioxidant properties that are responsible for therapeutic effects of plants/herbs [14, 32]. Evaluating antioxidant potential of herbal/plant extract is the easy and very useful way to validate extract's therapeutic efficacy toward oxidative stress.

2.2 Assessment of autophagy, lipophagy, and mitophagy using *in vitro* cell system

2.2.1 Analysis of autophagy by Western blotting

Introduction and principle: Autophagy is a catabolic cellular multistep process [38]. In autophagy, most commonly known as macroautophagy, targeted cytoplasmic constituents (such as damaged macromolecules and organelles) are isolated from the rest of the cell within a double-membraned vesicle called autophagosome. These autophagosome eventually fuses with lysosomes forming autolysosome and the contents inside are degraded under acidic environment and get recycled. Non-selective autophagy or macroautophagy is dissociated with other forms of specific autophagy processes such mitophagy (autophagic degradation of mitochondria), lipophagy (autophagic degradation of lipids), ERphagy (autophagic degradation of ER), etc.

Autophagy is executed by several autophagy-related (Atg) genes initially identified in Saccharomyces cerevisiae and later validated in mammalian systems [39]. During initiation of autophagy, two ubiquitin-like proteins Atg12 to Atg5 covalently conjugate and then bind to Atg16L1 to form an E3-like complex which functions as part of the second ubiquitin-like conjugation system. This complex binds and activates Atg3, which covalently attaches mammalian homologs of the ubiquitin-like protein Atg8/LC3B to the lipid phosphatidylethanolamine (PE) on the surface of autophagosomes after conversion from LC3B-I to LC3B-II. This lipidated LC3B (LC3B-II) closes the autophagosomes and enables docking of cargo-specific adaptor proteins such as Sequestosome-1/p62. The completed autophagosome then fuses with a lysosome through the actions of multiple proteins. After autophagosome-lysosome fusion, LC3B that is retained on the outer side is cleaved off and recycled whereas inside the vesicle along with Sequestosome-1/p62 gets degraded. Hence, Western blot analysis of the rate of accumulation of LC3B-II (autophagic flux) under lysosomal inhibition (using inhibitors such as chloroquine, Bafilomycin A1 etc.) is the most appropriate way to study autophagy and take as autophagic marker [40].

Methodology:

i. Seed the cells in appropriate density in multi-well plate and incubate at 370C and 5% CO2: 95% Air environment inside an incubator.

ii. At 70–80% confluency, make four groups as i. Untreated; ii. Lysosomal inhibitor; iii. Treatment X (extract or natural compound); and iv. Treatment X + Lysosomal inhibitor.

iii. Treat groups iii and iv with extract or natural compound for 24–72 h.

iv. After the treatment is over, treat the groups ii and iv with Lysosomal inhibitor for 4–6 h for acute accumulation of autophagosomes.

v. Harvest cells for protein extraction using standard RIPA buffer and run western blot to analyze LC3B-II (14 kDa) and internal control (GADH).

Analysis of LC3B-II western blot: As shown in the (**Figure 1A**), there will be 3 cases of LC3B-II western blot image.

Step-I: Take normalized densitometric ratios for basal autophagy (yellow box): i.e. lanes 2:1.

Step-II: Take normalized densitometric ratios for autophagy under treatment (green box): i.e. lanes 4:3.

Step-III: Compare the ratios from basal autophagy (step-I) vs. autophagy under treatment (step-II) for significance.

Interpretation of LC3B-II western blot:

As mentioned earlier that autophagy is a kinetic process. However, during the analysis, the expression of cellular LC3B-II at a given time point is a static and unable to represent the rate of autophagosome synthesis and degradation after lysosomal fusion. i.e. autophagy flux. Thus, to better analyze this, use of lysosomal inhibitors is important. This enables us to analyze any increase in the LC3B-II in the presence of a lysosomal inhibitor representing autophagy induction or autophagosome biogenesis before lysosomal activity was blocked. Hence, as shown in the **Figure 1A**, there will be 3 cases (**Figure 1B-D**) of autophagy flux according to LC3B-II expression before and after the treatment of lysosomal inhibitors.

- Case-1: Activation of autophagic flux by treatment X.

If relative LC3B-II protein expression is higher under treatment when compared to untreated group (as shown in **Figure 1B** by red dotted line), and it is further increased after lysosomal inhibition (as shown in Figure B by double-head arrows) in treated group compared to untreated group.

- Case-2: Inhibition of autophagic flux by treatment X at autophagosome formation step (i.e. early block).

If relative LC3B-II protein expression is lower under treatment when compared to untreated group (as shown in **Figure 1B** by red dotted line), and it is further decreased or unchanged after lysosomal inhibition (as shown in **Figure 2B** by double-head arrows) in treated group compared to untreated group.

- Case-3: Inhibition of autophagic flux by treatment X at autophagosome maturation/autolysosome step (i.e. late block).

- If relative LC3B-II protein expression is higher under treatment when compared to untreated group (as shown in **Figure 1D** by red dotted line; and this can be easily miss-interpreted to autophagy activation without flux analysis), and it is decreased or unchanged after lysosomal inhibition (as shown in **Figure 1D** by double-head arrows) in treated group compared to untreated group.

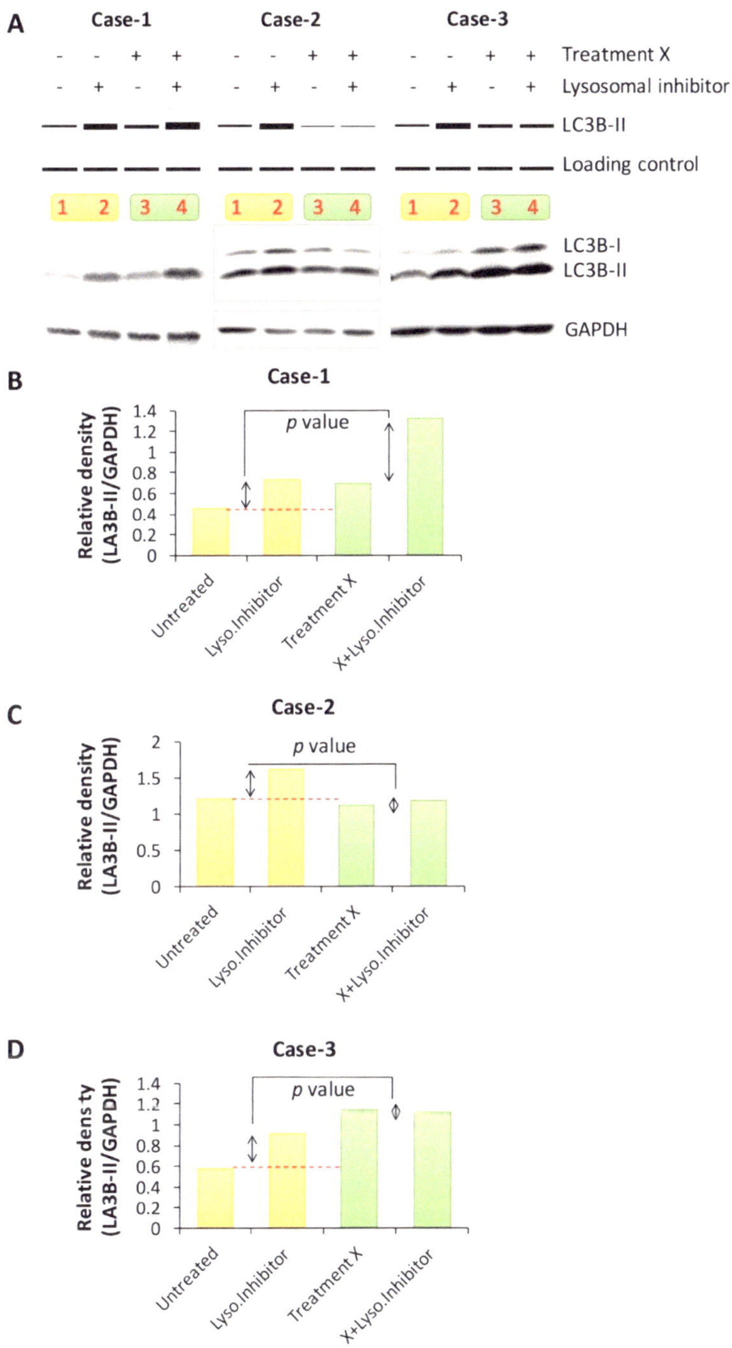

Figure 1.
Analysis of autophagic flux by Western blotting.

2.2.2 Analysis of autophagy and mitophagy by fluorescence microscopy

Introduction and principle: To analyze the progression of autophagy/mitophagy from autophagome/mitophagosomes to lysosomes, recently developed reporter

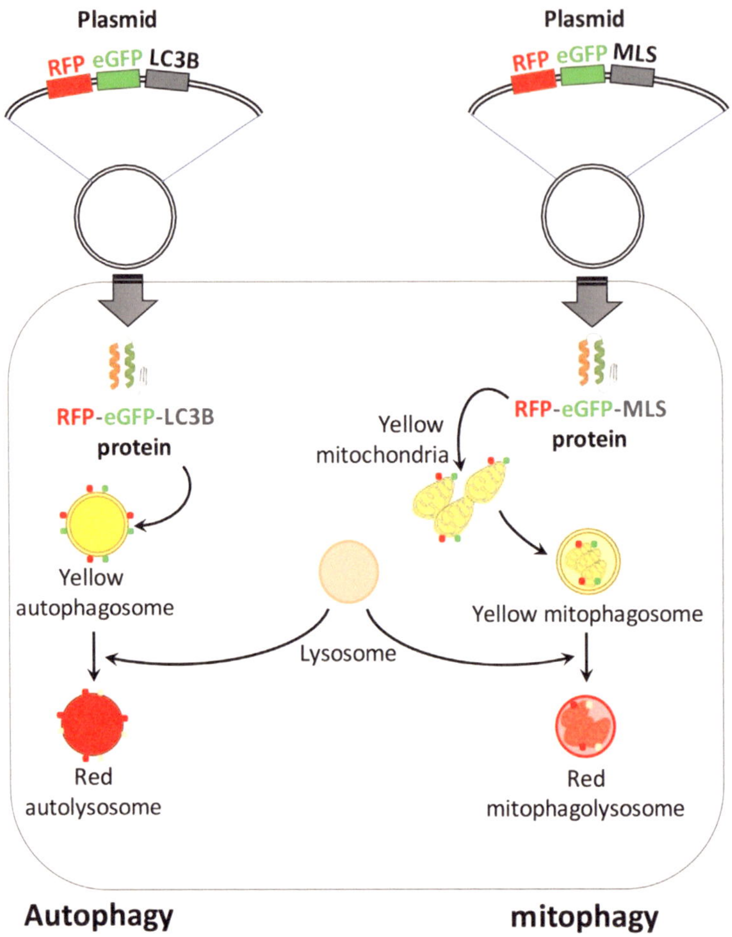

Figure 2.
RFP-GFP fusion plasmids to study autophagic/mitophagic flux under fluorescent microscope.

plasmids (**Figure 2**), that are tandem-tagged RFP-EGFP protein fused and cloned with LC3B (to analyze autophagy) or mitochondrial localization signal (MLS) typically COX-IV (also called mitoRFP-GFP plasmid; to analyze mitophagy), have been used, which exploit the differential stabilities of RFP and GFP in an acidic environment [40]. Green GFP signals are quenched in acidic pH, whereas RFP signals can be visualized in both autophagosome and acidic autophagolysosome, thus an increase in RFP fluorescence in the lysosomes indicates completion of the mitophagy process. However, at basal, yellow signals indicate merged images of GFP and RFP signals for only autophagosome containing mitochondria. Therefore, the formation of red puncta in treatment groups will indicate increased autophagic/mitophagic flux over untreated cells which will show yellow puncta.

Methodology:

i. Seed the cells in appropriate density in chambered slides and incubate at 370C and 5% CO2: 95% Air environment inside an incubator.

ii. Plasmid transfection: At 70–80% confluency, make two groups as i. untreated and ii. Treatment X (extract or natural compound), then use 0.5 μg plasmid and transfection reagent (usually Lipofectamin P3000 from Invitrogen, USA) to transfect DNA into the cells from both the groups, and incubate for 24 h at 370C and 5% CO2: 95% Air environment inside an incubator.

iii. Treat group ii with extract or natural compound for further 24–48 h.

iv. After the treatment is over, fix the cells in 4% paraformaldehyde for 15 min.

v. Stain cells with Hoechst 33342 or DAPI (Sigma) to stain nucleus.

vi. Wash the cells with 1x phosphate buffer saline (PBS) two times for 5 min each.

vii. Mount the slide using anti-fade reagent and visualize under the fluorescent microscope or confocal microscope at higher magnifications (40–60x is recommended).

Analysis of fluorescent images: Both RFP and GFP fluoresces on autophagosomes as LC3B and represents as yellow puncta in overlay under basal condition. However, in autolysosome, under acidic/low pH GFP (green) fluorescence get quenched and only RFP fluorescence can be visualized as red puncta (in overlay) of autolysosome suggesting increase in autophagy flux. Similarly, both RFP and GFP fluoresces on mitochondria and whole mitochondrial network can be visualized under basal condition. However, in mitophagolysosome, under acidic/low pH GFP (green) fluorescence get quenched and only RFP fluorescence will be visualized as red puncta (in overlay) of mitophagolysosome suggesting increase in mitophagy flux.

2.2.3 Analysis of lipophagy by fluorescence microscopy

Introduction and principle: To analyze lipophagy (autophagic degradation of lipids), cellular lipid droplets/triglycerides stain using BODIPY™ 493/503 dye (Invitrogen) and endogenous LC3B proteins stain using anti-LC3B antibody. Fluorescence microscopy analysis is performed to analyze co-localization of lipids/BODIPY with LC3B.

Methodology:

i. Seed the cells in appropriate density in chambered slides and incubate at 370C and 5% CO2: 95% Air environment inside an incubator.

ii. At 70–80% confluency, make two groups as i. untreated and ii. Treatment X (extract or natural compound), treat group ii with extract or natural compound for further 24–48 h, and incubate at 370C and 5% CO2: 95% Air environment inside an incubator.

iii. After the treatment is over, fix the cells in 4% paraformaldehyde for 15 min.

iv. Wash cells with 1x phosphate buffer saline (PBS) three times 5 min each.

v. Incubate cells with 1% BSA in PBST (PBS + 0.1% Tween 20) + 10% serum from the species the secondary antibody was raised in, for 1 h at room temperature to block unspecific binding of the primary antibody.

vi. Remove blocking reagent carefully and incubate cells in the diluted primary antibody (anti-LC3B) in 1% BSA in PBST overnight at 4°C.

vii. Wash cells with 1x phosphate buffer saline (PBS) three times for 5 min each.

viii. Incubate cells with the secondary antibody (Alexa Fluor® 594; Invitrogen) in 1% BSA in PBST for 1 h at room temperature in the dark.

ix. Remove reagent, wash one time in 1x PBS and incubate cells in BODIPY™ 493/503 stain (1:5000 dilution) for 15 min at room temperature.

x. Stain cells with Hoechst 33342 or DAPI (Sigma) to stain nucleus.

xi. Wash cells with 1x phosphate buffer saline (PBS) twice for 5 min each.

xii. Mount the slide using anti-fade reagent and visualize under the fluorescent microscope or confocal microscope at higher magnifications (40–60x is recommended).

Analysis of fluorescent images: Both BODIPY™ fluorescence is visualized using GFP/Annexin filter to observe green fluorescence of lipid droplets, however, RFP filter can be used to visualize LC3B red fluorescence. Lipophagy will be seen in overlay as yellow puncta formed by BODIPY-LC3B co-localization.

2.2.4 Analysis of autophagy, lipophagy, and mitophagy by electron microscopy

Introduction and principle: Electron microscopy is one of the best methods for the detection of autophagy and quantification of autophagic accumulation compared to fluorescence microscopy that has lower resolution than transmission electron microscopy [41]. During autophagy induction, the number of degradative compartments increases and the quantification of the number of autophagosome, autolysosomes, lysosomes, and amphisomes per cell section provides a simple morphological readout. But, it is very difficult to distinguish between autolysosomes, lysosomes, and amphisomes, hence, it is much simpler to classify all these organelles together, as degradative compartments or autophagic vesicles. However, autophagosome and autolysosome can be distinguished by observing their double membrane vs. single membrane walls respectively.

Methodology:

i. Seed the cells in appropriate density in chambered glass coverslips and incubate at 370C and 5% CO2: 95% Air environment inside an incubator.

ii. At 70–80% confluency, make two groups as i. untreated and ii. Treatment X (extract or natural compound), treat group ii with extract or natural compound for further 24–48 h, and incubate at 370C and 5% CO2: 95% Air environment inside an incubator.

iii. Fix the monolayers in 2% glutaraldehyde (EM grade) in 0.1 M phosphate buffer pH 7.4 for 1 h at room temperature.

iv. Wash with phosphate buffer twice for 3 min each.

v. Mix osmium tetroxide and NaCac in water to give final concentrations of 1% and 0.1 M respectively. Next, add potassium ferrocyanide (K4Fe(CN)6) at 15 mg/ml, add on to the cells and incubate for 1 h at room temperature.

vi. Wash with phosphate buffer twice for 3 min each, then with water once.

vii. Incubate the monolayers in 1% uranyl acetate and 0.3 M sucrose in water for 1 h at 4°C.

viii. Dehydrate cells with 70% ethanol for 1 min followed by 96% ethanol for 1 min, and 100% ethanol for 1 min three times.

ix. Dip cover slips into acetone and immediately drop some epoxy resin to cover the cells, before the acetone has completely evaporated.

x. Fill beem capsules with resin and place them upside down on the cover slips. Incubate for 2 h at room temperature and let the resin polymerize at 60°C overnight.

xi. Carefully remove the resin block from the cover slips and trim the block. Cut sections of 70- to 80-nm thickness.

xii. Post-stain the sections with uranyl acetate and lead citrate for 3 min for grid preparation [42].

xiii. Grids can be visualized under electron microscope and images can be acquired at 10000x as primary magnification, and 40,000x to 100,000x magnification can be used to visualized autophagic vesicles containing cargo (lipid droplets, mitochondria, etc.).

Application: Autophagy plays a key role in the cellular maintenance and development of disease [38, 43]. Impairment in autophagic removal of damaged mitochondria (mitophagy) and fatty acids/lipids (lipophagy) has been shown to be important in various pathological states including neurodegenerative diseases and metabolic disorders [38, 43]. Recently, phytochemicals have been shown to be an autophagy inducers and are beneficial in various diseases and metabolic disorders [44–47]. Hence, analyzing autophagy flux and their potential to remove damaged mitochondria and lipids can be an easy way to evaluate plant's therapeutic efficacy.

2.3 Assessment of mitochondrial turnover (synthesis of new and removal of damaged mitochondria) using MitoTimer

Introduction and principle: Fluorescent Timer, or DsRed, is a mutant of the red fluorescent protein, dsRed, in which fluorescence shifts over time from green to red as the protein matures or undergoes oxidation during the course of aging.

This molecular clock gives temporal and spatial information on protein turnover [48]. To monitor mitochondrial turnover, this Timer is engineered and tagged with mitochondrial-targeting sequence (called as "MitoTimer") and cloned in a plasmid. When cells get MitoTimer, it localizes in the mitochondria and disappearance of red fluorescence (for removal of damaged mitochondria) and appearance of green mitochondrial (for synthesis of new mitochondria) i.e. turnover, can be visualized.

Methodology:

i. Seed the cells in appropriate density in chambered slides and incubate at 370C and 5% CO2: 95% Air environment inside an incubator.

ii. Plasmid transfection: At 70–80% confluency, make two groups as i. untreated and ii. Treatment X (extract or natural compound), then use 0.5 μg plasmid and transfection reagent (usually Lipofectamin P3000 from Invitrogen, USA) to transfect DNA into the cells from both the groups, and incubate for 24 h at 370C and 5% CO2: 95% Air environment inside an incubator.

iii. Treat group ii with extract or natural compound for further 24–48 h.

iv. After the treatment is over, fix the cells in 4% paraformaldehyde for 15 min.

v. Stain cells with Hoechst 33342 or DAPI (Sigma) to stain nucleus.

vi. Wash the cells with 1x phosphate buffer saline (PBS) two times for 5 min each.

vii. Mount the slide using anti-fade reagent and visualize under the fluorescent microscope or confocal microscope at higher magnifications (40–60x is recommended).

Analysis of fluorescent images: As shown in the [49], both mitoRFP and mitoGFP fluoresces on mitochondria and represented as yellow mitochondrial network in overlay under basal condition, suggesting mitochondrial homeostasis. However, when there is an increase in mitochondrial turnover by mitophagy, mitoGFP (green) fluorescence increases and mitoRFP fluorescence decreases under (A) treatment X suggesting increase in new mitochondria via biogenesis and removal of old/damaged mitochondria by mitophagy. Similarly, when turnover is decreased, both mitoRFP increases and mitoGFP decreases suggesting decreased mitochondrial biogenesis and accumulation of damaged/aged mitochondria under (B) treatment X.

Application: Mitochondrial turnover is utmost important to maintain mitochondrial homeostasis that is required for cellular health [5, 50]. Dysregulated mitochondrial turnover leads to cause oxidative stress and cell death which is a major cause for metabolic disorders and cancer [24, 26, 51]. Study of mitochondrial turnover using MitoTIMER assay is a convenient method to evaluate phytochemicals efficacy.

2.4 Assessment of cell viability/cytotoxicity *in vitro*

2.4.1 MTT tetrazolium assay

Principle: Tetrazolium salts are widely used for detecting redox potential of cells for viability, cytotoxicity, and proliferation assays. This is a colorimetric assay that measures the reduction of yellow 3-(4,5-dimethythiazol2-yl)-2,5-diphenyl tetrazolium bromide (MTT) by mitochondrial succinate dehydrogenase, which demonstrates functional mitochondria [52]. The MTT is cell permeable and goes into the mitochondria and gets reduced to purple colored formazan crystals. These purple crystals are then solubilized using organic solvent. Lastly, this purple solution is measured spectrophotometrically. Notably, reduction of MTT only occurs in metabolically active cells, thus, the level of purple color is a measure of the viable cells.

Methodology:

i. Seed the cells in appropriate density in a 96-well plate and incubate at 370C and 5% CO2: 95% Air environment inside an incubator.

ii. At 70–80% confluency, make two groups as i. untreated and ii. Treatment X (extract or natural compound), treat group ii with extract or natural compound for further 24–48 h, and incubate at 370C and 5% CO2: 95% Air environment inside an incubator.

iii. After treatment, add 20 μl /well of MTT solution (Sigma) and incubate for 4–5 h in 37oC in dark.

iv. Aspirate MTT containing media and add 200 μl Di methyl Sulphoxide (DMSO, Sigma), mix gently and incubate for 1 h in 37oC in dark.

v. Measure absorbance at 570 nm (and background wavelength is 630 nm) using 96-well plate spectrophotometer.

vi. The linear relationship between absorbance and cell number should be taken into account and the data should be expressed as a percentage of viability in untreated group.

2.4.2 MTS assay

Principle: Colorimetric method for sensitive quantification of viable cells in proliferation and cytotoxicity assay. This assay is also based on the same principle as MTT assay, where we read colored formazan product by metabolically active cells is soluble in cell culture media. The MTS assay offers a number of advantages over the MTT assay. First, it eliminates the need for washing or solubilization steps, making it a more efficient and streamlined process. This high-throughput assay can be performed directly in the cell culture media, and can be conducted using a 96-well microtiter plate. The MTS assay is versatile and can be used to measure cell proliferation in response to a variety of stimuli, such as growth factors, cytokines, mitogens, and nutrients. Additionally, it is useful for analyzing the cytotoxic effects of compounds such as anticancer drugs, as well as other toxic agents, and pharmaceuticals.

Methodology:

i. Seed the cells in appropriate density in a 96-well plate and incubate at 370C and 5% CO2: 95% Air environment inside an incubator.

ii. At 70–80% confluency, make two groups as i. untreated and ii. Treatment X (extract or natural compound), treat group ii with extract or natural compound for further 24–48 h, and incubate at 370C and 5% CO2: 95% Air environment inside an incubator.

iii. Add 20 µl/well MTS Reagent into each well and incubate for 4 h at 37°C in standard culture conditions.

iv. Shake the plate briefly on a shaker and measure absorbance at 490 nm using 96-well plate spectrophotometer.

v. The linear relationship between absorbance and cell number should be taken into account and the data should be expressed as a percentage of viability of untreated group.

2.4.3 Luciferase-based assay for real-time cytotoxicity analysis

Principle: The latest innovation in viable cell number measurement is a real-time approach using a pro-substrate and luciferase. This method involves adding the reagent directly to the culture medium, where viable cells with active metabolism reduce the pro-substrate into a Luc-substrate. The Luc-substrate then diffuses into the culture medium, and luciferase utilizes it to generate a luminescent signal.

The pro-substrate is non-reactive with luciferase and is well-tolerated by cells, remaining stable in complete cell culture medium at 37°C for at least 72 hours. This long-term stability allows for multiple measurements from the same sample over a period of days without the need for pro-substrate replenishment.

The assay can be performed in both continuous-read and endpoint formats. In continuous-read mode, the luminescent signal is recorded repeatedly from the same sample wells over an extended period to measure the number of viable cells in real time. This new approach provides a reliable and efficient way to monitor viable cell numbers and metabolic activity in a variety of research applications [53].

Methodology:

RealTime-GloTM MT Cell Viability Assay (Promega) for continuous read mode:

i. Equilibrate the MT Cell Viability Substrate and the NanoLuc® Enzyme to 37°C.

ii. Seed the cells in appropriate density in a white colored 96-well plate optimize for luminescence.

iii. Add MT Cell Viability Substrate and the NanoLuc® Enzyme to 1X final concentration to the cell suspension and incubate at 370C and 5% CO2: 95% Air environment inside an incubator.

iv. After 24 h, make two groups as i. untreated and ii. Treatment X (extract or natural compound), treat group ii with extract or natural compound.

v. Measure luminescence using a 96-well plate luminometer, and return cells to the cell culture incubator between measurements to maintain a consistent temperature at each measurement. To determine the linear range of cell densities for the continuous-read format, continue measuring luminescence at various times over the desired time course (e.g. 12 h, 24 h, 72 h). The data should be expressed as a percentage of initial viability (at 0 h) of respective groups.

Application: Analysis of cell viability plays a diverse role in establishing the relationship between drug safety and its toxicity. High-throughput methods are very well embraced in the scientific community for drug's safety evaluation tastings. Hence, above-mentioned cytotoxicity/cell viability assays are best fit to titrate the safe dosage of phytochemicals.

3. Discussion

One of the important and attractive uses of the *in vitro* methods is to evaluate therapeutic efficacy and safety of active compounds/phytochemicals and/or crude extracts in a high-throughput manner, and that also limits the use of animals which can be expensive, labor-intensive, and time taking. The most common assays (as mentioned above in Section 2.1) are to evaluate antioxidant potential of herbal extracts or active compounds in a cell-free system or by using animal cell culture system. Prof. Y. Osumi has received Noble prize for re-stating the role of autophagy in cell health and disease [54]. Technological advancements in scientific field further made it possible to evaluate autophagic flux and related processes, such as mitophagy and lipophagy, in a high-throughput way by using fusion proteins (as discussed above in Sections 2.2 and 2.4) in animal cell culture system. Mitochondrial dysfunction and de-regulation of mitochondrial turnover and homeostasis play key roles in may diseases including cancer. Analysis of OCR that is an indicator of mitochondrial respiration and mitochondrial activity is utmost important for safety evaluation plant-based therapeutic compounds. Recently developed engineered mitochondria-localized fluorescent proteins [48], that aged or oxidized in time-dependent manner and fluoresce differentially (as discussed above in Section 2.3) are boon for new age research, can be used to evaluate mitochondrial turnover in a high-throughput way. Cytotoxicity is desirable in the cancer-targeted drugs; however, it can be detrimental for normal cells, so analysis of cytotoxicity is important for successful use of plant-based chemo-drugs or to target metabolic diseases. Although, there are several high-throughput assays available for assessing the potential cell viability and cytotoxicity of plant extracts and phytochemicals, this chapter (in Section 2.5) provides a summary of both commonly used and recently developed cell-free and cell-based assays. These assays serve as important tools for evaluating the safety and therapeutic effectiveness of plant-based drugs and compounds.

Acknowledgements

The authors are thankful to the Ministry of Health (MOH), and National Medical Research Council (NMRC), Singapore, for financial assistance grant number NMRC/OFYIRG/0002/2016 and MOH-000319 (MOH-OFIRG19may-0002) to BKS; NMRC/OFYIRG/077/2018 to MT; and CSAI19may-0002 to PMY.

Conflict of interest

The authors declare no conflict of interest.

Notes/thanks/other declarations

None.

Author details

Madhulika Tripathi[1*], Manish Kumar Verma[2] and Brijesh Kumar Singh[1*]

1 Cardiovascular and Metabolic Disorders Program, Duke-NUS Medical School, Singapore

2 Raghuveer Singh Government (P.G.) College, Lalitpur, Uttar Pradesh, India

*Address all correspondence to: madhulika.tripathi@duke-nus.edu.sg and singhbrijeshk@duke-nus.edu.sg

References

[1] Molaei E et al. Nephroprotective activity of natural products against chemical toxicants: The role of Nrf2/ ARE signaling pathway. Food Science & Nutrition. 2021;**9**(6):3362-3384

[2] Nejabati HR, Roshangar L. Kaempferol as a potential neuroprotector in Alzheimer's disease. Journal of Food Biochemistry. 2022;**46**(12):e14375

[3] Rolt A, Cox LS. Structural basis of the anti-ageing effects of polyphenolics: Mitigation of oxidative stress. BMC Chemistry. 2020;**14**(1):50

[4] Sapian S et al. Role of polyphenol in regulating oxidative stress, inflammation, fibrosis, and apoptosis in diabetic nephropathy. Endocrine, Metabolic & Immune Disorders Drug Targets. 2022;**22**(5):453-470

[5] Kakkar P, Singh BK. Mitochondria: A hub of redox activities and cellular distress control. Molecular and Cellular Biochemistry. 2007;**305**(1-2):235-253

[6] Singh S et al. Nrf2-ARE stress response mechanism: A control point in oxidative stress-mediated dysfunctions and chronic inflammatory diseases. Free Radical Research. 2010;**44**(11):1267-1288

[7] Ying M, Hu X. Tracing the electron flow in redox metabolism: The appropriate distribution of electrons is essential to maintain redox balance in cancer cells. Seminars in Cancer Biology. 2022;**87**:32-47

[8] Andres CMC et al. Superoxide anion chemistry-its role at the Core of the innate immunity. International Journal of Molecular Sciences. 2023;**24**(3):1841

[9] Trachootham D, Alexandre J, Huang P. Targeting cancer cells by ROS-mediated mechanisms: A radical therapeutic approach? Nature Reviews. Drug Discovery. 2009;**8**(7):579-591

[10] Ascensao A et al. Modulation of hepatic redox status and mitochondrial metabolism by exercise: Therapeutic strategy for liver diseases. Mitochondrion. 2013;**13**(6):862-870

[11] Baffy G. Mitochondrial uncoupling in cancer cells: Liabilities and opportunities. Biochimica et Biophysica Acta - Bioenergetics. 2017;**1858**(8):655-664

[12] Georgieva D, Egli D. Tying genetic stability to cell identity. Cell Cycle. 2017;**16**(12):1139-1140

[13] Akbari B et al. The role of plant-derived natural antioxidants in reduction of oxidative stress. BioFactors. 2022;**48**(3):611-633

[14] Antonio Pereira I et al. Traditional plants used in southern Brazil as a source to wound healing therapies. Chemistry & Biodiversity. 2023:e202201021

[15] Jalali J, Ghasemzadeh Rahbardar M. Ameliorative effects of Portulaca oleracea L. (purslane) and its active constituents on nervous system disorders: A review. Iranian Journal of Basic Medical Sciences. 2023;**26**(1):2-12

[16] Kamkin V et al. Comparative analysis of the efficiency of medicinal plants for the treatment and prevention of COVID-19. International Journal of Biomaterials. 2022;**2022**:5943649

[17] Tarmizi NM et al. Genus Lepisanthes: Unravelling its botany, traditional uses, phytochemistry, and pharmacological properties. Pharmaceuticals (Basel). 2022;**15**(10):1261

[18] Huang PL. A comprehensive definition for metabolic syndrome. Disease Models & Mechanisms. 2009;**2**(5-6):231-237

[19] Lemieux I, Despres JP. Metabolic syndrome: Past, present and future. Nutrients. 2020;**12**(11):3501

[20] Pavlova NN, Zhu J, Thompson CB. The hallmarks of cancer metabolism: Still emerging. Cell Metabolism. 2022;**34**(3):355-377

[21] Ali M et al. Selected hepatoprotective herbal medicines: Evidence from ethnomedicinal applications, animal models, and possible mechanism of actions. Phytotherapy Research. 2018;**32**(2):199-215

[22] Martinez-Outschoorn UE et al. Cancer metabolism: A therapeutic perspective. Nature Reviews. Clinical Oncology. 2017;**14**(1):11-31

[23] Rodic S, Vincent MD. Reactive oxygen species (ROS) are a key determinant of cancer's metabolic phenotype. International Journal of Cancer. 2018;**142**(3):440-448

[24] Bhatti JS, Bhatti GK, Reddy PH. Mitochondrial dysfunction and oxidative stress in metabolic disorders - A step towards mitochondria based therapeutic strategies. Biochimica et Biophysica Acta - Molecular Basis of Disease. 2017;**1863**(5):1066-1077

[25] Chan DC. Mitochondrial dynamics and its involvement in disease. Annual Review of Pathology. 2020;**15**:235-259

[26] Gaude E, Frezza C. Defects in mitochondrial metabolism and cancer. Cancer & Metabolism. 2014;**2**:10

[27] Picca A et al. Mitochondrial dysfunction, oxidative stress, and neuroinflammation: Intertwined roads to neurodegeneration. Antioxidants (Basel). 2020;**9**(8):647

[28] Picca A et al. Mitochondrial dysfunction and aging: Insights from the analysis of extracellular vesicles. International Journal of Molecular Sciences. 2019;**20**(4):805

[29] Falzon CC, Balabanova A. Phytotherapy: An introduction to herbal medicine. Primary Care. 2017;**44**(2): 217-227

[30] Tillotson AK. Essential concepts and vocabulary in herbal medicine. Journal of Dietary Supplements. 2008;**5**(4):411-421

[31] Padmavathi G et al. Butein in health and disease: A comprehensive review. Phytomedicine. 2017;**25**:118-127

[32] Waltenberger B et al. Natural products to counteract the epidemic of cardiovascular and metabolic disorders. Molecules. 2016;**21**(6):807

[33] Rice-Evans CA. Measurement of total antioxidant activity as a marker of antioxidant status in vivo: Procedures and limitations. Free Radical Research. 2000;**33**(Suppl):S59-S66

[34] Feelisch M, Noack EA. Correlation between nitric oxide formation during degradation of organic nitrates and activation of guanylate cyclase. European Journal of Pharmacology. 1987;**139**(1):19-30

[35] El Assar M, Angulo J, Rodriguez-Manas L. Oxidative stress and vascular inflammation in aging. Free Radical Biology & Medicine. 2013;**65**:380-401

[36] Guzik TJ, Touyz RM. Oxidative stress, inflammation, and vascular aging in hypertension. Hypertension. 2017;**70**(4):660-667

[37] Sies H, Jones DP. Reactive oxygen species (ROS) as pleiotropic physiological signalling agents. Nature Reviews. Molecular Cell Biology. 2020;**21**(7):363-383

[38] Sinha RA, Singh BK, Yen PM. Reciprocal crosstalk between autophagic and endocrine Signaling in metabolic homeostasis. Endocrine Reviews. 2017;**38**(1):69-102

[39] Harnett MM et al. From Christian de Duve to Yoshinori Ohsumi: More to autophagy than just dining at home. Biomedical Journal. 2017;**40**(1):9-22

[40] Klionsky DJ et al. Guidelines for the use and interpretation of assays for monitoring autophagy (4th edition)(1). Autophagy. 2021;**17**(1):1-382

[41] Yla-Anttila P et al. Monitoring autophagy by electron microscopy in mammalian cells. Methods in Enzymology. 2009;**452**:143-164

[42] Wyffels, J.T., Principles and Techniques of Electron Microscopy: Biological Applications, Fourth edition, by M. A. Hayat. Microscopy and Microanalysis, 2001. 7(1): p. 66

[43] Galluzzi L et al. Molecular definitions of autophagy and related processes. The EMBO Journal. 2017;**36**(13):1811-1836

[44] Moosavi MA et al. Phytochemicals as potent modulators of autophagy for cancer therapy. Cancer Letters. 2018;**424**:46-69

[45] Nabavi SF et al. Regulation of autophagy by polyphenols: Paving the road for treatment of neurodegeneration. Biotechnology Advances. 2018;**36**(6):1768-1778

[46] Sinha RA et al. Caffeine stimulates hepatic lipid metabolism by the autophagy-lysosomal pathway in mice. Hepatology. 2014;**59**(4):1366-1380

[47] Zhou J et al. Epigallocatechin-3-gallate (EGCG), a green tea polyphenol, stimulates hepatic autophagy and lipid clearance. PLoS One. 2014;**9**(1):e87161

[48] Hernandez G et al. MitoTimer: A novel tool for monitoring mitochondrial turnover. Autophagy. 2013;**9**(11):1852-1861

[49] Singh BK et al. Thyroid hormone receptor and ERRalpha coordinately regulate mitochondrial fission, mitophagy, biogenesis, and function. Science Signaling. 2018;**11**(536):eaam5855

[50] Peinado JR et al. Mitochondria in metabolic disease: Getting clues from proteomic studies. Proteomics. 2014;**14**(4-5):452-466

[51] Amorim JA et al. Mitochondrial and metabolic dysfunction in ageing and age-related diseases. Nature Reviews. Endocrinology. 2022;**18**(4):243-258

[52] Gerlier D, Thomasset N. Use of MTT colorimetric assay to measure cell activation. Journal of Immunological Methods. 1986;**94**(1-2):57-63

[53] Riss TL, Moravec RA, Niles AL. Cytotoxicity testing: Measuring viable cells, dead cells, and detecting mechanism of cell death. Methods in Molecular Biology. 2011;**740**:103-114

[54] Fullgrabe J et al. Transcriptional regulation of mammalian autophagy at a glance. Journal of Cell Science. 2016;**129**(16):3059-3066

Section 3

Application of Cell Culture in Enhancing Drug Discovery

Chapter 6

Perspective Chapter: Evaluating New Drugs against K-Ras4B/PDE6δ Using an *In Vitro* Approach

Dayan A. Carrion-Estrada, Paola Briseño-Diaz, Sandra Delfín-Azuara, Arturo Aguilar-Rojas and Miguel Vargas

Abstract

Cancer represents the leading cause of global mortality worldwide. Recent estimates have shown that approximately 25% of all cancer types exhibiting *KRAS* mutations, making these mutations one of the most reported so far. Given the important role played by KRas during the progression of different tumors, the search for new therapeutic compounds that can reduce the adverse effects of this oncogene becomes evident. However, discovering effective anticancer compounds is a complex and time-consuming task. These compounds should ideally exhibit potent anticancer properties at low concentrations, with minimal impact on healthy cells. The validation of potential candidates involves several stages and methods, including *in vitro* techniques such as cell lines or primary cell cultures grown under 2D and 3D conditions. This chapter provides a comprehensive review of methods to support the effectiveness of two compounds, C14 and P8, specifically targeting mutant KRas as potential antitumor agents. Cytotoxicity assays were employed on breast and pancreatic cancer cell lines and primary cell cultures grown in 2D and 3D conditions to evaluate the effectiveness of these compounds. The use of multiple cell culture systems provides more pertinent data, enhancing our understanding and assessment of the potential benefits of new therapeutic molecules.

Keywords: human primary cell, 3D cell cultures, K-Ras4B, PDE6δ, therapy

1. Introduction

Cancer is the leading cause of death worldwide. According to GLOBOCAN, every year, over 20 million people are diagnosed with cancer globally, which leads to nearly 10 million deaths reported in 2020 [1]. The increasing incidences of cancer worldwide have created a demand for identifying and developing effective and tolerable therapies for treating different types of cancer [2].

Among the most common causes of cancer death in 2020 were: lung, colorectal, stomach, breast and pancreatic cancer [1]. It is important to highlight that, from these six pathologies, they share in common the presence and dependence of *KRAS* gene mutations [3].

KRAS (Kirsten rat sarcoma viral oncogene homolog) was first reported to be a transforming gene related to oncogenesis in humans in the early 1980's [4, 5]. Since then, *KRAS* has emerged as one of the extensively studied proto-oncogenes, owing to its established mutation prevalence of approximately 90% in pancreatic adenocarcinomas, 45% of colorectal adenocarcinomas, 22% of lung adenocarcinomas and 5% of breast cancers [6]. KRAS mutations are found in 30% of all cancers, making it the most frequently mutated oncogene across all cancer types and a promising drug target due to its participation in the progression and maintenance of several types of tumors [7].

KRAS belongs to the RAS family of small GTPases and behaves as a molecular switch by cycling between a GTP-bound active state and a GDP-bound inactive state. Wild-type KRAS is a tumor suppressor that is usually activated via specific receptors to regulate different signaling pathways involved in, cell cycle progression, cell growth, proliferation, migration, and apoptosis (**Figure 1**). For K-Ras protein to be active it needs to be transported to the cell membrane, through phosphodiesterase 6 delta protein (PDE6δ) [10, 11]. The interaction between K-Ras and PDE6δ is performed by the recognition of a post-translational modification (farnesyl group) present in the GTPase (**Figure 1**).

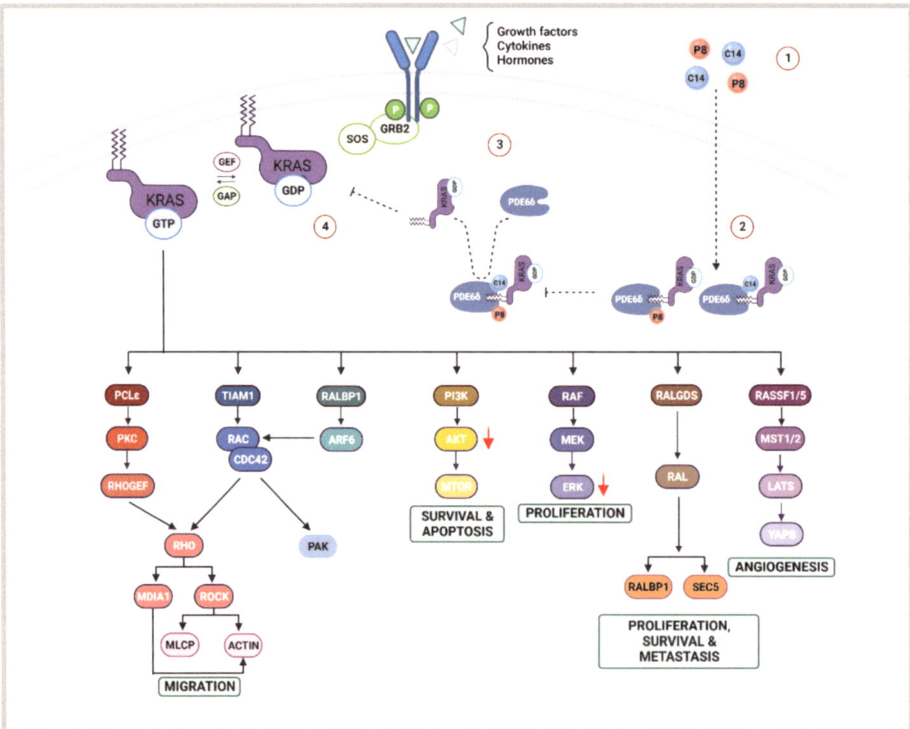

Figure 1.
RAS effector pathways. Plasma membrane-associated Ras-GTP can directly interact with multiple different effectors to activate different signaling pathways such as cell proliferation, migration, survival/death, differentiation, migration, and adhesion. Previous studies have shown that the administration of C14 and P8 compounds [1] cause a strong binding of the mutant K-Ras4B/PDE6δ complex [2] inhibiting either it dissociation [3], or, and binding to the membrane [4] resulting in a diminution in the activation of its principal effectors Akt and Erk (red arrows). Modified from Refs. [8, 9].

However, its function has been shown to be frequently lost during tumor progression in many types of cancer. The disruption of KRAS regulation is triggered by specific point mutations occurring at codons 12, 13, 61, and 146 within the *KRAS* gene. These mutations hinder GTP hydrolysis, resulting in continuous activation of the K-Ras protein and abnormal activation of K-Ras signaling pathways. Consequently, this abnormal activation promotes cell proliferation, differentiation, and survival (**Figure 1**) [10].

For decades, several groups made unsuccessful attempts to target K-Ras in cancer treatment due to the protein's unique characteristics, specifically the absence of binding grooves on its surface for small molecules. As a result, K-Ras was considered undruggable [12–14]. However, in 2013, an innovative strategy emerged with the discovery that PDE6δ plays a role in regulating the subcellular location of K-Ras. In this research, inhibiting the interaction between both proteins resulted in K-Ras being retained in the cytoplasm, which impacted its ability to transmit signals. Consequently, the K-Ras-PDE6*δ* complex was identified as a potential pharmacological target [15]. However, these strategies have shown be limited to a single K-Ras mutant, and toxic to non-cancerous cells due to their mechanism affecting both K-Ras WT and other proteins recognized by PDE6*δ* [15, 16].

In 2018, our research group presented a new strategy. We proposed the stabilization of the K-Ras/PDE6*δ* complex by small molecules instead of preventing its interaction (**Figure 1**). Based on this, a series of compounds capable of binding and firmly fixing the interaction between the K-Ras and PDE6*δ* proteins were selected *in silico* (**Figure 2**) [17]. These molecules showed excellent results against *KRAS*-dependent cancers, with elevated affinity towards mutated variants of K-Ras, thereby preserving the signaling activity of K-Ras WT [8, 18–20]. After several assays carried out in a big library of compounds, the molecules denominated as C14 and P8 showed the best ability to specifically bind to and stabilize the mutated K-Ras4B/PDE6δ complex (**Figure 2**) [17].

The antitumoral efficacy of these molecules has been assessed in various *in silico* models, *in vitro* models (cell lines and primary cultures), and *in vivo* models of pancreatic and colon cancer [8, 18]. These have resulted in the decreased activity of the principal K-Ras signaling pathways (e.g., MAPK and PI3K/Akt pathways) (**Figures 1** and **2**), apoptosis-mediated cancer cell death, reduction in clonogenic capacity and decreased tumor volume [8, 17].

Given the high prevalence of cancer, there is a need to find novel drugs to treat this complex disease. These new treatments should have the minimal side effects on the patients, while also being more effective in eliminating malignant cells and showing low rates of resistance to them. For these reasons, it is necessary to evaluate new therapeutic options in many models to achieve their validation.

In vitro models play a vital role in drug discovery for cancer, as they afford a high degree of control over numerous variables within a research study, and the experimental design is less complex. Moreover, these models offer a cost-effective approach and serve as alternatives to animal use, thereby addressing ethical considerations [21]. In cancer drug discovery, *in vitro* models include various cell culture techniques, with three primary types demonstrating progressive degrees of complexity: 2D, primary, and 3D cultures. These advancements have significantly contributed to our understanding of cancer cells growth and migration, disease response, and drug interactions within the cell [22]. Currently, drug discovery and testing involve the use of both *in vitro* models and *in vivo*-animal models.

This chapter reviews the most common *in vitro* models used to characterize and validate the efficacy of C14 and P8 as the main drugs against the K-Ras/PDE6δ complex. An overview of cell cultures is provided, including cell lines in 2D conditions, 3D systems, and primary cell cultures.

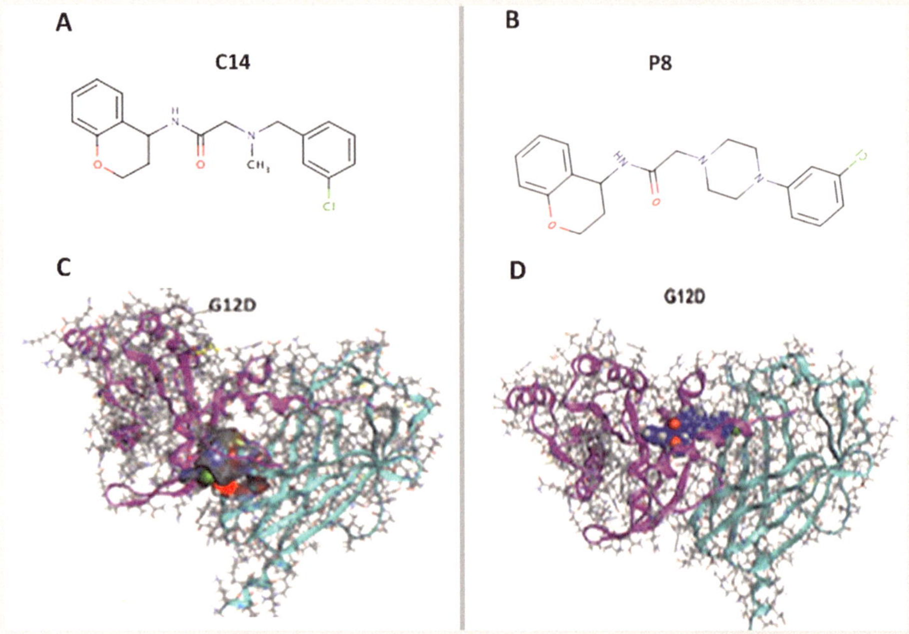

Figure 2.
Anti-cancer compounds identified in-silico search. (A) 2-[(3-chlorophenyl) methyl-methyl-amino]-N-chroman-4-yl-acetamide structure known as C14. (B) 2-[4-(3-chlorophenyl)piperazin-1-yl]-N-[(4R)-chroman-4-yl] acetamide structure known as P8. (C and D) By molecular dynamic assays, both molecules demonstrated strong binding and stabilization of the most common mutant K-Ras4B G12D/PDE6δ complex. The dissociation of this complex is impaired and the GTPase function of K-Ras4B inhibited. Both compounds bind to the complex in different positions, which could have a synergic effect [8].

2. *In vitro* models for anti-cancer drug discovery

The process of drug development and its evaluation can indeed be costly. As a result, *in vitro* cell culture models have been extensively utilized for these purposes due to their numerous benefits, including reduced expenses and time compared to animal studies. These *in vitro* models can be classified into two types: 2D and 3D models, each with its own set of advantages and disadvantages, as described in **Table 1**. Furthermore, the source from which the cells used as models are obtained can be adapted to the study's characteristics. Therefore, the combination of cell lineage or culture conditions provides great versatility to *in vitro* models. For instance, when screening and selecting leads from hundreds of compounds, a 2D assay would be the best option over 3D models, as it allows for the lowest feasible cost due to the reduced requirement of cells and time for these assays. However, for fewer compounds requiring more in-depth characterization, a 3D model would be more suitable [23].

2.1 2D cell culture for anticancer drug testing

2D cultures are the most employed cell models to test drugs due to two major advantages: the methods to cultivate and preserve the cells under these conditions are uncomplicated and importantly; maintenance in terms of materials that are needed is low-cost and being a simpler model interpretation of results is easier [24].

	Cell lines	**Primary cultures**	**3D models**
Accessibility	Can be purchased from a company (derived from established cell lines)	Derived directly from isolated tissues.	Significant time must be invested in developing these models (generated from cell line or primary culture cells)
Proliferation	Unlimited	Limited	Limited
Characterization	Well characterized	Must be characterized	Must be characterized
Complexity and genetic variation	Simpler, with fewer cell types. Minimal genetic variation.	Presents higher heterogeneity, with various cell types. High genetic variation that reflects the genetic makeup of the tissue of origin.	Higher complexity than 2D cultures. Resembles tumor structure. Higher genetic variation.
Biological relevance	May not fully represent *in vivo* conditions	Closer to *in vivo* physiological conditions	Closer to *in vivo* physiological conditions and structures 3D cell culture models more closely mimic *in vivo* behavior and better represents drug metabolism. Additionally, protein and gene expression levels in culture resembling those found *in vivo*.
Ease of result interpretation	Results are easier to interpret	More challenging to interpret due to a higher number of variables involved	More challenging to interpret due to a higher number of variables involved
Ease of handling	Uncomplicated to handle	More challenging to handle	More challenging to handle
Cost	Less expensive to maintain	Cost is higher at the beginning when the cell line is being established.	More costly to maintain.

Table 1.
Advantages and disadvantages of in vitro *culture methods.*

In this classic model, cells grow as a monolayer in a polystyrene culture flask or petri dish [24]. By definition, a cell line is a collection of cells originated from one cell. These cell lines are *in vitro* systems typically established and propagated in a growth medium in tubes, flasks, or dishes as a monolayer culture. Here, cells can continue to divide indefinitely. Using a cell line allows each cell in an experiment to have a similar response because genetic variation is minimal. This can help researchers to observe differences between test groups more clearly [25].

As previously stated, our research group has evaluated the antineoplastic effects of a group of compounds capable of stabilizing the K-Ras-PDE6δ complex. To validate their effects in 2D conditions, we utilized various pancreatic and breast cancer cell lines (**Table 2**). Additionally, antitumoral effects were determinate by viability assays were used.

2.1.1 Pancreatic cancer cell lines

About 90% of pancreatic cancers are histologically classified as pancreatic ductal adenocarcinoma (PDAC) [26], and although it ranks as the 12th most

Pancreatic cancer-derived cell lines
hTERT-HPNE (ATCC, CRL-4023). Cultured in 75% DMEM without glucose (Biowest) with additional 2 mM L-glutamine (Gibco) and 1.5 g/L sodium bicarbonate plus, 25% Medium M3 Base (Gibco). The media is supplemented with fetal bovine serum 5%, 10 ng/ml human recombinant EGF (Sigma-Aldrich), 5.5 mM D-glucose (1 g/L) (Sigma-Aldrich) and 750 ng/ml puromycin (Sigma-Aldrich) at 37°C in an atmosphere with 5% CO2.
BxPC-3 (ATCC, CRL-1687). Cultured in RPMI medium (Biowest, USA) supplemented with 10% fetal bovine serum and 1% antibiotic 100 U/mL (penicillin/streptomycin) and 1.5 g/L sodium bicarbonate at 37°C in an atmosphere with 5% CO2.
Panc-1 (ATCC, CRL-1469). Cultured in DMEM medium (Biowest, USA) supplemented with 10% fetal bovine serum and 1% antibiotic 100 U/mL (penicillin/streptomycin) and 1.5 g/L sodium bicarbonate at 37°C in an atmosphere with 5% CO2.
MIA PaCa-2 (ATCC, CRL-1420). Cultured in DMEM medium (Biowest, USA) supplemented with 10% fetal bovine serum and 1% antibiotic 100 U/mL (penicillin/streptomycin) and 1.5 g/L sodium bicarbonate at 37°C in an atmosphere with 5% CO2.
Capan-1 (ATCC, HTB-79). Cultured in Iscove's Modified Dulbecco's Medium (ATCC) supplemented with 20% fetal bovine at 37°C in an atmosphere with 5% CO2.
Breast cancer-derived cell lines
MCF 10A (ATCC, CRL-10317): Cultured in Dulbecco's modified Eagle's medium F12 (DMEM F12) (Gibco, New York, USA) supplemented with 10% fetal bovine serum (Gibco) and 1% antibiotic 100 U/mL (penicillin/streptomycin) (Sigma-Aldrich) and 1.5 g/L sodium bicarbonate (Sigma-Aldrich) at 37°C in an atmosphere with 5% CO2.
MDA-MB-231 (ATCC, HTC-26): Cultured in Lebovitz (L-15) medium (Gibco) supplemented with 10% fetal bovine serum and 1% antibiotic 100 U/mL (penicillin/streptomycin) and 1.5 g/L sodium bicarbonate at 37°C in an atmosphere with 5% CO2.
MDA-MB-231 RR (ATCC, HTC-26): Cultured in DMEM F12 medium (Gibco) supplemented with 10% fetal bovine serum and 1% antibiotic 100 U/mL (penicillin/streptomycin) and 1.5 g/L sodium bicarbonate at 37°C in an atmosphere with 5% CO2.
MCF 7 (ATCC, HTB-22): Cultured in DMEM medium (Biowest, USA) supplemented with 10% fetal bovine serum and 1% antibiotic 100 U/mL (penicillin/streptomycin) and 1.5 g/L sodium bicarbonate at 37°C in an atmosphere with 5% CO2.
MCF 7 RR (ATCC, HTB-22): Cultured in DMEM F12 medium (Biowest, USA) supplemented with 10% fetal bovine serum and 1% antibiotic 100 U/mL (penicillin/streptomycin) and 1.5 g/L sodium bicarbonate at 37°C in an atmosphere with 5% CO2.

Table 2.
Cell lines employed for drug screening.

common cancer, for decades it has firmly held the first position as the cancer type with the lowest 5-year survival rate [27]. Although a general mutational profile has been well determined for PDAC, the standardly gene reported as mutated is *KRAS* (>90%) [28]. Among the muted forms that are present in PDAC, the most frequently are G12D (~51%), G12V (~32%) and G12R (~12%) [29]. These mutations have been associated with a poor prognosis due to their ability to induce alterations in cancer cells, enabling them to evade immune response, reprogramming their metabolism, and developing resistance to therapy [30]. Given the importance of *KRAS* mutations in the carcinogenesis and tumor progression of PDAC, mutant pancreatic cell lines are an excellent *in vitro* model for anti K-Ras/PDE6δ drug testing. In our research group, we evaluated the antitumoral effects of the candidates molecules in the pancreatic cell lines (ATCC, USA) as describe in **Figure 3**.

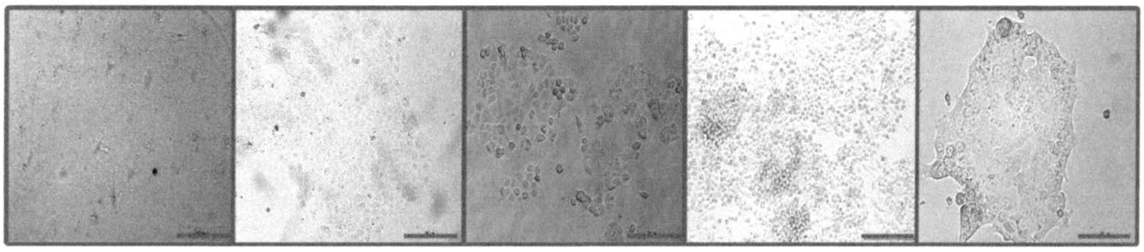

Figure 3.
*Representative images of principal pancreatic cancer cell lines that present most of the KRAS mutations. In brief, BxPc-3 is a pancreatic cell line, which does not express the mutant form of K-Ras4B (**Figure 1B**). Panc-1 expresses the mutation in exon 2 of the KRAS gene, K-Ras4B^{G12D} (**Figure 1C**). MIA-PaCa expresses the mutation in exon 2 of the KRAS gene, K-Ras4B^{G12C} (**Figure 1D**) and CAPAN-1 expresses the mutation in exon 2 of the KRAS gene, K-Ras4B^{G12V}. The non-tumoral pancreatic cell line hTERT HPNE, was employed as control (**Figure 1A**). Scale bar 200 μM.*

2.1.2 Breast cancer cell lines

According to Globocan, breast cancer (BC) ranks first in incidence and second in mortality among all oncologic malignancies worldwide [1]. This is a heterogeneous group of diseases that can be grouped into four subtypes into four subtypes (luminal A, luminal B, HER2 and Triple negative Breast Cancer (TNBC)) [31, 32]. Regarding the genomic landscape of BC, various alterations have been identified within each subtype. Among them, mutations in *KRAS* play an important role in carcinogenesis or progression of BC [33, 34] and, this gene is mutated in about 7%-12% of the patients (2% luminal A, 20% luminal B, 17% HER2, 0–8% TNBC) [3]. Specifically, the presence of oncogenic K-Ras is associated with the more aggressive luminal B and HER2 subtypes compared to the luminal A type [35]. Significantly, this gene is amplified in approximately 32% of TNBC cases [36] leading to an upregulation of *KRAS* signaling pathways and an enhanced tumor microenvironment that promotes tumor progression [33]. Breast cancer cell lines are described in **Figure 4**.

For initial drug screening it is suggested to determine the proper concentration to use in the desired assays through dose-esponse curve and IC50 determination. This process is often performed in 2D assay of the cell line of interest. The cells are plated and treated with various concentrations of the test compounds, typically in an

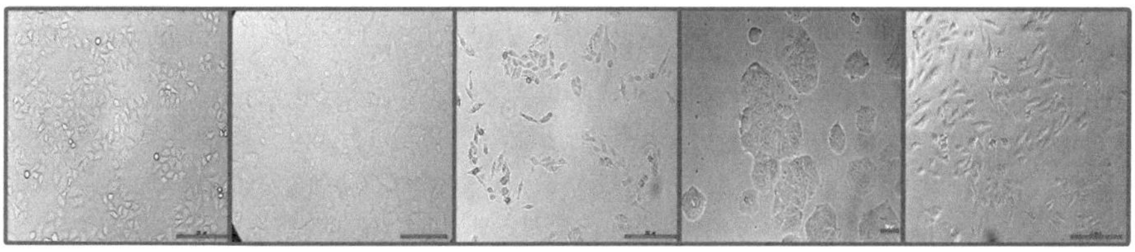

Figure 4.
*Representative images of breast cancer cell lines. Breast cancer cell lines and derived radio-resistant cells growth in monolayer. Scale bar 200 μM. In brief, MCF-7 is an estrogen receptor (ER)-positive cell line, which does not express the mutant form of K-Ras4B (**Figure 3B**). MDA-MB-231 is a TNBC that expresses the mutation in exon 2 of the KRAS gene, K-Ras4B^{G13D} (**Figure 5C**) [37]. To explore the effect of C14 and P8 over radioresistant cell lines, MCF 7RR which does not contain the mutation in K-Ras4B (**Figure 3D**) and MDA-MB-231RR, that possess the mutation K-Ras4BG13D as was observed in its parental cell line were used (**Figure 5D**) and MDA-MB-231RR, that possess the mutation K-Ras4BG13D as was observed in its parental cell line (**Figure 5E**) [38]. The non-tumoral breast cell line MCF 10A, was employed as control (**Figure 5A**).*

eight-point dose range, along with vehicle and media only controls The methods are described in detail in Ref. [39].

2.2 Methods for viability/cytotoxicity methods determination

Cell viability or cytotoxicity in monolayer cultures can be determined by different methods. Cell viability assays are frequently employed, with a variety of markers as indicators of metabolically active (living) cells, such as measuring ATP levels (e.g., CellTiter-Glo® Luminescent assay), measuring the ability to reduce a substrate (e.g., tetrazolium reduction and resazurin reduction), and detecting enzymatic/protease (e.g., CellTiter-Fluor assay) activities present only in living cells.

The most widely used technique is Tetrazolium Reduction Cell Viability Assays, which measures cellular metabolism as an indicator of cell viability, proliferation, and cytotoxicity. In this method, the positively charged dye, 3-(4,5-dimethylthiazol-2-yl)-2,5-diphenyltetrazolium bromide (MTT), penetrates viable cells and is metabolized into a purple-colored product known as formazan. As a result, color formation can be a helpful indicator of viable cells. However, the incubation time for this method is long (more than 4 hours). To address this limitation, other negatively charged compounds such as (2,3-bis(2-methoxy-4-nitro-5-sulfophenyl)-2H-tetrazolium-5-carboxanilide) (XTT), which cannot penetrate cells, must be combined with intermediate electron coupling reagents that help them enter into the viable cells. Once inside the cells, XTT is reduced to tetrazolium, a soluble formazan product. The incubation time for this method is 1–4 hours, making it more convenient [40]. For our experiments, we used MTT and XTT assays to build a dose-response curves and determine the effect of different concentrations of compounds C14 and P8 in pancreatic and breast cancer cell lines viability.

Anti-cancer drugs C14 and P8 should demonstrate inhibition of cancer cells growth or proliferation, and these is most often determined by IC50. In this work, IC50 is defined as the particular inhibitory drug concentration required to reduce the percentage of viable cells by 50% compared to cells grown without drug exposure [41].

The IC50 value (molar concentration) is important for *in vitro* models because it indicates the amount of a drug is required to inhibit a biological process by half, indicating changes in the population due to increased cell death or decreased cell proliferation. In cancer, using the IC50 concentration of a compounds means killing cancer cells and preventing cancer cell growth while having a less damage on healthy cells in the body or inhibiting tumor growth by half. A low IC50 suggests that the drugs will be effective at lower concentrations and will consequently cause less systemic toxicity when administered to patients for therapy [42].

2.3 Evaluation of compound targeting K-Ras4B/PDE6δ in pancreatic cancer cell lines

A dose-response curve to evaluate the antitumoral effects of compounds of interest was performed using MTT assays. Pancreatic cancer cells were plated in 96-well plate and treated with various concentrations of C14 and P8 compounds, along with vehicle (DMSO) and media-only as controls. The IC50 concentration was determined using Prism 8 software (GraphPad, USA).

As observed below, C14 possess a high cytotoxic property in eliminating cancer cells (**Figure 5B–E**, blue line, and **Figure 5F**). On the other hand, P8, an analogue of C14, demonstrates a more potent cytotoxic effect on cancer cells (**Figure 5B–E**, red line, and **Figure 5F**) compared to the vehicle (**Figure 5B–E** black line).

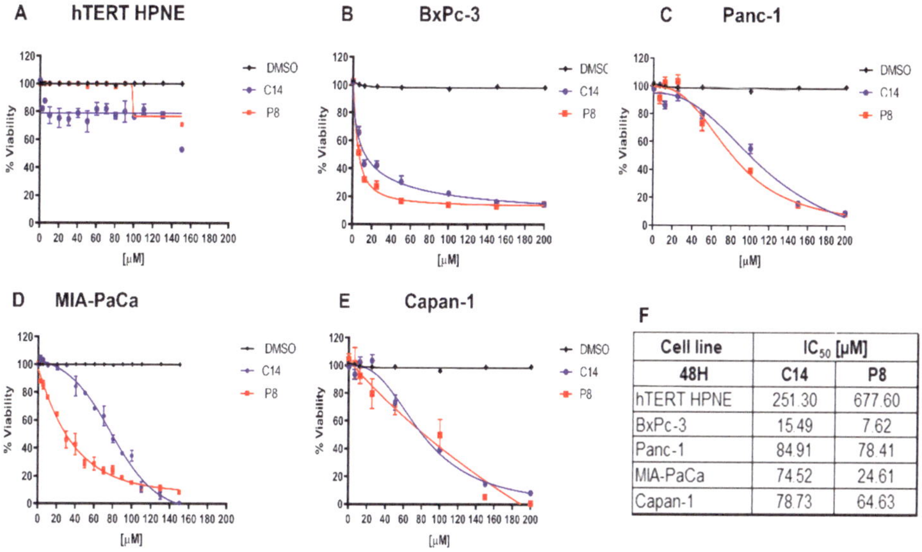

Cell line	IC_{50} [μM]	
48H	C14	P8
hTERT HPNE	251.30	677.60
BxPc-3	15.49	7.62
Panc-1	84.91	78.41
MIA-PaCa	74.52	24.61
Capan-1	78.73	64.63

Figure 5.
Dose-response curves of pancreatic cell lines treated with compounds C14 and P8. Dose-response curves using 0–200 μM concentrations of compound C14 and P8 tested at 48 hours post incubation measured with MTT. Each curve was done by three independent replicates.

Another crucial point is the specific effect of C14 and P8 over malignant cells. As the results showed above, *KRAS* mutant cell lines Panc-1-G12D, MIA-PaCa-G12C and Capan-1-G12V exhibit a considerable cytotoxic effect for both compounds compared to the vehicle (DMSO) (**Figure 5C,D** and **E**). This behavior was also observed in BxPC-3WT cell line, although BXPC-3 cells are WT for *KRAS*, the present several other mutations, which contribute to the activation of K-Ras and the dependence to other effectors downstream K-Ras signaling pathway, potentially explaining their susceptibility to the treatment [43].

Remarkably no inhibitory effect was observed in the viability of the control cell line hTERT-HPNE, even at concentrations higher than 200 μM after 48 hours of exposure (**Figure 5A**, red and blue line and **Figure 5F**).

These IC50 values of C14 and P8 in cancer cell lines are lower than 90 μM, which is 2 and 6 times less than the concentrations required to affect the viability of control cells (**Figure 5A** and **F**). This suggests that these compounds could be an excellent treatment alternative for pancreatic cancer, as they exhibit cytotoxicity to cancer cells while sparing normal pancreatic cells from side effects.

As a result, this assay allows to observe the specific cytotoxic effect of compounds on the viability mainly in *KRAS* mutant cancer cells and highly *KRAS*-dependent cells without affecting healthy cells, making them a promising treatment alternative.

2.4 Evaluation of compound targeting K-Ras4B/PDE6δ in breast cancer cell lines

As mentioned earlier, dose-response curves were generated to determine the cytotoxic effect of compounds C14 and P8 on the breast cancer cell lines using XTT assays. The breast cancer cells were plated in 96-well plate and treated with various concentrations of the compounds, with DMSO used as a vehicle and media only as control. The IC50 concentration was determined using Prism 8 software (GraphPad, USA). These data are presents in **Figure 6**.

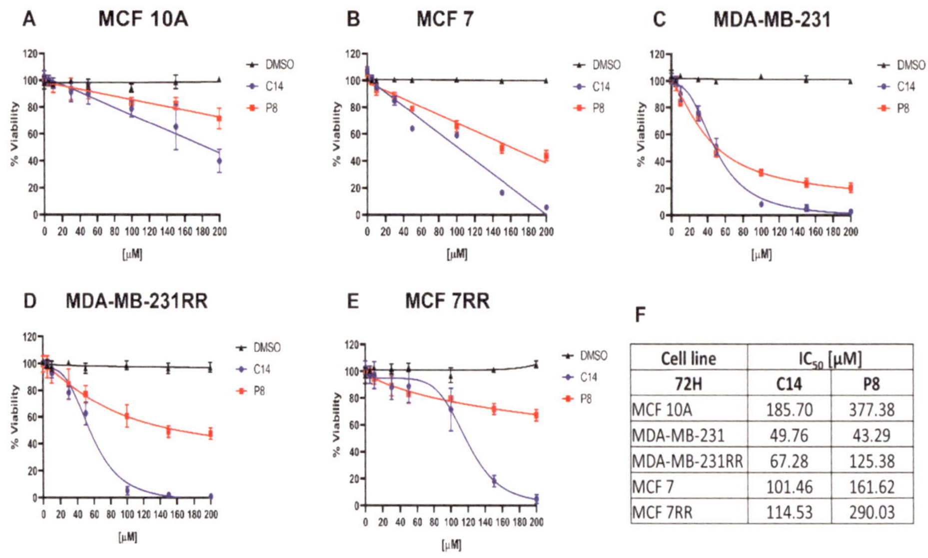

Cell line	IC_{50} [μM]	
72H	C14	P8
MCF 10A	185.70	377.38
MDA-MB-231	49.76	43.29
MDA-MB-231RR	67.28	125.38
MCF 7	101.46	161.62
MCF 7RR	114.53	290.03

Figure 6.
Dose-response curves of the breast cell lines treated with compounds C14 and P8. Dose-response curves using 0–200 μM concentrations of compound C14 and P8 tested at 72 hours post incubation measured with XTT each curve was done by three independent replicates.

Similar to observed in pancreatic cancer cells, no noticeable effect on the viability of the control cell line MCF 10A was observed after its exposure to C14 and P8 (**Figure 6A**, lines red and blue and **Figure 6F**). However, both compounds exhibited a potent cytotoxic effect over the rest of the tested breast cancer cell lines tested (**Figure 6B–E**, lines red and blue and **Figure 6F**). According with the IC50 values, even at concentrations higher than 100 μM after 72 hours of exposure, both compounds did not have impact over the viability of non-tumoral cells (**Figure 6A** and **F**). This finding supports the notion that C14 and P8 have a specific effect over *KRAS*-mutant cells such as, MDA-MB-231 (**Figure 6C**). This discovery was also observed in the radioresistant cell line MDA-MB-231RR, where the compounds also had a significant inhibitory effect at 72 hours (**Figure 6D**). This finding suggests an alternative course of treatment for patients who have experienced a cancer recurrence after radiation therapy. Finally, in the case of MCF 7 and MCF 7RR, the IC50 values in them were higher than those found in *KRAS* mutant lines (**Figure 6B** and **E**). It is important to note that the compounds have an inhibitory effect on the viability of luminal and radioresistant breast cancer cells, even though that these cell lines do not carry the *KRAS* gene mutation (**Figure 6A, E** and **F**).

Taken together, these findings indicate that the conventional 2D cell-based cancer drug screening on pancreatic and breast cancer cell yields reproducible results of the specific cytotoxic effect over cancer cells without affecting non-tumor control cells. This makes them suitable for further experiment and for evaluating their effects in a more complex cell culture.

Primary cultures generated directly from tumors are a viable alternative to employing cell lines. In comparison to cell lines, in which all cells are genetically identical, primary culture cells gene expression can vary from cell to cell. Primary cultures offer several advantages. Not only are cells isolated directly from the tumor site but provide the complete pathology which allows to compare the properties of the culture to those of the original. In general, such cultures can be generated as explants, in which mixed cell populations develop out of small fragments of tissue, or as enriched populations of particular cell types, such as cancer cell or cancer stem cells, being ideal to evaluate the efficacy of a new anticancer drug [44].

3. Primary cell culture for anticancer drug testing

Primary cultures consist of cell cultures derived directly from isolated tissue cells (**Figure 7**). These cultures can proliferate under 2D and 3D conditions without undergoing any genetic modifications. However, the number of passages for their use is highly limited [45]. Primary cultures derived from malignant tumors are particularly of great interest in research. Various protocols are currently available for obtaining primary cell cultures, and the selection of a specific method depends on factors such as tissue type and sample size. In **Table 3**, summarizes the different methods for obtaining primary cell cultures in 2D from pancreatic cancer tumor samples [46–48]. Once the primary cultures are obtained, they must be characterized to determine the presence of the required cell lineage. Currently, various methods facilitate their characterization including, the evaluation of specific markers by immunofluorescence microscopy or Fluorescent Activated Cell Sorter (FACS) (e.g., **Figures 8** and **9**, respectively). The most commonly used antibodies for malignancy markers are: E-Cadherin, Ki-67, Vimentin, B-Catenin, MUC-1, MUC-4, MUC-16, EGFR, CEA, lineage markers: CK19 and CK7, and for cancer stem cells (CSCs) marker: CD24, CD44 and CD133 [8, 49–52].

3.1 Isolation and characterization of primary cultures derived from pancreatic cancer

Primary PDAC cultures were obtain in the laboratory using an enzymatic method involving collagenase and trypsin (1:5 trypsin/EDTA solution, 15 min until the tissue was loose) under sterile conditions (**Table 3** and **Figure 7**). The requirements to maintain the primary culture vary depending on the aggressiveness of the cancer and the stage of the cancer being studied. The percentage of serum and growth factors is gradually decreased until the cells can survive in standard conditions (10% serum and 1% antibiotic) (**Table 4**).

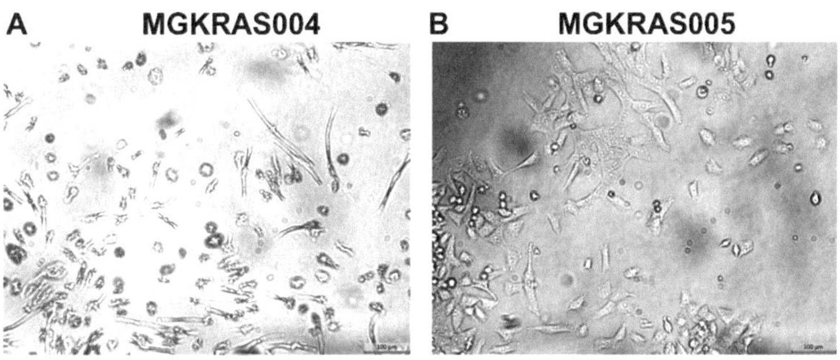

Figure 7.
Representative image of the primary cell cultures after three passages denominated as MGKRAS004 (A) and MGKRAS005 (B), obtained from malignant pancreatic tumors. Scale bar (red line) 100 μM.

Method	Mechanism	Advantages	Disadvantages
Explant	The tissue is cut into several fragments with the aid of a scalpel or dissecting scissors.	The cells coming from the niche leave and give rise to the exit of the tumors	Culture is very slow as there is no cell dissociation
Chemical	A tissue fragment is placed in a 1.5 mL tube and 10 volumes of 0.5% EDTA are added.	Intercellular bonds are broken.	Cells do not dissociate.
Mechanic	Cut, scratch, the tissues into small pieces to separate the cells and wash them with gentle agitation.	It is a fast technique for soft tissues.	Cells are not dissociated, and cell viability may be decreased.
Enzymatic	Enzymes for cutting or digesting pieces of tissue into free cells, collagenase and trypsin are commonly used.	Cells dissociate more easily.	Enzymatic cleavage can modify proteins on the surface of cells.
Combination	Combination of the 4 previous methods.	A greater number of cells are obtained, and a suitable niche can be generated.	Caution with the times.

Table 3.
Different methods to obtain primary cultures from pancreatic cancer.

The characterization of those cultures previously was previously performed using immunofluorescence microscopy employing the specific markers, CD44+ (Ambion, USA) (**Figure 8A**), CD24+ (Ambion) (**Figure 8B**) and CD133+ (Ambion)(**Figure 8C**). These assays strongly suggested the presence of cancer stem cells (CSCs), in both cultures. To verify this information, flow cytometry assays were conducted which revealed the presence of CSC in 2.4% of total population of MGKRAS004 cells and in 2.18% of MGKRAS005 population (**Figure 9A**).

Based on these results, there is a high abundance of CSCs in both primary cultures. This finding is significant due CSCs have the ability to contribute to the development of a more aggressive tumor characterized by high rates of proliferation, invasion, and resistance to conventional therapies [53]. Keeping this in mind, C14 and P8 compounds were evaluated on these primary cell cultures.

3.2 Evaluation of compounds targeting K-Ras4B/PDE6δ on primary cultures from pancreatic tumors

After demonstrating the heterogenicity and complexity of pancreatic primary cultures, the efficacy of C14 and P8 compounds was evaluated on 2D conditions using primary cultures obtained from samples of pancreatic malignant tumors, with the same conditions used for cell lines and DMSO as control (**Figure 10A** and **B**). According to our evaluation, both molecules were more effective against the MGKRAS004 cell line. In this case, C14 exhibited an IC50 of 38.72 μM, while P8 showed an IC50 of 80.88 μM with (**Figure 10A** and **C**). These IC50 values are consistent with those obtained for the common pancreatic cell lines (**Figure 5F**). However, in the case of MGKRAS005, the effect of C14 was reduced, and an IC50 of 268.2 μM

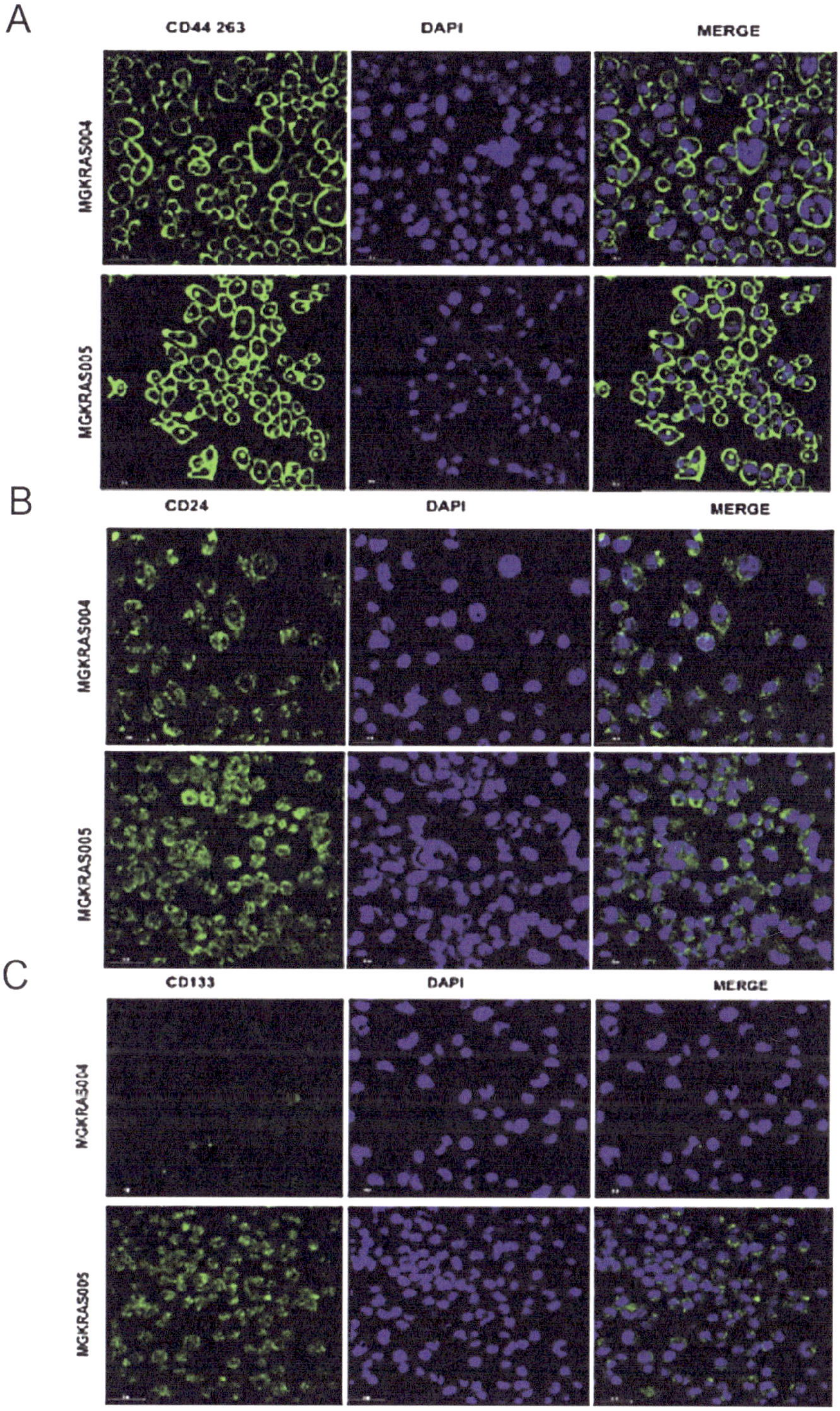

Figure 8.
Expression of CSC markers in primary pancreatic cultures. (A) CD44 expression in MGKRAS004 and MGKRAS005 cells. (B) CD24 expression in MGKRAS004 and MGKRAS005 cells. (C) CD133 expression in MGKRAS004 and MGKRAS005 cells.

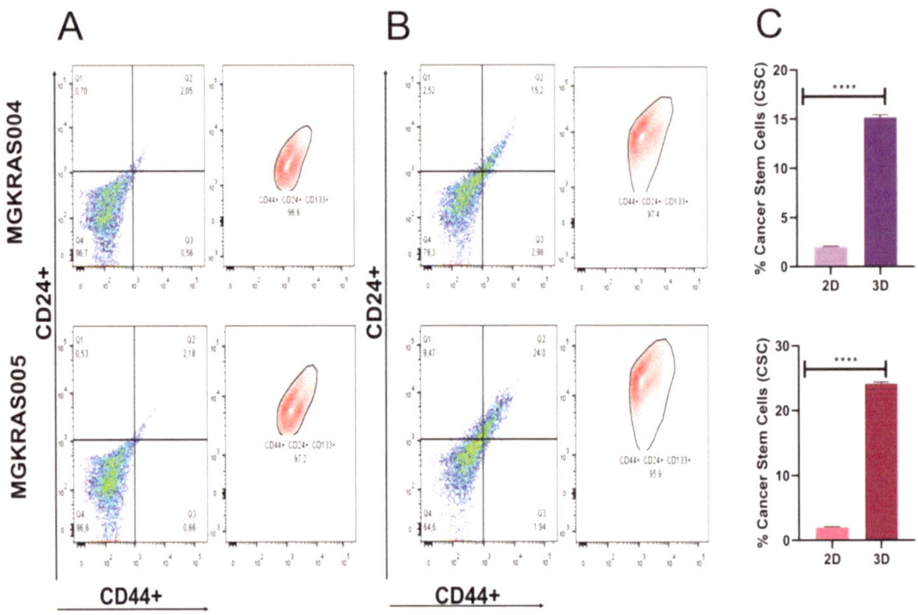

Figure 9.
Cancer stem cell proportion in pancreatic primary culture. (A) Quantification by FACS of CD44+, CD24+ and CD133+ in 2D conditions of pancreatic cancer primary cultures. (B) Quantification by FACS of CD44+, CD24+ and CD133+ in 3D conditions of pancreatic cancer primary cultures. (C) Comparison of % of CSC in 2D and 3D conditions.

was observed (**Figure 10B** and **C**). A similar trend was observed with P8, where the IC50 observed was 148.9 μM (**Figure 10B** and **C**). The explanation for this data could be related to the amount of CSC present in each population. As mentioned earlier, both primary cultures contain a similar proportion of CSC. However, in the case of MGKRAS005, the population of cells expressing CD133 is greater (**Figure 8C**). This higher expression of CD133 in MGKRAS005 cells may contribute to their reduced sensitivity to C14 and P8, leading to higher IC50 values observed in the treatment response. These data are in agreement with previously reports that associated this molecule with chemoresistance in various cancer cells [54].

Requirements	P0	P1–3	P4–10
Medium	DMEM High Glucose or Advance DMEM F12 Medium	DMEM High Glucose or Advance DMEM F12 Medium	DMEM High Glucose or Advance DMEM F12 Medium
FBS	70%	50%	10%
Matrigel	0.1%	—	—
Antibiotic/ antimycotic	5%	3%	1%
EGF	10 ng/mL	10 ng/mL	—
Other	Ciprofloxacin 10ug/mL	Ciprofloxacin 10ug/mL	—

Table 4.
Nutritional requirements for primary pancreatic cultures since its insolation and placed culture passage 0 (P0) until passages 4 to 10 (P4–10).

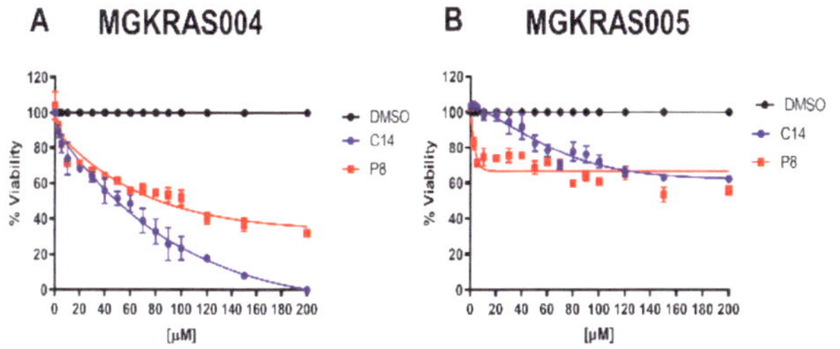

C

Cell Lines	IC_{50} [μM]	
48H	C14	P8
MGKRAS004	38.72	80.88
MGKRAS005	268.2	148.9

Figure 10.
Effect on the viability of primary cultures in 2D. (A) MGKRAS004 in 2D treated with C14 and P8. (B) MGKRAS005 in 2D treated with C14 and P8.

It is important to highlight that many drug candidates will fail during clinical trials, leading to the loss of money and time invested in research. *In vitro* models, plays an important role in drug discovery, especially in the early and preclinical stages. Considering this, despite the increase in the IC50 value in pancreatic primary cultures these compounds shown a significant cytotoxic effect in cancer cells, making them candidates for more in-depth characterization.

Although the most used cell culture growth conditions in cancer research are 2D models, they do not accurately reflect the native environment and general physiology of malignant cells and tumors such as, mechanical and biochemical signaling or intercellular communications [55], cell invasion [56] and expression of pathological markers [56]. To address these limitations, three-dimensional (3D) *in vitro* models have been developed. In this models, it has been possible to reproduce a more detailed aspects of tumoral behavior including interactions among tumor cells, the extracellular matrix, and stromal cells [57]. Additionally, cells growing in 3D, conditions have been exhibited a greater resistance to chemotherapeutics agents and mayor invasiveness compared to 2D models [58, 59]. As a result, the use of 3D models has been gaining more popularity every day. Considering this, we present the advances obtained in 3D models to evaluate the effect of C14 and P8 in one more realistic pancreatic cancer cells systems.

4. 3D models for anticancer drug testing

Currently, various types of 3D models are available for cancer research. The most commonly used ones include spheroid models, organ-on-a-chip models, hydrogel models, and bio-printed models [57]. Each of these approaches has its own specific advantages and disadvantages. However, despite the significant time and cost involved in their production, spheroids are the most widely utilized due to their unique characteristics [60].

Spheroids are small-scale 3D models that could self-assemble into spherical cell aggregates (with radii ranging from 100 to 600 μm) [60–62]. These models exhibit various tumoral characteristics, such as, a central necrotic core, surrounded by quiescent cells and an outer layer consisting of actively proliferating cells [60]. Other important features observed are, pH, oxygen, and metabolic compound gradients [63], and in some cases, micrometastases [64].

In this section, the focus is on spheroids, generated from cell lines or cells obtained from primary cultures of tumoral tissues. The following will describe the procedure for creating spheroids using cell lines (refer to **Table 2**) as well as spheroids from primary cultures, along with the method for characterizing their cellular composition. Subsequently, the cytotoxic effects of compounds C14 and P8 on the 3D models will be discussed.

4.1 Spheroids derived from cell lines and primary cell cultures

The following method outlines the protocol utilized by our research group to generate spheroids from cell lines or primary cultures (**Figure 11**). Firstly, the process begins with expanding of cultures under 2D conditions. Cell harvesting is conducted using a standard procedure. Briefly, when the cells reach 70-80% confluency in a plate, the medium is removed, and the cells are collected using trypsinization (0.05% trypsin, 0.53 mM EDTA, Gibco, USA). Subsequently, the cells are centrifuged and resuspended in 1–5 ml of spheroid medium, consisting of DMEM/F12 (Gibco), 2 mM L-glutamine, 100 U/ml penicillin, and 100 U/ml streptomycin (Sigma-Aldrich). The spheroid medium is further supplemented with 20 ng/ml recombinant human epidermal growth factor (Sigma-Aldrich), 10 ng/ml recombinant human basic fibroblast growth factor (R&D Systems, USA), and 1X B27 supplement (Sigma-Aldrich).

From this cellular suspension, approximately 3000–5000 cells are seeded per well in 96-well ultra-low adherence plates (Gibco). The plates are then maintained at 37°C and 5% CO2 for a period of 4 to 7 days, depending on the cell line and desired spheroid size. During the initial 5-day growth phase, it is crucial to minimize disturbances to the plates, particularly during the initial 5-day growth phase [65]. At the end of the culture period, spheroids are collected through gentle centrifugation at 800 rpm and utilized in functional assays.

Alternatively, enzymatic dissociation of the 3D models can be achieved by incubating the culture for 10 minutes with 0.05% trypsin (Invitrogen), followed by mechanical pipetting using a flame-polished Pasteur pipette. The dissociated cells are then passed through a 40 μm pore size filter (Corning). Subsequently, the cells can

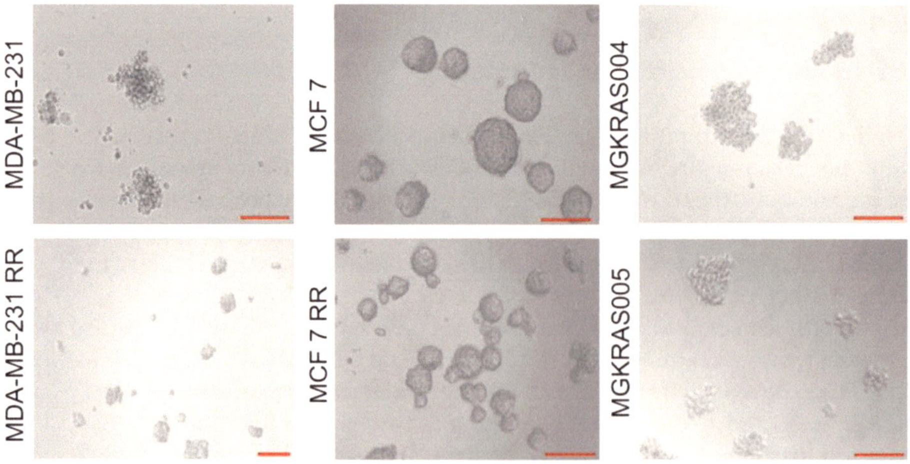

Figure 11.
Representative images of spheroids from cell lines (MDA-MB-231, MDA-MB-231 RR, MCF7 and MCF 7RR) and primary cell cultures (MGKRAS004 and MGKRAS005) after 7 days of culture. Scale bar 200 μM.

be analyzed for specific molecule expression using techniques such as FACS [66] or conventional immunofluorescence microscopy.

Figure 11 shows all the cell lines and cells from primary cell culture, that were able to develop spheroids. In the case of MDA-MB-231, this cell line is able to generate spheroids in ultra-low adherence plates, although they present a less compact appearance and less well-defined edges (**Figure 11**) [61]. Something similar is observed in its derivative radioresistant cell line (**Figure 11**). On the other hand, MCF 7 and MCF 7RR cells, quickly formed created three-dimensional spheroids with a homogenous inner structure when placed on plates with extremely low attachment properties as previously shown [67, 68]. Finally, although the structure of 3D models derived from primary cell cultures MGKRAS004 and MGKRAS005 is looser and more challenging to obtain compared to cell lines, these models provide a closer representation of the *in vivo* microenvironment of the patient's tumor [69].

4.2 Calculation of spheres formation efficiency (%)

To verify the effectiveness of the aforementioned method, the percentage of spheroids can be determined. This involves counting the number of spheroids (larger than 40 μm) after the culture period using a microscope at 40X magnification. Digital images of 5 random fields are captured using a digital camera connected to an optical microscope, and the size of the spheroids is determined using acquisition software. The spheroid formation efficiency (MFE%) is calculated using Eq. (1).

$$MFE(\%) = \frac{(\text{number of mammospheres } \textit{per} \text{ well})}{(\text{number of cells seeded } \textit{per} \text{ well})} \times 100 \qquad (1)$$

4.3 Fluorescent activated cell sorter (FACS)

As mentioned earlier, the characterization of spheroids involves various methods, including FACS. According to previous reports [70, 71], the generation of spheroids leads to the enrichment of a subpopulation known as Cancer Stem Cell (CSC). These CSCs population, are cell that are distinguished by their high tumorigenic potential, which includes self-renewal, pluripotency, and proliferative abilities. Additionally, they exhibit resistance to conventional treatments such as chemotherapy and radiation [72, 73]. Given these characteristics, CSCs are widely used as therapeutic targets for new drugs.

The following section outlines, the method employed by our research group to evaluate the expression of specific CSC markers using FACS in spheroids. This approach is employed to determine the antitumoral effect of the compounds of our interest. The cells derived from the spheroids are dissociated, as described above, and then stained with the following antibodies: anti-CD24-phycoerythrin (PE) (Abcam, USA), anti-CD44-allophycocyanin (APC) (Abcam), and anti-ALDH1-fluorescein (FITC) (Abcam). The cellular staining is conducted according to a procedure previously reported [66]. In brief, the dissociated cells are passed through a 40 μm cell filter (Corning) and counted to adjust the concentration to 100,000 cells/ml. Each antibody is diluted at (1:100), added and incubated at 37°C for 1 hour. After incubation, the cells are washed three times with PBS. Subsequently, flow cytometry analysis is performed.

4.4 Enrichment of CSC in spheroids derived from primary cell cultures

In the following section, representative data of cells derived from primary cultures of pancreatic tumors are presented. FACS assays demonstrated the presence of CD44+, CD24+, and CD133+ markers in spheroids derived from primary cultures (**Figure 9**) [74]. These findings support the existence of CSCs in spheroids obtained from MGKRAS004 and MGKRAS005 samples [75]. As mentioned earlier, the generation of spheroids selectively enriches the CSC population. The enrichment of CD44+, CD24+, and CD133+ marker expressions in spheroids, compared to 2D cultures, further confirms the selective effect of 3D systems on this cell population (**Figure 9C**). In the case of MGKRAS004, 2.4% of CSCs were found in 2D cultures, whereas this percentage increased to 6.3% in 3D conditions (**Figure 8A** and **B**). Similar results were observed for MGKRAS005 cells, with 2.18% of CSCs detected in 2D cultures and a significant increase to 24% in 3D cultures (**Figure 9A** and **B**). These findings also support the utilization of these 3D models as a more realistic system to evaluate the effects of new therapeutic compounds. It is worth noting that the levels of CD44+ and CD24+ in patients have recently been associated as predictors of mortality [76].

4.5 3D cell viability assay and IC50 determination

Using the 3D models described above, the cell viability effects of compounds C14 and P8 were determined. For this purpose, the CellTiter-Glo® 3D Cell Viability Assay Reagent from Promega (USA) was employed. This method measures ATP as an indicator of viability and provides a luminescent readout. Briefly, spheroids were generated as mentioned previously, and then exposed to both compounds for 48 hours. Cell lysis and luminescent signal recording were performed according to the manufacturer's instructions, and IC50 values were calculated using Prism 8 software (GraphPad, USA).

In the 3D models, although C14 and P8 reduced the cell viability of spheroids, their IC50 values were higher compared to the 2D conditions. For MGKRAS004, the IC50 for C14 in 3D was 395.7 μM, whereas it was 38.72 μM in 2D (**Figure 12A,C** and **Figure 10A,C**). P8 had an IC50 104.4 μM in 3D and 80.88 μM in 2D (**Figure 9A**). In the case of MGKRAS005, C14 exhibited an IC50 of 470.4 μM in 3D and 268.2 μM in 2D (**Figure 12B,C** and **Figure 10B,C**). On the other hand, P8 showed an IC50 of 96.71 μM in 3D and 148.9 μM in 2D (**Figure 12B, C** and **Figure 10B,C**). This increase in IC50 values in 3D models is a phenomenon that has been previously demonstrated [59] and is attributed to the greater complexity observed in these systems. In conclusion, despite the increased IC50 values observed 3D culture conditions, both

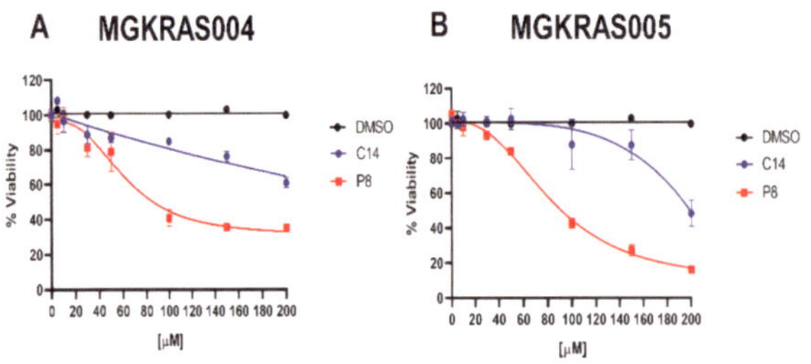

C

Cell Lines	IC_{50} [μM]	
48H	C14	P8
MGKRAS004	395.7	470.4
MGKRAS005	104.4	96.71

Figure 12.
Effect on the viability of primary cultures in 3D, treated with C14 and P8. (A) MGKRAS004 in 3D treated with C14 and P8. (B) MGKRAS005 in 3D treated with C14 and P8.

compounds remain active. In this regard, it is evident that compound P8 exhibited better activity, which aligns with its chemical attributes as a compound selected as a C14 analog with enhanced activity. Thus, P8 demonstrates antitumor properties on CSC-enriched populations.

5. Drug screening: Challenges and troubleshooting

The search for new antitumor compounds is a long and challenging process. Demonstrating the effectiveness of candidate compounds must go through different stages. Firstly, they should be highly effective in exerting their effects at low concentrations, while also desirably minimizing their adverse effects on non-tumor targets.

Ensuring the effectiveness of novel antineoplastic candidates requires extensive evaluations in both *in vitro* and *in vivo* models prior to patient administration [24]. This chapter presents the assessment of compounds C14 and P8 in diverse models. The data presented supports the efficacy of both molecules in mitigating the adverse effects induced by various mutant forms of K-Ras in both pancreatic and breast models. However, we encountered several challenges in stablishing the effectiveness of these molecules. One significant challenge was selecting the appropriate method to determine the IC (inhibitory concentration) of each compound. Conventional methods using XTT and/or MTT as cell proliferation quantification techniques have limitations since their response relies on the metabolic rate of the specific cell lines used, which may not accurately reflect the true growth rate of the models being studied. To address this issue, we recognize the importance of implementing models that measure cell growth by quantifying elements independent of cellular metabolism, such as DNA/RNA or ATP [77].

Another crucial aspect that we deem essential during the implementation of the various methods presented in this chapter is their strict standardization to ensure the precision obtained results. This becomes particularly important when working with complex models like 3D cultures. Based on our own experience, incorporating procedures that enhance the homogeneity of mammospheres, such as gently centrifuging cells during seeding, has proven to be highly advantageous.

Finally, we emphasize the significant importance of developing methodologies to assess the physiology of 3D models. Although several effective systems exist for evaluating different aspects in 2D conditions, corresponding tools for 3D cultures are often scarce or nonexistent. Therefore, research groups must take the initiative to develop these methods themselves.

6. Conclusion

In vitro models have been of great importance in cancer research and drug screening. The use of these models allows the study of numerous cell process, the effect of drugs on cells, and the processes that trigger apoptosis.

2D models are the standard for drug evaluation keeping the cost lower and requiring fewer cells and less time compared to animal models or human clinical trials. However, this model does not closely mimic the behavior and drug metabolism seen in *in vivo*, resulting in many drug candidates failing during clinical trials.

For more reliable inhibitory effects and better characterization of drugs, primary and 3D models are preferable for analyzing pathologies with the highest mortality, such as cancer. Primary cells provide broader spectrum of cell types from a greater number of patients to be studied without inducing artificial genetic mutations, and maintaining the same phenotype throughout the culture, while keeping the advantages of a 2D cell line-base assay in terms of time and cost. Nevertheless, management of primary cell culture poses challenges. Obtaining the patient consent, quality of samples, practicing sterility, and maintaining the culture are some of the challenges faced when working with primary cultures. To overcome these difficulties, proper collection with the help of pathologist and selection of appropriate isolation methods and culture media based on tissue type can help to increase the cell viability.

On the other hand, 3D cell culture models offer a number of advantages as they better represent the microenvironment of *in vivo* conditions, with protein and gene expression similar to those found *in vivo*. Due to these advantages, 3D primary cell culture has gained popularity and holds the potential to replace animal *in vivo* models in future, eventually leading to direct human clinical trials.

In conclusion, this chapter reviews several *in vitro* models for assessing of anticancer compounds, demonstrating their selective effectiveness in reducing viability of pancreatic and breast cancer cells in different conditions. These findings are of vital importance, as current chemotherapies often result in severe side effects. The pursuit of alternative therapies with fewer adverse effects could greatly benefit patients with these pathologies. Considering the evidence presented, we propose that compounds C14 and P8 represent novel therapeutic alternatives for the treatment of tumors dependent on oncogenic forms of K-Ras.

Acknowledgements

Dayan Andrea Carrion Estrada and Sandra Delfin Azuara received a predoctoral scholarship from CONACyT (935796 and 1078870). Dr. Paola Briseño Díaz, researcher, thanks to the Faculty of Medicine of UNAM and CONAHCYT for their support in carrying out this work. We thank Dr. Maria del Rocio Thompson Bonilla from the ISSSTE Mexico Hospital 1° de Octubre, since without her support the primary cultures would not have been performed. We thank Dr. Elena Arechega from the UAM Cuajimalpa Mexico, for the donation of radioresistant breast cancer cell lines.

Conflict of interest

The authors declare no conflict of interest.

DOI: http://dx.doi.org/10.5772/intechopen.113019

Author details

Dayan A. Carrion-Estrada[1], Paola Briseño-Diaz[2], Sandra Delfín-Azuara[1], Arturo Aguilar-Rojas[3] and Miguel Vargas[1*]

1 Department of Molecular Biomedicine, Center for Research and Advanced Studies of the National Polytechnic Institute (CINVESTAV-IPN), Mexico City, Mexico

2 Department of Biochemistry of the Faculty of Medicine of the National Autonomous University of Mexico (UNAM), Mexico City, Mexico

3 Medical Research Unit in Reproductive Medicine, Mexican Institute of Social Security (IMMS), Mexico City, Mexico

*Address all correspondence to: mavargas@cinvestav.mx

References

[1] Chhikara BS, Parang K. Global cancer statistics 2022: The trends projection analysis. Chemical Biology Letters. 2023;**10**(1):451

[2] Benson JD, Chen Y-NP, Cornell-Kennon SA, Dorsch M, Kim S, Leszczyniecka M, et al. Validating cancer drug targets. Nature. 2006;**441**(7092):451-456

[3] Kodaz H, Kostek O, Hacioglu MB, Erdogan B, Kodaz CE, Hacibekiroglu I, et al. Frequency of Ras mutations (Kras, Nras, Hras) in human solid cancer. Breast Cancer. 2017;**7**(12):1-7

[4] Capon DJ, Seeburg PH, McGrath JP, Hayflick JS, Edman U, Levinson AD, et al. Activation of Ki-ras 2 gene in human colon and lung carcinomas by two different point mutations. Nature. 1983;**304**(5926):507-513

[5] Barbacid MMM, Barbacid M. RAS oncogenes: The first 30 years. Nature Reviews Cancer. 2003;**3**:459-465

[6] Mukhopadhyay S, Vander Heiden MG, McCormick F. The metabolic landscape of RAS-driven cancers from biology to therapy. Nature Cancer. 2021;**2**(3):271-283

[7] Timar J, Kashofer KJC, Reviews M. Molecular epidemiology and diagnostics of KRAS mutations in human cancer. Cancer and Metastasis Reviews. 2020;**39**:1029-1038

[8] Briseño-DíazPSMBM,Correa-BasurtoJ, Rojo-Domínguez A, Arregui L, Vega-Loyo L, Núñez-González E, et al. Synergistic effect of antagonists to KRas4B/PDE6 molecular complex kill tumor pancreatic cancer cells. Life Science Alliance. Forthcoming. 2023. p. 1-23

[9] Soriano O, Alcón-Pérez M, Vicente-Manzanares M, Castellano E. The crossroads between RAS and RHO signaling pathways in cellular transformation, motility and contraction. Genes. 2021;**12**(6):819

[10] Jančík S, Drábek J, Radzioch D, Hajdúch M. Clinical relevance of KRAS in human cancers. Journal of Biomedicine and Biotechnology. Hindawi Publishing Corporation; 2010;**2010**:1-13

[11] Leung ELH, Luo LX, Li Y, Liu ZQ, Li LL, Shi DF, et al. Identification of a new inhibitor of KRAS-PDEδ interaction targeting KRAS mutant nonsmall cell lung cancer. International Journal of Cancer. 2019;**145**(5):1334-1345

[12] Lamei H, Zhixing G, Wang F, Liwu FJST, Therapy T. KRAS mutation: From undruggable to druggable in cancer. Signal Transduction and Targeted Therapy. 2021;**6**:1, 386

[13] Lanman BA, Allen JR, Allen JG, Amegadzie AK, Ashton KS, Booker SK, et al. Discovery of a Covalent Inhibitor of KRASG12C (AMG 510) for the Treatment of Solid Tumors. Journal of Medicinal Chemistry. 2019;**63**(1):52-65

[14] Nakajima EC, Drezner N, Li X, Mishra-Kalyani PS, Liu Y, Zhao H, et al. FDA approval summary: Sotorasib for KRAS G12C-mutated metastatic NSCLC. Clinical Cancer Research. 2022;**28**(8):1482-1486

[15] Lam KK, Wong SH, Cheah PY. Targeting the 'Undruggable'Driver protein, KRAS, in epithelial cancers: Current perspective. Cell. 2023;**12**(4):631

[16] Zimmermann G, Papke B, Ismail S, Vartak N, Chandra A,

Hoffmann M, et al. Small molecule inhibition of the KRAS–PDEδ interaction impairs oncogenic KRAS signalling. Nature. 2013;**497**(7451):638-642

[17] Casique-Aguirre D, Briseno-Diaz P, Garcia-Gutierrez P, la Rosa CHG, Quintero-Barceinas RS, Rojo-Dominguez A, et al. KRas4B-PDE6delta complex stabilization by small molecules obtained by virtual screening affects Ras signaling in pancreatic cancer. BMC Cancer. 2018;**18**(1):1299

[18] Cruz-Nova P, Schnoor M, Correa-Basurto J, Bello M, Briseno-Diaz P, Rojo-Dominguez A, et al. The small organic molecule C19 binds and strengthens the KRAS4b-PDEdelta complex and inhibits growth of colorectal cancer cells in vitro and in vivo. BMC Cancer. 2018;**18**(1):1056

[19] Cruz-Nova P, Ocampo-García B, Carrión-Estrada DA, Briseño-Diaz P, Ferro-Flores G, Jiménez-Mancilla N, et al. 131I-C19 iodide radioisotope and synthetic I-C19 compounds as K-Ras4B–PDE6δ inhibitors: A novel approach against colorectal cancer—Biological characterization, biokinetics and dosimetry. Molecules. 2022;**27**(17):5446

[20] Briseño-Díaz P, Velez-Uriza DE, Cruz-Nova P, Ramirez MB, Correa-Basurto J, Hernandez-Rivas R, et al. KRas4BG12C/D/PDE6δ Heterodimeric Molecular Complex: A Target Molecular Multicomplex for the Identification and Evaluation of Nontoxic Pharmacological Compounds for the Treatment of Pancreatic Cancer. Challenges in Pancreatic Cancer. Mexico City, Mexico: IntechOpen; 2020. p. 1-15

[21] Polli JE. In vitro studies are sometimes better than conventional human pharmacokinetic in vivo studies in assessing bioequivalence of immediate-release solid oral dosage forms. The AAPS Journal. 2008;**10**:289-299

[22] Duval K, Grover H, Han L-H, Mou Y, Pegoraro AF, Fredberg J, et al. Modeling physiological events in 2D vs. 3D cell culture. Physiology. 2017;**32**(4):266-277

[23] Lovitt CJ, Shelper TB, Avery VM. Advanced cell culture techniques for cancer drug discovery. Biology. 2014;**3**(2):345-367

[24] Kapałczyńska M, Kolenda T, Przybyła W, Zajączkowska M, Teresiak A, Filas V, et al. 2D and 3D cell cultures–a comparison of different types of cancer cell cultures. Archives of Medical Science. 2018;**14**(4):910-919

[25] Gillet J-P, Varma S, Gottesman MM. The clinical relevance of cancer cell lines. Journal of the National Cancer Institute. 2013;**105**(7):452-458

[26] Park W, Chawla A, O'Reilly EM. Pancreatic cancer: A review. JAMA. 2021;**326**(9):851-862

[27] Liot S, Balas J, Aubert A, Prigent L, Mercier-Gouy P, Verrier B, et al. Stroma involvement in pancreatic ductal adenocarcinoma: An overview focusing on extracellular matrix proteins. Frontiers in Immunology. 2021;**12**:612271

[28] Wang S, Zheng Y, Yang F, Zhu L, Zhu X-Q, Wang Z-F, et al. The molecular biology of pancreatic adenocarcinoma: Translational challenges and clinical perspectives. Signal Transduction and Targeted Therapy. 2021;**6**(1):249

[29] Dai M, Chen S, Teng X, Chen K, Cheng WJ. KRAS as a key oncogene in the clinical precision diagnosis and treatment of pancreatic cancer. Journal of Cancer. 2022;**13**(11):3209-3220

[30] Zhang Z, Zhang H, Liao X, Tsai HI. KRAS mutation: The booster of pancreatic ductal adenocarcinoma transformation and progression. Frontiers in Cell and Developmental Biology. 2023;**11**:1147676

[31] Orrantia-Borunda E, Anchondo-Nuñez P, Acuña-Aguilar LE, Gómez-Valles FO, Ramírez-Valdespino CA. Subtypes of breast cancer. In: Breast Cancer. Brisbane, Australia: Exon Publications; 2022. p. 31-42

[32] Yersal O, Barutca SJ. Biological subtypes of breast cancer: Prognostic and therapeutic implications. World Journal of Clinical Oncology. 2014;**5**(3):412

[33] Tokumaru Y, Oshi M, Katsuta E, Yan L, Satyananda V, Matsuhashi N, et al. KRAS signaling enriched triple negative breast cancer is associated with favorable tumor immune microenvironment and better survival. American Journal of Cancer Research. 2020;**10**(3):897

[34] Koffa M, Malamou-Mitsi V, Agnantis N, Spandidos D. Mutational activation of k-ras oncogene in human breast-tumors. International Journal of Oncology. 1994;**4**(3):573-576

[35] Pereira CB, Leal MF, de Souza CR, Montenegro RC, Rey JA, Carvalho AA, et al. Prognostic and predictive significance of MYC and KRAS alterations in breast cancer from women treated with neoadjuvant chemotherapy, PLoS one. 2013;**8**(3):e60576

[36] Chin L, Park PJ, Kucherlapa R, et al. Comprehensive molecular portraits of human breast tumours. Nature. 2012;**490**(7418):61-70

[37] Fitzpatrick SL, LaChance MP, Schultz GS. Characterization of epidermal growth factor receptor and action on human breast cancer cells in culture. Cancer Research. 1984;**44**(8):3442-3447

[38] Perez-Añorve IX, Gonzalez-De la Rosa CH, Soto-Reyes E, Beltran-AnayaFO,DelMoral-HernandezO, Salgado-Albarran M, et al. New insights into radioresistance in breast cancer identify a dual function of miR-122 as a tumor suppressor and oncomiR. Molecular Oncology. 2019;**13**(5):1249-1267

[39] Larsson P, Engqvist H, Biermann J, Werner Rönnerman E, Forssell-Aronsson E, Kovács A, et al. Optimization of cell viability assays to improve replicability and reproducibility of cancer drug sensitivity screens. Scientific Reports. 2020;**10**(1):5798

[40] Zhang J-H, Chung TD, Oldenburg KR. A simple statistical parameter for use in evaluation and validation of high throughput screening assays. Journal of Biomolecular Screening. 1999;**4**(2):67-73

[41] Swinney DC. Molecular Mechanism of Action (MMoA) in Drug Discovery. In: Annual Reports in Medicinal Chemistry. 46. Belmont, United States: Elsevier; 2011. pp. 301-317

[42] Damiani E, Solorio JA, Doyle AP, Wallace HM. How reliable are in vitro IC50 values? Values vary with cytotoxicity assays in human glioblastoma cells. Toxicology Letters. 2019;**302**:28-34

[43] Kota S, Hou S, Guerrant W, Madoux F, Troutman S, Fernandez-Vega V, et al. A novel three-dimensional high-throughput screening approach identifies inducers of a mutant KRAS selective lethal phenotype. Oncogene. 2018;**37**(32):4372-4384

[44] Burdall SE, Hanby AM, Lansdown MR, Speirs V. Breast cancer cell lines: Friend or foe? Breast Cancer Research. 2003;**5**(2):1-7

[45] Ishiwata T, Matsuda Y, Yoshimura H, Sasaki N, Ishiwata S, Ishikawa N, et al. Pancreatic cancer stem cells: Features and detection methods. Pathology & Oncology Research. 2018;**24**:797-805

[46] April-Monn SL, Wiedmer T, Skowronska M, Maire R, Schiavo Lena M, Trippel M, et al. Three-dimensional primary cell culture: a novel preclinical model for pancreatic neuroendocrine tumors. Neuroendocrinology. 2021;**111**(3):273-287

[47] Bryja A, Popis M, Borowiec B, Dyszkiewicz-Konwińska M, Kocherova I, Angelova-Volponi A, et al. Overview of the different methods used in the primary culture of oral mucosa cells. Journal of Biological Regulators and Homeostatic Agents. 2019;**33**(2):397-401

[48] Jagadeeshaprasad MG, Govindappa PK, Nelson AM, Elfar JC. Isolation, culture, and characterization of primary Schwann cells, keratinocytes, and fibroblasts from human foreskin. JoVE (Journal of Visualized Experiments). 2022;**181**:e63776

[49] Chang JC, Kundranda M. Novel diagnostic and predictive biomarkers in pancreatic adenocarcinoma. International Journal of Molecular Sciences. 2017;**18**(3):667

[50] Carraway KL, Theodoropoulos G, Kozloski GA, Carothers Carraway CA. Muc4/MUC4 functions and regulation in cancer. Future Oncology. 2009;**5**(10):1631-1640

[51] Roy LD, Sahraei M, Subramani DB, Besmer D, Nath S, Tinder TL, et al. MUC1 enhances invasiveness of pancreatic cancer cells by inducing epithelial to mesenchymal transition. Oncogene. 2011;**30**(12):1449-1459

[52] Liang C, Qin Y, Zhang B, Ji S, Shi S, Xu W, et al. Oncogenic KRAS targets MUC16/CA125 in pancreatic ductal adenocarcinoma. Molecular Cancer Research. 2017;**15**(2):201-212

[53] Sumbly V, Landry I. Understanding pancreatic cancer stem cells and their role in carcinogenesis: A narrative review Stem Cell Investigation. 2022;**9**:1

[54] Zhu Y, Yu J, Wang S, Lu R, Wu J, Jiang B. Overexpression of CD133 enhances chemoresistance to 5-fluorouracil by activating the PI3K/Akt/p70S6K pathway in gastric cancer cells. Oncology Reports. 2014;**32**(6):2437-2444

[55] Pampaloni E, Reynaud EG, Stelzer EHK. The third dimension bridges the gap between cell culture and live tissue. Nature Reviews Molecular Cell Biology. 2007;**8**(10):839-845

[56] Hadjipanayi E, Mudera V, Brown RA. Guiding cell migration in 3D: A collagen matrix with graded directional stiffness. Cell Motility and the Cytoskeleton. 2009;**66**(3):121-128

[57] Huerta-Reyes M, Aguilar-Rojas A. Three-dimensional models to study breast cancer. International Journal of Oncology. 2021;**58**(3):331-343

[58] Fautc MA, Laurent L, Ploton D, Poupon M-F, Jardillier J-C, Bobichon HJC, et al. Distinctive alterations of invasiveness, drug resistance and cell–cell organization in 3D-cultures of MCF-7, a human breast cancer cell line, and its multidrug resistant variant. Clinical & Experimental Metastasis. 2002;**19**:161-167

[59] Lovitt CJ, Shelper TB, Avery VM. Evaluation of chemotherapeutics in a three-dimensional breast cancer model. Journal of Cancer Research and Clinical Oncology. 2015;**141**:951-959

[60] Mehta G, Hsiao AY, Ingram M, Luker GD, Takayama S. Opportunities and challenges for use of tumor spheroids as models to test drug delivery and efficacy. Journal of Controlled Release. 2012;**164**(2):192-204

[61] Froehlich K, Haeger J-D, Heger J, Pastuschek J, Photini SM, Yan Y, et al. Generation of multicellular breast cancer tumor spheroids: Comparison of different protocols. Journal of Mammary Gland Biology and Neoplasia. 2016;**21**:89-98

[62] Bahcecioglu G, Basara G, Ellis BW, Ren X, Zorlutuna P. Breast cancer models: Engineering the tumor microenvironment. Acta Biomaterialia. 2020;**106**:1-21

[63] Nagelkerke A, Bussink J, Sweep FC, Span PN. Generation of multicellular tumor spheroids of breast cancer cells: How to go three-dimensional. Analytical Biochemistry. 2013;**437**(1):17-19

[64] Singh M, Mukundan S, Jaramillo M, Oesterreich S, Sant S. Three-dimensional breast cancer models mimic hallmarks of size-induced tumor ProgressionThree-dimensional tumor progression model. Cancer Research. 2016;**76**(13):3732-3743

[65] Lombardo Y, de Giorgio A, Coombes CR, Stebbing J, Castellano L. Mammosphere formation assay from human breast cancer tissues and cell lines. JoVE (Journal of Visualized Experiments). 2015;**97**:e52671

[66] Dontu G, Abdallah WM, Foley JM, Jackson KW, Clarke MF, Kawamura MJ, et al. In vitro propagation and transcriptional profiling of human mammary stem/progenitor cells. Genes & Development. 2003;**17**(10):1253-1270

[67] Chen G, Liu W, Yan B. Breast cancer MCF-7 cell spheroid culture for drug discovery and development. Journal of Cancer Therapy. 2022;**13**(3):117-130

[68] Pulze L, Congiu T, Brevini TA, Grimaldi A, Tettamanti G, D'Antona P, et al. MCF7 spheroid development: New insight about spatio/temporal arrangements of TNTs, amyloid fibrils, cell connections, and cellular bridges. International Journal of Molecular Sciences. 2020;**21**(15):5400

[69] El Harane S, Zidi B, El Harane N, Krause K-H, Matthes T, Preynat-Seauve O. Cancer spheroids and organoids as novel tools for research and therapy: State of the art and challenges to guide precision medicine. Cell. 2023;**12**(7):1001

[70] Ishtiah AA, Yahaya BH. The enrichment of breast cancer stem cells from MCF7 breast cancer cell line using spheroid culture technique. In: Stem Cell Assays: Methods and Protocols. New York, NY: Springer; 2022. pp. 475-484

[71] Dhandapani H, Siddiqui A, Karadkar S, Tayalia P. In vitro 3D spheroid model preserves tumor microenvironment of hot and cold breast cancer subtypes. Advanced Healthcare Materials. 2023;**12**(21):2300164

[72] Shackleton M, Quintana E, Fearon ER, Morrison SJ. Heterogeneity in cancer: Cancer stem cells versus clonal evolution. Cell. 2009;**138**(5):822-829

[73] Yu Z, Pestell TG, Lisanti MP, et al. Cancer stem cells. The International Journal of Biochemistry and Cell Biology. 2012;**44**(12):2144-2151

[74] Sumbly V, Landry I. Understanding pancreatic cancer stem cells and their role in carcinogenesis: A narrative review. Stem Cell Investigation. 2022;**9**:1

[75] Xia P, Liu DH. Cancer stem cell markers for liver cancer and

pancreatic cancer. Stem Cell Research. 2022;**60**:102701

[76] Ferrara B, Dugnani E, Sordi V, Pasquale V, Pellegrini S, Reni M, et al. A comprehensive characterization of Stemness in cell lines and primary cells of pancreatic ductal adenocarcinoma. International Journal of Molecular Sciences. 2022;**23**(18):10663

[77] Braissant O, Astasov-Frauenhoffer M, Waltimo T, Bonkat GJ. A review of methods to determine viability, vitality, and metabolic rates in microbiology. Frontiers in Microbiology. 2020;**11**:547458

Chapter 7

Using Tumor-Like Spheroids to Study the Effect of Anti-Cancer Drugs *In Vitro*

Yousuf Alam, Pawel Borowicz, Stefan W. Vetter and Estelle Leclerc

Abstract

Cell culture techniques have evolved in the last decades and allow now testing anti-cancer drugs using tumor-like spheroids. We describe here issues and troubleshooting solutions when generating spheroids from three human melanoma cell lines (A375, WM115 and WM266). A375 cells generated irregular shape spheroids that were difficult to study due to their fragility. Spheroids generated from all cell lines initially reduced their diameter and increased compacity before increasing in size overtime. Cells present at the periphery of the spheroids showed higher metabolic activity than cells present in the core of the spheroids. When grown as spheroids, a smaller fraction of the A375 and WM115 cells was sensitive to the chemotherapeutic agent temozolomide as compared to cells grown on flat surface. However, this difference was not observed with WM266 cells. Although the presence of spheroids resulted in a smaller fraction of WM155 cells sensitive to the anti-cancer agent vemurafenib, the opposite was observed with A375 cells. Among the cells, WM266 cells were the most resistant to vemurafenib. In conclusion, our study suggests that cell lines behave differently in terms of spheroid formation, and that the effect of the 3D cellular architecture on drug effect is cell type and drug dependent.

Keywords: spheroids, cancer, melanoma, vemurafenib, temozolomide

1. Introduction

For many decades, cell culture experiments have been performed with cells attached to the flat surface of plastic wells. These experiments have provided valuable information in many aspects of cell biology, from the identification of new signaling pathways to the screening of new therapeutic anti-cancer agents [1–3]. However, the outcomes of many experiments performed in these *in vitro* conditions could not be reproduced in pre-clinical studies or in clinical trials [4]. One of the reasons for this failure is that two-dimensional (2D) cell culture growth conditions do not mimic the complex three-dimensional (3D) architecture of tumors [5, 6]. Indeed, when cells are grown on flat surfaces, they form only limited contacts with other cells, and the cellular architecture is different from the complex 3D structures observed in tumors

(**Figure 1**). For instance, in 2D growth conditions all cells receive similar levels of oxygen from the environment. However, when grow as spheroids, diffusion of oxygen and nutrients vary according to the position of the cells in the spheroid: cells at the center of the spheroid do not receive the same levels of nutrient and oxygen and are often subject to hypoxia [7, 8].

To overcome the limitations of 2D cell culture conditions, researchers have developed 3D *in vitro* cell culture systems [9–11]. Experimental conditions have been optimized to allow the formation of monoculture or co-culture of 3D assemblies (spheroids) of cells that mimic tumors. Studies have shown that spheroids are valuable 3D cellular models that possess features shared with tumors [12–15]. These features include the presence of gradients of nutrients, waste and gazes (oxygen and CO_2) throughout the spheroids, the presence of layers of cells with different metabolic activities (proliferative, quiescent and necrotic cells), as well as stronger deposition of extracellular matrix proteins as compared to 2D cultures of cells [5, 6, 16, 17].

Two main approaches are used for the generation of spheroids: scaffold-based and scaffold-free approaches [12–15, 18]. Each approach has its advantages and disadvantages and the choice of the approach to use is guided by the type of assays the investigator will perform with these spheroids. In the scaffold-based approach, the cells are seeded onto artificial matrices with different porosities, resulting in the growth of cells both around and inside the matrices. This approach is not the most appropriate for drug studies because of the low reproducibility between batches of spheroids, the possible adsorption of the drugs with the matrices, the presence of non-transparent matrices that disturb spectroscopic analysis and the difficulties in detaching and extracting the cells from the scaffold when performing Western blot or flow

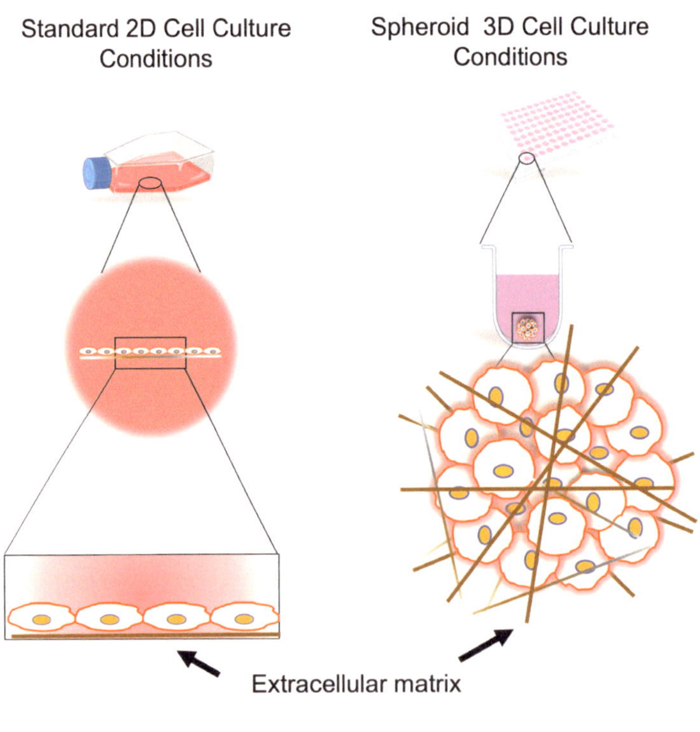

Figure 1.
Schematic representation of cells grown in standard 2D growth conditions (left) or as spheroids (right). A denser extracellular matrix is present around the cells in 3D growth conditions.

cytometry analysis [18]. The second approach and the one that was chosen in this study is the generation of scaffold-free spheroids. Several methods deriving from the scaffold-free approaches have been developed and include the hanging-drop method, the agitation-based approach and the liquid overlay technique [9–11, 18]. This later method is one of the simplest, reproducible and cheapest methods to produce spheroids, and was chosen in the present study for these reasons [19, 20]. In this study, we report the generation of 3D spheroids from three different human melanoma cell-lines (WM115, WM266-4 and A375) using the liquid overlay technique, with the purpose of determining how the growth of cells in 3D conditions affect their sensitivity towards two different types of anti-cancer agents: temozolomide and vemurafenib. To assess the effect of the anti-cancer drugs, we used the Alamar Blue assay that uses changes of oxidative state of resazurin as an indication of changes in metabolic activities. We discuss issues and trouble-shooting solutions throughout our study.

2. Materials and methods

2.1 Cell lines

The three human melanoma cell-lines, WM115, WM266.4 and A375 were purchased from ATCC (Manas, VA) and were grown in Dulbecco's Modified Eagle's Medium (DMEM, ATCC) supplemented with 10% FBS and penicillin (100 U/ml) and streptomycin (100 μg/ml). The absence of mycoplasma contamination in the cells prior to all experiments was confirmed by PCR. Note that cells in our laboratory are typically tested every 2 months for possible contamination with mycoplasma by PCR using the method described in Tang et al. [21]. For simplicity, WM226.4 cells will be referred to as WM266 throughout the rest of the manuscript.

2.2 Cell culture plates

For the generation of spheroids using the liquid overlay technique, cells are typically seeded in round-bottom 96-well plates. These plates are made of plastic polymer. Standard tissue culture plates are made from two types of plastic polymer: polystyrene or polypropylene [22]. Because these polymers are hydrophobic in nature, they need to be modified or "treated" by surface functionalization to become more hydrophilic, therefore allowing cells to adhere to the surface of the plates [22]. The generation of spheroids using round-bottom plates requires the use of non-tissue culture treated plastic plates, because cells should not adhere to the plastic but rather form multi-dimensional assemblies between them. However, it is still possible to use tissue-culture treated multi-well plates for the generation of spheroids. In this case, the surface of the wells should be coated with a polymer, such as poly 2-hydroxyethyl methacrylate (poly-HEMA), to prevent attachment of the cells to the plastic [23]. However, multi-well tissue culture plates with low adhesive properties designed for the generation of spheroids are now available from many suppliers.

In the present study, we compared the generation of spheroids from four different types of round-bottom 96-well plates provided by different manufacturers (**Table 1**). One type of plate was made of tissue-culture treated polystyrene (Plate 1). To prevent cell adhesion to this specific plastic surface, this plate was coated with poly-HEMA as described below. The three other plates (Plates 2, 3 and 4) were not tissue-culture treated. Two plates were made of polystyrene (Plates 1 and 2) and one plate of

Number of the plate referred to in this paper	#1	#2	#3	#4
Round-bottom 96-well plate. Cat. #	#353917	#650185	#3879	#174929
Manufactu-rer	BD Falcon	Greiner Bio-One Cellstar	Corning	ThermoScientific
Polymer type	Polystyrene	Polystyrene	Polypropylene	Polystyrene Nunclon-sphera treated*
Tissue-culture treatment	Yes	No	No	No
Poly-HEMA coating requirement	Yes	No	No	No

**Nucleon-sphera treated plates have low adhesive properties following a non-disclosed proprietary process.*

Table 1.
Characteristics of the round-bottom plates used for the generation of spheroids.

polypropylene (Plate 3). One type of plate (Plate 4) was specifically marketed for the generation of spheroids.

2.3 Poly-HEMA coating

The wells of the tissue-culture treated 96-well plate 1 (**Table 1**) were coated with poly-HEMA by adding 200 µl of 5 mg/ml poly-HEMA (Sigma-Aldrich, Saint-Louis, MO), prepared in 95% ethanol, to each well of a round bottom 96-well plate [23]. After addition of the poly-HEMA solution to the wells, the plates were left in a Biosafety cabinet for 72 h, allowing ethanol to evaporate completely.

2.4 Methylcellulose preparation

Methylcellulose was added to the cells to promote spheroid formation by acting as a cell repellent additive [24]. Methylcellulose is an inert polymer that forms a viscous and gel-like solution that promotes cell aggregation [25–27]. The methylcellulose solution was prepared after modification of the procedure described in Longati et al. [26]. Briefly, a 1.2% (w/v) methylcellulose stock solution was prepared by resuspending 6 g autoclaved methylcellulose (M1314, Spectrum) in 50 ml DMEM containing 10% FBS, penicillin (100 U/ml) and streptomycin (100 µg/ml). The solution was stirred overnight at 4° C and centrifuged for 2 h at 7500 RPM. After centrifugation, 20 ml of the clear viscous solution was added to 80 ml of DMEM containing 10% FBS, penicillin (100 U/ml) and streptomycin (100 µg/ml), resulting in a methylcellulose final concentration of 0.24%.

2.5 Cell seeding

Cells grown to 80–90% confluence in T-75 cell culture flasks were detached with 0.05% trypsin, centrifuged for 2 minutes at 500 g and resuspended in DMEM media supplemented with 10% FBS, penicillin (100 U/ml), streptomycin (100 µg /ml), and 0.24% methylcellulose. Cells were then counted using an hematocytometer and added to the wells of the four different 96 well-plates (Plates 1, 2, 3 and 4) at two different seeding densities: 2500 and 5000 cells per well. The wells were observed every day (0, 24, 48, 72,

96, 120, 144 and 168 h) by microscopy, images of spheroids were taken from 12 different wells, and the total number of formed spheroids per plate was counted after 8 days.

2.6 Resazurin assay

Resazurin or 7-hydroxy-10-oxidophenoxazin-10-ium-3-one is a water soluble and permeable dye [28]. After entering cells, resazurin is reduced by accepting electrons from multiple intracellular electrons donors such as NADPH, FMNH, NADH or cytochromes, without altering the normal transfer of electrons in cells [28]. Non-viable cells will not reduce resazurin. Resazurin is generally non-toxic for many cell-types, but cellular toxicity has been reported [29]. Oxidized resazurin has a dark blue color and has little intrinsic fluorescence, however, reduced resazurin or resofurin, is pink and highly fluorescent with measured excitation and emission fluorescence wavelengths typically of 540 nm and 590 nm, respectively.

Resazurin or Alamar Blue (AB) (Amresco/VWR/Avantor, Radnor, PA.) was prepared at 0.1 mg/ml in 50 mM Phosphate buffer pH 7.4 containing 150 mM NaCl (PBS), sterile filtered using a 0.2 um filter, and stored at 4° C in the absence of light. When added to cells, 10% (vol./vol.) resazurin was added to each well of the plate. Control wells contained resazurin and media only. When measuring changes in metabolic activity of cells grown in flat surface, the AB fluorescence signals were detected after 6 h incubation with the dye. However, after adding AB to the spheroids, we waited a 24 h incubation period before measuring the AB signal.

2.7 Staining of spheroids by immunofluorescence

Ten-days old spheroids generated from A375, WM115 and WM266 cells were fixed with 4% paraformaldehyde (PFA) for 30 minutes and then stained with 0.1% eosin Y (Sigma-Aldrich, Saint Louis, MO) in 80% ethanol to facilitate the visualization of spheroids in the paraffin block. The spheroids were then paraffin embedded using a Leica ASP300S processor. Briefly, the spheroids were washed with a series of ethanol solutions, starting with 95% ethanol followed by two washes with 100% ethanol. The spheroids were then cleared by two 100% xylene baths and finally embedded in paraffin blocks (Tissue-Tek® Base Mold 4162, 15 x 15 x 5 mm) using a Sakura Tissue-Tek embedding station. The paraffin blocks were sectioned at 5 μm using Thermo Microm 355S microtome. The sections were then floated on a water bath, placed on VWR VistaVision TM Histobond © charged slides and air dried. The spheroid sections were treated with three 100% xylene washes to remove the wax from slides and rehydrated through a series of 100–70% ethanol washed followed by a last wash in deionized (DI) water. An antigen retrieval step was then performed for 20 minutes at 120°C in citrate buffer (at pH 6.0) using a Retriever 2100 (Diagnostic Technology). For the staining procedure, the spheroid sections were first washed with TBS-T for 3 min and blocked for 1 h at room temperature (RT) with 5% NGS in TBS blocking buffer (Vector laboratories, Newark, CA). The sections were then incubated with a Ki67 rabbit polyclonal primary antibody (AbCam # ab15580, Waltham, MA) at a 1:500 dilution in 1% BSA overnight at 4°C. After 3 min washes with TBS-T, the sections were further incubated with CF 633 labeled Goat anti-rabbit secondary antibody (Biotium #20122, Fremon, CA) at a 1:250 dilution in 1% Bovine Serum Albumin (BSA) in TBS for an hour at room temperature. Finally, the sections were washed for 1 min with DI water and stained with 4,6-diamidino-2-phenylindole (DAPI) (Biotium #40043) before being mounted using the EverBrite™ Hardset Mounting Medium

(Vector laboratories #23003). The slides were imaged using a Zeiss Axio Observer Z1 LSM 700 microscope using 20x 0.8NA lens. (Carl Zeiss Microscopy LLC, White Plains, NY).

In a different experiment, 72 h old spheroids were stained for the cell viability marker calcein acetoxymethyl (AM) ester (2 μM; Sigma-Aldrich), the dead cell marker Ethidium Homodimer-1 (EthD-1), (3 μM; Invitrogen/Thermo Fisher Scientific, Waltham, MA) and the nuclear marker Hoescht 33,342 (33 μM; Thermo Fisher Scientific) according to a procedure described in [30]. The spheroids were stained simultaneously with the three dyes in the 96 well plate for 30 min at 37°C. After incubation, the spheroids were carefully transferred to a glass bottom dish (Ibidi USA, Fitchburg, WI) for imaging with a Zeiss Axio Observer Z1 epifluorescence microscope equipped with a LSM 700 confocal scanning head, and using a 20×, 0.8 NA objective (Carl Zeiss, Thornwood, NY). Images were captured after excitation/emission of 494/517 nm for the detection of calcein-AM positive cells, 528/617 nm for the detection of EthD-1-positive cells and 360/461 nm for the detection of Hoescht 33,342 positive cells.

2.8 Statistical analysis of the data

Twelve spheroids were imaged for each cell line in each plate. **Figure 2** shows only a representative image from one of these 12 wells. The diameter of 12 independent spheroids were measured each day, for a period of 8 days, using ImageJ software [31]. The mean and the standard deviation were used for the graphs shown in **Figure 3**. A student's t-test was performed to indicate the significance between spheroids'

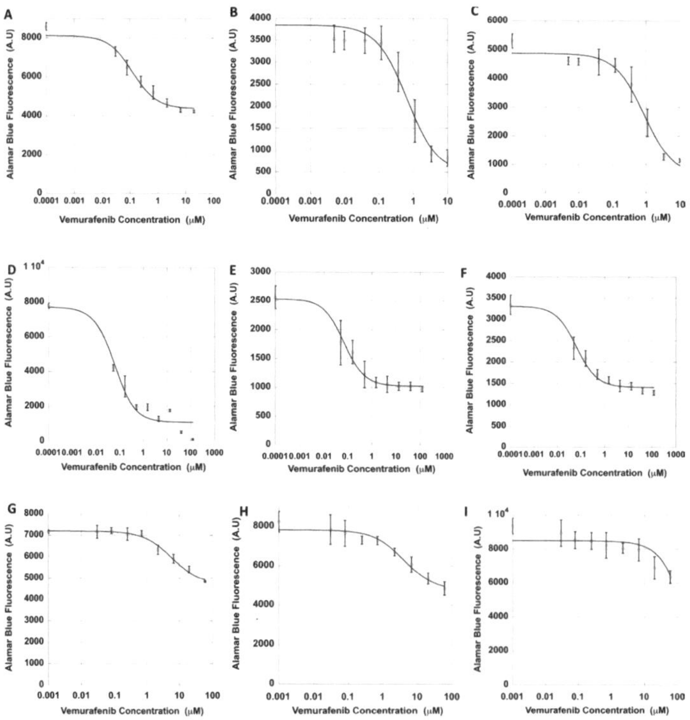

Figure 2.
Images of spheroids formed from A375 cells (A), WM115 cells (B) and WM266 cells (C) at different time points: 0, 24, 48, 72, 96, 120, 144 and 168 h scale bar: 250 μm.

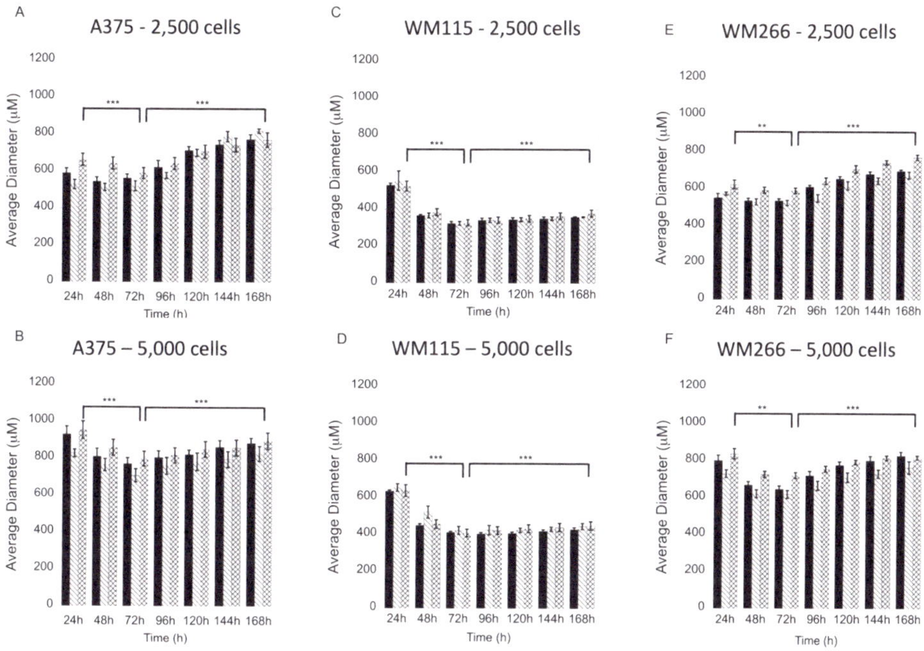

Figure 3.
*Diameters of spheroids formed from 2500 (A) and 5000 (B) A375 cells, from 2500 (C) and 5000 (D) WM115 cells, and from 2500 (E) and 5000 (F) WM266 cells at different time points (0 h, 24 h, 48 h, 72 h, 96 h, 120 h, 144 h and 168 h). ** p < 0.001; *** p < 0.001.*

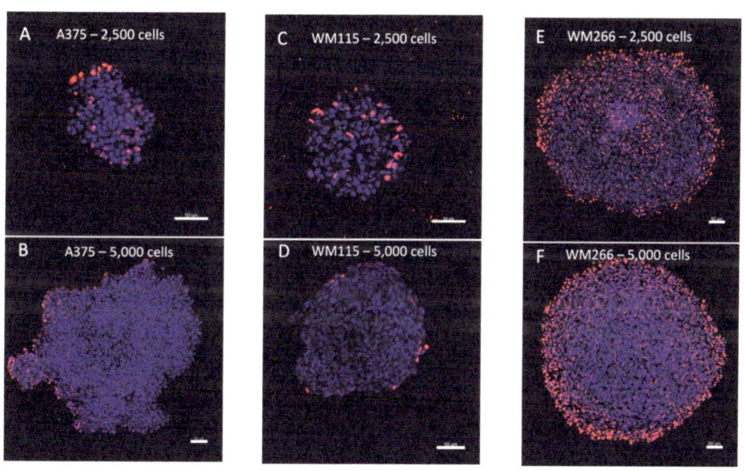

Figure 4.
Staining of spheroids formed from 2500 (A) and 5000 (B) A375 cells, 2500 (C) and 5000 (D) WM115 cells, and 2500 (E) and 5000 (F) WM266 cells with the proliferative marker Ki-67 (pink). The nuclei are stained in blue with Hoescht 33,342. Shown on the figure are the merged images. Scale bar: 50 µm.

diameters at different time points with $^{*}p < 0.05$; $^{**}p < 0.001$ and $^{***}p < 0.001$. **Figures 4** and **5** show representative images of stained spheroids. The cell-based assays were repeated at least three independent time. The data points from the titration with vemurafenib and temozolomide were fit using Kaleidagraph version 5.0 software (Synergy Software).

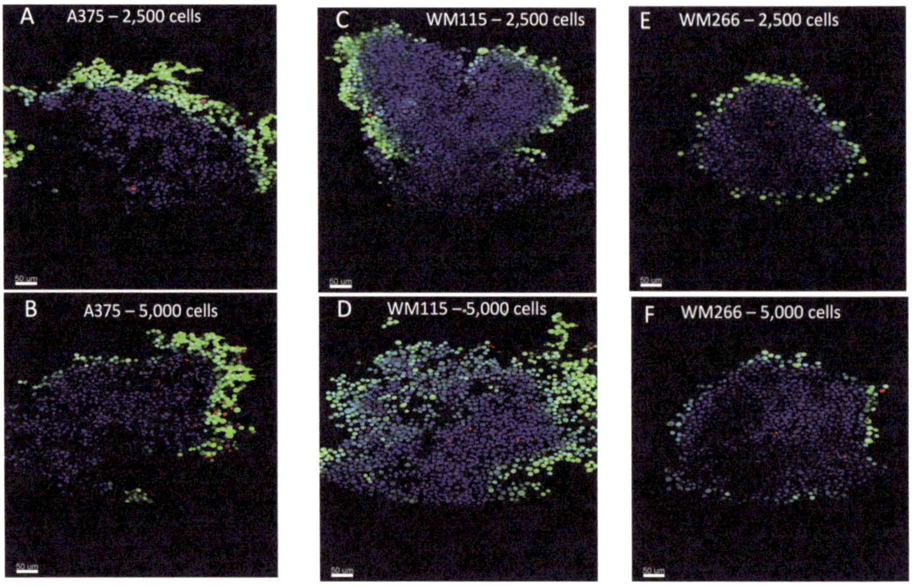

Figure 5.
Staining of spheroids formed from 2500 (A) and 5000 (B) A375 cells, 2500 (C) and 5000 (D) WM115 cells, and 2500 (E) and 5000 (F) WM266 cells with calcein-AM (green) and EthD-1 (pink). The nuclei are stained in blue with Hoescht 33,342. Shown on the figure are the merged images. Scale bar: 50 μm.

3. Results and discussion

3.1 The morphology, shape and growth rate vary among cell lines

Our first goal was to determine if the three different cell lines could form spheroids using the scaffold-free liquid overlay technique. For each cell line, we tested the formation of spheroids in four commercial plates. These plates were made of polystyrene (Plates 1, 2 and 4) or polypropylene (Plate 3) and were either tissue-culture treated (Plate 1) or not (Plates 2, 3 and 4). In case of plate 1, to avoid cell adhesion to the plastic surface, the plate was coated with poly-HEMA prior to seeding the cells. The surface of plate 4 had already been pre-treated by the manufacturer using a non-disclosed proprietary process to enhance spheroid formation. Two different cell densities (2500 and 5000 cells) were seeded for each type of plate. The presence of spheroid was assessed every day for 7 days (0, 24, 48, 72, 96, 120, 144 and 168 h) by microscopy.

We first compared the yield of spheroids produced on the different plates for each cell line. We noticed that in our experimental conditions, a larger number of spheroids was generated using the plates P2, P3 and P4, whereas the poly-HEMA coated plate produced lower yields of spheroid for each of the three cell lines examined (**Table 2**).

Observation of the spheroids by microscopy every day during the time course of 8 days (168 h), revealed that the spheroids generated with the different cell lines had different morphologies and shapes (**Figure 2**). Differences in the ability of cancer cells to form spheroids has been previously reported [32]. A375 cells did not produce spherical spheroids but rather irregular-shape spheroids with small assemblies of cells protruding in different directions. WM155 cells produced

Cell line	Cells/well	Spheroid yield			
		Plate 1 (%)	Plate 2 (%)	Plate 3 (%)	Plate-4 (%)
A375	2500	68.3	100	100	100
	5000	48.1	100	100	100
WM115	2500	35	100	100	100
	5000	77.7	100	100	100
WM266	2500	78.3	100	100	100
	5000	54	100	100	100

Table 2.
Number of spheroids formed per number of examined wells (54 or 60). Poly-HEMA coated plate 1 produced the smallest yield of spheroids. All other plates generated spheroids in every well.

spheroids with the most regular shape (spheres) whereas WM266 cells generated more oval-shaped spheroids. WM115 cells produced the smallest size spheroids from all three cell lines (**Figure 2**). Note that the WM115 and WM266 cell lines originate from the same patient: whereas WM115 was established from a primary tumor, WM266 was established from a metastatic tumor [33]. Our observation suggests that the choice of the cell line for the generation of spheroid is an important aspect for assuring the success of the study. The size of regular-shape spheroids is more easily measured than that of irregular-shape spheroids. We indeed experienced issues when measuring the diameter of many A375 spheroids as well as some WM266 spheroids whose shapes were not spherical. We also noticed that it was very difficult observing spheroids generated in the P3 polypropylene plate. The plastic presented some opacity that hindered the microscopic observation of the spheroids. If a study aims to follow changes in spheroids' size overtime by brightfield microscopy, these aspects are important to consider, as previously reported by other groups [34].

To assess the changes in spheroids' size overtime, we measured the diameter of 12 spheroids every day for a period of 7 days, in plates 2, 3 and 4. Because the yield of spheroid formation was very variable from plate 1, we chose not to study the change in spheroids' size overtime from this plate. The first measurement was performed 24 h after seeding the cells in the round-bottom wells. The average diameter for these 12 spheroids over time is presented in **Figure 3**. We observed similar growth changes in the size of the spheroids in all three plates. Overall, the size of the spheroids tended to decrease for the first 48 h to 72 h, before slowly increasing over the next 4 days. Observation of the spheroids by microscopy showed that immediately after seeding and centrifuging the cells in the wells, the spheroids initially consisted of an assembly of cells loosely connected. The space between cells seemed then to reduce within the first 3 days, resulting in an apparent shrinkage of the spheroids. The size of the spheroids then increased in the next 4 days, probably because of cell division (**Figures 2** and **3**). This observation was similar for each cell line in each of the three plates. For conciseness, we give below numerical data of the changes in diameter observed for each cell-line in plate 4. As mentioned above, the changes were similar in plates 2 and 3.

From all three cell lines, WM115 cells showed the largest initial size reduction between 24 h and 72 h, with a reduction of 37.5% and 36.1% in the average diameter for spheroids generated in Plate 4 with 2500 and 5000 cells respectively. Between

72 h and 168 h, these spheroids grew and increased their size by 15.5 and 9.5%, respectively. WM115 generated spheroids with the smallest size from the three cell lines examined. After reaching their lowest size at 72 h with a decrease of 5.3 and 13.9% for spheroids generated with 2500 and 5000 WM266 cells respectively, these spheroids later increased their size between 72 h and 168 h by 30.7 and 13.6% respectively (Plate 4). A375 spheroids also saw a decrease in their diameters between 24 h and 72 h 0f 10.9% and 16.3% for spheroids generated from 2500 and 5000 cells respectively, and an increase between 72 h and 168 h by 30.7% and 12.1% respectively. For all three cell lines examined, the growth increase between 72 h and 168 h was larger with 2500 cells than with 5000 cells (**Figure 3**). This suggests that cell seeding density is an important factor to consider if the study aims to use changes of spheroid size when investigating the effect of drugs on spheroid growth. Another important point to consider when performing a drug study is the choice of the age of the spheroids for initiating the drug study. As we just showed, the spheroids generated from the three cell lines all "shrank" in the first 72 h. Starting a drug study earlier than 72 h could hamper the analysis of the data, if one would aim to investigate changes in spheroid size as an effect of the drug. Initial changes of spheroid size have been recently explained by the effect of cell-cell adhesion forces that pull cells towards each other's resulting in more compact structures [35]. In a different study, Thakuri et al. did not observe the initial size reduction in spheroids when dispensing dextran-respuspended HT-29 cells in a poly-ethylene glycol solution [36], suggesting that experimental conditions can greatly modulate the behavior of cells within spheroids.

3.2 Spheroids possess proliferative and metabolically active cells at their periphery and edges

We imaged 8–10 days old spheroids for the presence of proliferative cells using antibodies against the proliferative marker Ki-67. Ki-67 is a nuclear protein that is present in all stages of the cell cycle but is absent in quiescent cells and has been previously used for the staining of spheroids [37]. We first observed differences in abundancy in Ki-67 positive cells in the spheroids generated from the three different cell types. WM266 spheroids showed the highest amount of Ki-67 positive cells (**Figure 4E** and **F**). WM115 showed a lower percentage of Ki-67 positive cells, both with spheroids generated from 2500 and 5000 cells (**Figure 4C** and **D**). The presence of Ki-67 positive cells at the edges of the spheroid was clearly evidenced with the irregular-shape A375 spheroids generated from 5000 cells (**Figure 4A** and **B**).

To further evaluate the viability of cells within the spheroids, we stained 72 h old spheroids with calcein-AM, ethidium homodimers and Hoescht 33324 as described in [30, 38]. In these conditions, nuclei from both viable and dead cells are labeled with Hoescht 33342, the viable cells are labeled with calcein-AM (green) and the dead cells with ethidium homodimer-1 (pink). Imaging of the stained spheroids revealed that the calcein-AM positive cells were at the periphery of the spheroids (**Figure 5**). After the staining procedure and during the transfer of the spheroids from the round-bottom well of the 96 well plate to the imaging dish, many spheroids from A375 and WM115 cells broke, and only incomplete fragments could be imaged (**Figure 5**). However, these fragments do show that the metabolically active and calcein-AM positive cells are located at the edge of the spheroids in agreement with the observation of Ki-67 positive proliferative cells present at the edge of the spheroids as well (**Figure 5**). Furthermore, the non-fragmented WM266 spheroids also showed the presence of metabolically more active cells at the periphery of the spheroids. Staining of the spheroids with EthD-1 revealed only few dead cells (pink) in the 72 h old spheroids (**Figure 5**). The presence of the external layer of proliferative and metabolically active cells in the spheroids generated from all melanoma three cell lines (**Figures 4** and 5) is in agreement with previous studies (reviewed in [39]).

3.3 Sensitivity of melanoma cells towards temozolomide and vemurafenib in 2D and spheroid 3D growth conditions

After identifying the optimal properties for the generation of the spheroids from the three different melanoma cell lines, our goal was to compare the sensitivity of the three melanoma cell types, grown in standard 2D growth conditions and as spheroids, towards two cancer drugs, temozolomide (TMZ) and vemurafenib. TMZ is an alkylating agent that alkylate guanine bases in DNA leading to DNA cross-linkages and impaired cell division, thus reducing melanoma cell growth [40].

When comparing the effect of TMZ on cells either grown in 2D conditions in cell culture flasks and as spheroids, we observed similar patterns of sensitivity. In general, A375, WM115 and WM266 all showed some resistance to TMZ as the highest concentration tested (1000 μM) only reduced the signal by 50% with WM115 cells (**Figure 6D**). For A375 and WM266, the signal observed at 1000 μM was only 37.5% and 30% lower than that in the absence of TMZ (**Figure 6A** and **G**). In our experimental conditions, we were not able to increase the TMZ concentration at higher levels due to the toxic effect of DMSO. For A375 and WM115, when comparing the effect of TMZ on the spheroids to that on cells, we observed that TMZ effected a smaller proportion of cells grown as spheroids than on flat surfaces (**Figure 6B** and **C** for A375 spheroids and E and F for WM115 spheroids). Only WM266 spheroids appeared to be slightly more sensitive towards TMZ with only 30% signal reduction by 1000 μM TMZ in cells and about 50% and 45% reduction by the same concentration of TMZ in the spheroids generated from 2500 and 5000 cells, respectively (**Figure 6G, H** and **I**). These results reflect the pattern and distribution of the proliferative marker Ki-67 in the A375, WM115 and WM266 spheroids. Indeed, because TMZ affects cell division, it is only effective in cells that divide, thus being most effective in proliferative cells, such as those staining positive for Ki-67. As observed in **Figure 5**, the spheroids generated from WM266 are those that possess the largest percentage of proliferative cells, from the three cell lines examined, and therefore, should be the spheroids that respond the most to TMZ. However, we also observed that a larger fraction of the WM266 cells responded to TMZ when assembled as spheroids than as monolayers (**Figure 6G**, **H** and **I**).

The second drug that we tested in our study was vemurafenib. Vemurafenib is a BRAF kinase mutant inhibitor [41]. This small molecule drug only inhibits the mutant form of the kinase and is therefore only effective when the melanoma cells carry the appropriate BRAF V600E mutation. By blocking the BRAF kinase, vemurafenib inhibits the growth of the melanoma cells by inhibiting the BRAF/Map kinase pathway. All three cell lines used in our study carry the BRAF V600E mutation and were thus suitable for the study [33, 42].

We first observed that when testing vemurafenib on cells grown on 2D flat surfaces, WM115 showed the strongest IC50 for vemurafenib with a calculated IC50 = 0.057 ± 0.021. Based on the shape of the curve, it appeared that most cells in the dish were affected by vemurafenib (**Figure 7D**). This was not the case for A375 (**Figure 7A**) and WM266 (**Figure 7G**) where only a portion (less than 50%) of the

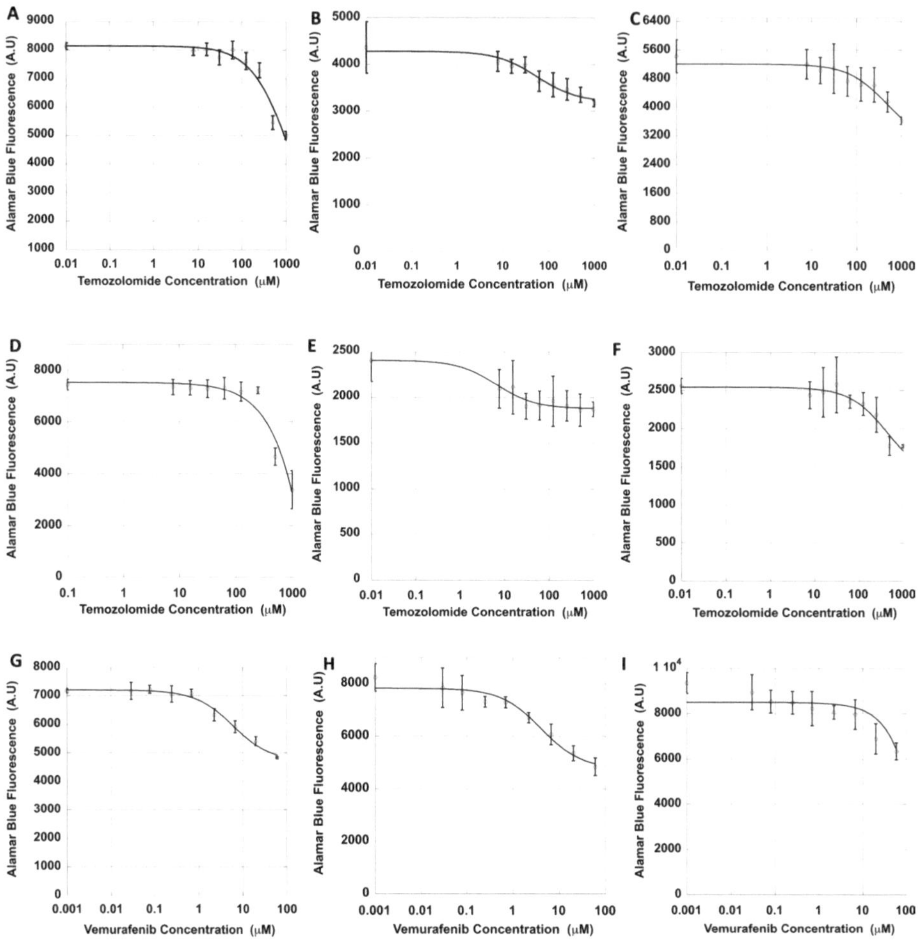

Figure 6.
Effect of temozolomide on: 2D grown A375 cells (A) and 3D A375 spheroids formed from 2500 (B) and 5000 (C) cells; 2D grown WM115 cells (D) and 3D WM115 spheroids formed from 2500 (E) and 5000 (F) cells; and 2D grown WM266 cells (G) and 3D WM266 spheroids formed from 2500 (H) and 5000 (I) cells.

cells seemed to be affected by the BRAF inhibitor. A similar effect of vemurafenib on only a portion of the cells was also observed in spheroids generated from WM115 and WM266 cells (**Figure 7E**, **F**, **H** and **I**). Interestingly, vemurafenib appeared to affect most of the cells within the A375 spheroids, with the fluorescence signal decreasing by up to 80% at the highest vemurafenib concentration used (**Figure 7B** and **C**).

Vemurafenib inhibits the BRAF/Map kinase pathway, and by doing so, affects cell proliferation [41]. Based on the larger population of proliferative cells in the WM266 spheroids (**Figure 5E** and **F**) when compared to the A375 and WM115 spheroids, it was expected that these spheroids would respond the best towards vemurafenib. However, we observed that for these spheroids, the fluorescence signal decreased by about 37 and 25% in the presence of the highest concentration of vemurafenib, for spheroids generated from 2500 and 5000 cells respectively. We are currently investigating the reasons of this discrepancy. As future studies, it would be particularly interesting to stain the vemurafenib treated spheroids with Ki-67, as well as with calcein-AM and ETHD-1 to identify the regions of the spheroids affected by the drug.

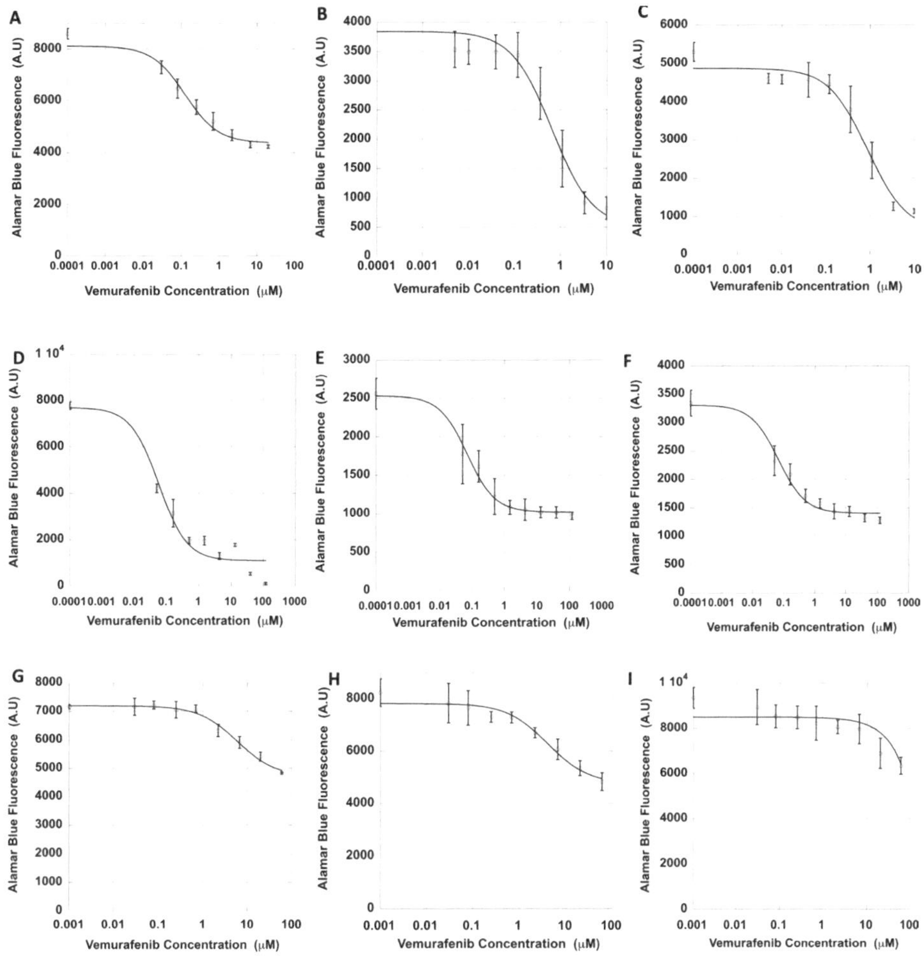

Figure 7.
Effect of vemurafenib on: 2D grown A375 cells (A) and 3D A375 spheroids formed from 2500 (B) and 5000 (C) cells; 2D grown WM115 cells (D) and 3D WM115 spheroids formed from 2500 (E) and 5000 (F) cells; and 2D grown WM266 cells (G) and 3D WM266 spheroids formed from 2500 (H) and 5000 (I) cells.

Our study showed that the response of the cancer cells towards the two cancer agents TMZ and vemurafenib was affected by the architecture of the cells (3D spheroids versus 2D monolayers), but that the response to the drugs depended on both the cell type and the drug itself. Indeed, although the A375 cells appeared to be more resistant towards TMZ when present as 3D spheroids than as 2D monolayers (**Figure 6A–C**), they were more sensitive to vemurafenib when present as 3D spheroids than as monolayers (**Figure 7A–C**). In an earlier study, Filipiak-Duliban et al. had also shown that the drug sensitivity of cancer cells was affected by the mode of culture of the cells [43]. In their study, the authors showed that the B16F10 murine melanoma cells were more resistant to the three anti-cancer drugs everolimus, doxorubicin and cisplatin when grown as spheroids than as 2D monolayers [43]. However, the authors also showed that the effect of the 3D structure on drug sensitivity was not observed when studying a renal cancer cell line [43]. These data, in agreement with our own observations, strongly suggest that the effect of the 3D structure of the cell assemblies depends on both the cell type and the drug that is studied. In addition, the size of the spheroids has also been shown to affect drug resistance, with spheroids of larger sizes being in general more resistant than spheroids of smaller sizes [44]. In our study, we also observed that spheroids generated from 5000 WM115 cells were more resistant towards TMZ than spheroids generated from 2500 cells. Similarly, Spheroids generated from 5000 WM266 cells were more resistant to vemurafenib than spheroids generated from 2500 cells. However, the size of the spheroids did not seem to affect the sensitivity of WM115 cells towards vemurafenib, or the sensitivity of WM266 cells towards TMZ, making the relationship between each cell line and each drug unique.

4. Conclusions and perspectives

Research performed in the last decade has clearly demonstrated that cells grown as 3D spheroids represent a more accurate model of tumors than cancer cells grown as monolayers in cell culture flasks [11, 39]. As shown in our study, the spheroids generated from the three human melanoma cells all showed features of tumors. For instance, all three types of spheroids exhibited a layer of proliferative and metabolically active cells at the edge of the spheroids, as shown by their positive staining with Ki67 and calcein-AM. However, we are also aware that our spheroid-based model presents weaknesses and needs to be improved to better mimic tumors. One of the weaknesses is that our spheroids were generated from cancer cells only. Tumors are formed not only from cancer cells, but also from other cell types such as fibroblasts, macrophages, pericytes and endothelial cells. All these cells communicate with each other's in tumors [45]. Therefore, in tumors, responses to anti-cancer drugs can be very different than the response to the same drugs in spheroids generated from a single type of cells. To solve this issue, new generation of multicellular spheroids are being developed and studied [10, 46–49]. For example, Klicks et al. recently generated spheroids composed of a mixture of fibroblasts, keratinocytes and melanoma cells [48]. However, challenges remain when using multicellular spheroids for the study of anticancer drugs. One of those challenges reside in the generation of consistent spheroids of homogeneous size and shape [50–52]. As we have shown in our study, certain cell lines such as A375, generate irregular shape spheroids that are difficult to study. In addition, in an effort to better mimic the complexity of tumors' architecture and cellular heterogeneity, spheroids are now combined with bioprinting technologies, resulting in additional challenges due to, among other factors, the mechanical instability of spheroids [53, 54]. However, with the fast advancement of new technologies, these challenges will be surely overcome in a near future [55–59].

Acknowledgments

We would like to thank Virginia Montgomery from the Advanced Imaging Microscopy (AIM) Core Facility at NDSU for her assistance in staining the spheroids. This study was supported by NIH-NIGMS Award number U54GM128729 (DaCCoTA Scholar Award to E.L.), the National Science Foundation under NSF EPSCoR Track-1 Cooperative Agreement OIA#1946202 and the College of Health Professions at NDSU.

Conflict of interest

"The authors declare no conflict of interest."

Author details

Yousuf Alam, Pawel Borowicz, Stefan W. Vetter and Estelle Leclerc*
North Dakota State University, Fargo, NC, United States

*Address all correspondence to: estelle.leclerc@ndsu.edu

References

[1] Aggarwal BB, Danda D, Gupta S, Gehlot P. Models for prevention and treatment of cancer: Problems vs promises. Biochemical Pharmacology. 2009;**78**(9):1083-1094

[2] Hait WN. Anticancer drug development: The grand challenges. Nature Reviews. Drug Discovery. 2010;**9**(4):253-254

[3] Edmondson R, Broglie JJ, Adcock AF, Yang L. Three-dimensional cell culture systems and their applications in drug discovery and cell-based biosensors. Assay and Drug Development Technologies. 2014;**12**(4):207-218

[4] Hutchinson L, Kirk R. High drug attrition rates—Where are we going wrong? Nature Reviews. Clinical Oncology. 2011;**8**(4):189-190

[5] Chitcholtan K, Asselin E, Parent S, Sykes PH, Evans JJ. Differences in growth properties of endometrial cancer in three dimensional (3D) culture and 2D cell monolayer. Experimental Cell Research. 2013;**319**(1):75-87

[6] Baker BM, Chen CS. Deconstructing the third dimension: How 3D culture microenvironments alter cellular cues. Journal of Cell Science. 2012;**125**(Pt 13):3015-3024

[7] Grimes DR, Kelly C, Bloch K, Partridge M. A method for estimating the oxygen consumption rate in multicellular tumour spheroids. Journal of the Royal Society Interface. 2014;**11**(92):20131124

[8] Grimes DR, Currell FJ. Oxygen diffusion in ellipsoidal tumour spheroids. Journal of the Royal Society Interface. 2018;**15**(145):20180256

[9] Guillaume L, Rigal L, Fehrenbach J, Severac C, Ducommun B, Lobjois V. Characterization of the physical properties of tumor-derived spheroids reveals critical insights for pre-clinical studies. Scientific Reports. 2019;**9**(1):6597

[10]] Lee KH, Kim TH. Recent advances in multicellular tumor spheroid generation for drug screening. Biosensors (Basel). 2021;**11**(11). DOI: 10.3390/bios11110445. PMID: 34821661; PMCID: PMC8615712

[11] Kitaeva KV, Rutland CS, Rizvanov AA, Solovyeva VV. Cell culture based in vitro test systems for anticancer drug screening. Frontiers in Bioengineering and Biotechnology. 2020;**8**:322

[12] Turetta M, Ben FD, Brisotto G, Biscontin E, Bulfoni M, Cesselli D, et al. Emerging technologies for cancer research: Towards personalized medicine with microfluidic platforms and 3D tumor models. Current Medicinal Chemistry. 2018;**25**(35):4616-4637

[13] Polini A, Del Mercato LL, Barra A, Zhang YS, Calabi F, Gigli G. Towards the development of human immune-system-on-a-chip platforms. Drug Discovery Today. 2019;**24**(2):517-525

[14] Pampaloni F, Reynaud EG, Stelzer EH. The third dimension bridges the gap between cell culture and live tissue. Nature Reviews. Molecular Cell Biology. 2007;**8**(10):839-845

[15] Abolhassani H, Zaer M, Shojaosadati SA, Hashemi-Najafabadi S. Rapid generation of homogenous tumor spheroid microtissues in a scaffold-free platform for high-throughput screening of a novel combination nanomedicine. PLoS One. 2023;**18**(2):e0282064

[16] Francia G, Kerbel RS. Raising the bar for cancer therapy models. Nature Biotechnology. 2010;**28**(6):561-562

[17] Mekhileri NV, Major G, Lim K, Mutreja I, Chitcholtan K, Phillips E, et al. Biofabrication of modular spheroids as tumor-scale microenvironments for drug screening. Advanced Healthcare Materials. 2023;**12**(14):e2201581

[18] Costa EC, de Melo-Diogo D, Moreira AF, Carvalho MP, Correia IJ. Spheroids formation on non-adhesive surfaces by liquid overlay technique: Considerations and practical approaches. Biotechnology Journal. 2018;**13**(1). DOI: 10.1002/biot.201700417. Epub 2017 Nov 15. PMID: 29058365

[19] Costa EC, Moreira AF, de Melo-Diogo D, Gaspar VM, Carvalho MP, Correia IJ. 3D tumor spheroids: An overview on the tools and techniques used for their analysis. Biotechnology Advances. 2016;**34**(8):1427-1441

[20] Amaral RLF, Miranda M, Marcato PD, Swiech K. Comparative analysis of 3D bladder tumor spheroids obtained by forced floating and hanging drop methods for drug screening. Frontiers in Physiology. 2017;**8**:605

[21] Tang J, Hu M, Lee S, Roblin R. A polymerase chain reaction based method for detecting mycoplasma/Acholeplasma contaminants in cell culture. Journal of Microbiological Methods. 2000;**39**(2):121-126

[22] Lerman MJ, Lembong J, Muramoto S, Gillen G, Fisher JP. The evolution of polystyrene as a cell culture material. Tissue Engineering. Part B, Reviews. 2018;**24**(5):359-372

[23] Phung YT, Barbone D, Broaddus VC, Ho M. Rapid generation of in vitro multicellular spheroids for the study of monoclonal antibody therapy. Journal of Cancer. 2011;**2**:507-514

[24] Leung BM, Lesher-Perez SC, Matsuoka T, Moraes C, Takayama S. Media additives to promote spheroid circularity and compactness in hanging drop platform. Biomaterials Science. 2015;**3**(2):336-344

[25] Maritan SM, Lian EY, Mulligan LM. An efficient and flexible cell aggregation method for 3D spheroid production. Journal of Visualized Experiments: JoVE. 2017;**121**. DOI: 10.3791/55544. PMID: 28448014; PMCID: PMC5564434

[26] Longati P, Jia X, Eimer J, Wagman A, Witt MR, Rehnmark S, et al. 3D pancreatic carcinoma spheroids induce a matrix-rich, chemoresistant phenotype offering a better model for drug testing. BMC Cancer. 2013;**13**:95

[27] Urich E, Patsch C, Aigner S, Graf M, Iacone R, Freskgard PO. Multicellular self-assembled spheroidal model of the blood brain barrier. Scientific Reports. 2013;**3**:1500

[28] Rampersad SN. Multiple applications of Alamar blue as an indicator of metabolic function and cellular health in cell viability bioassays. Sensors (Basel). 2012;**12**(9):12347-12360

[29] Gloeckner H, Jonuleit T, Lemke HD. Monitoring of cell viability and cell growth in a hollow-fiber bioreactor by use of the dye Alamar blue. Journal of Immunological Methods. 2001;**252**(1-2):131-138

[30] Singh SK, Abbas S, Saxena AK, Tiwari S, Sharma LK, Tiwari M. Critical role of three-dimensional tumorsphere size on experimental outcome. BioTechniques. 2020;**69**(5):333-338

[31] Schneider CA, Rasband WS, Eliceiri KW. NIH image to ImageJ: 25 years of image analysis. Nature Methods. 2012;**9**(7):671-675

[32] Chitcholtan K, Sykes PH, Evans JJ. The resistance of intracellular mediators to doxorubicin and cisplatin are distinct in 3D and 2D endometrial cancer. Journal of Translational Medicine. 2012;**10**:38

[33] Herlyn M, Balaban G, Bennicelli J, Dt G, Halaban R, Herlyn D, et al. Primary melanoma cells of the vertical growth phase: Similarities to metastatic cells. Journal of the National Cancer Institute. 1985;**74**(2):283-289

[34] Zanoni M, Piccinini F, Arienti C, Zamagni A, Santi S, Polico R, et al. 3D tumor spheroid models for in vitro therapeutic screening: A systematic approach to enhance the biological relevance of data obtained. Scientific Reports. 2016;**6**:19103

[35] Amereh M, Edwards R, Akbari M, Nadler B. In-Silico Modeling of tumor spheroid formation and growth. Micromachines (Basel). 2021;**12**(7). DOI: 10.3390/mi12070749

[36] Thakuri PS, Gupta M, Plaster M, Tavana H. Quantitative size-based analysis of tumor spheroids and responses to therapeutics. Assay and Drug Development Technologies. 2019;**17**(3):140-149

[37] Laurent J, Frongia C, Cazales M, Mondesert O, Ducommun B, Lobjois V. Multicellular tumor spheroid models to explore cell cycle checkpoints in 3D. BMC Cancer. 2013;**13**:73

[38] Sirenko O, Mitlo T, Hesley J, Luke S, Owens W, Cromwell EF. High-content assays for characterizing the viability and morphology of 3D cancer spheroid cultures. Assay and Drug Development Technologies. 2015;**13**(7):402-414

[39] Pinto B, Henriques AC, Silva PMA, Bousbaa H. Three-dimensional spheroids as in vitro preclinical models for cancer research. Pharmaceutics. 2020;**12**(12). DOI: 10.3390/pharmaceutics12121186. PMID: 33291351; PMCID: PMC7762220

[40] Wesolowski JR, Rajdev P, Mukherji SK. Temozolomide (Temodar). AJNR. American Journal of Neuroradiology. 2010;**31**(8):1383-1384

[41] Luke JJ, Hodi FS. Vemurafenib and BRAF inhibition: A new class of treatment for metastatic melanoma. Clinical Cancer Research. 2012;**18**(1):9-14

[42] Al Hashmi M, Sastry KS, Silcock L, Chouchane L, Mattei V, James N, et al. Differential responsiveness to BRAF inhibitors of melanoma cell lines BRAF V600E-mutated. Journal of Translational Medicine. 2020;**18**(1):192

[43] Filipiak-Duliban A, Brodaczewska K, Kajdasz A, Kieda C. Spheroid culture differentially affects cancer cell sensitivity to drugs in melanoma and RCC models. International Journal of Molecular Sciences. 2022;**23**(3). DOI: 10.3390/ijms23031166. PMID: 35163092; PMCID: PMC8835769

[44] Gong X, Lin C, Cheng J, Su J, Zhao H, Liu T, et al. Generation of multicellular tumor spheroids with microwell-based agarose scaffolds for drug testing. PLoS One. 2015;**10**(6):e0130348

[45] Ma L, Hernandez MO, Zhao Y, Mehta M, Tran B, Kelly M, et al. Tumor cell biodiversity drives microenvironmental reprogramming in liver cancer. Cancer Cell. 2019;**36**(4):418-430e6. DOI: 10.1016/j.ccell.2019.08.007. Epub 2019 Oct 3. PMID: 31588021; PMCID: PMC6801104

[46] Saraiva DP, Matias AT, Braga S, Jacinto A, Cabral MG. Establishment

of a 3D co-culture with MDA-MB-231 breast cancer cell line and patient-derived immune cells for application in the development of immunotherapies. Frontiers in Oncology. 2020;**10**:1543

[47] Lazzari G, Nicolas V, Matsusaki M, Akashi M, Couvreur P, Mura S. Multicellular spheroid based on a triple co-culture: A novel 3D model to mimic pancreatic tumor complexity. Acta Biomaterialia. 2018;**78**:296-307

[48] Klicks J, Masslo C, Kluth A, Rudolf R, Hafner M. A novel spheroid-based co-culture model mimics loss of keratinocyte differentiation, melanoma cell invasion, and drug-induced selection of ABCB5-expressing cells. BMC Cancer. 2019;**19**(1):402

[49] Saleh NA, Rode MP, Sierra JA, Silva AH, Miyake JA, Filippin-Monteiro FB, et al. Three-dimensional multicellular cell culture for anti-melanoma drug screening: Focus on tumor microenvironment. Cytotechnology. 2021;**73**(1):35-48

[50] Han SJ, Kwon S, Kim KS. Challenges of applying multicellular tumor spheroids in preclinical phase. Cancer Cell International. 2021;**21**(1):152

[51] Mitrakas AG, Tsolou A, Didaskalou S, Karkaletsou L, Efstathiou C, Eftalitsidis E, et al. Applications and advances of multicellular tumor spheroids: Challenges in their development and analysis. International Journal of Molecular Sciences. 2023;**24**(8). DOI: 10.3390/ijms24086949

[52] Rossi M, Blasi P. Multicellular tumor spheroids in nanomedicine research: A perspective. Frontiers in Medical Technology. 2022;**4**:909943

[53] Zhuang P, Chiang YH, Fernanda MS, He M. Using spheroids as building blocks towards 3D bioprinting of tumor microenvironment. International Journal of Bioprint. 2021;**7**(4):444

[54] Banerjee D, Singh YP, Datta P, Ozbolat V, O'Donnell A, Yeo M, et al. Strategies for 3D bioprinting of spheroids: A comprehensive review. Biomaterials. 2022;**291**:121881

[55] Minaeva ED, Antoshin AA, Kosheleva NV, Koteneva PI, Gonchukov SA, Tsypina SI, et al. Laser bioprinting with cell spheroids: Accurate and gentle. Micromachines (Basel). 2023;**14**(6). DOI: 10.3390/mi14061152

[56] Decarli MC, Seijas-Gamardo A, Morgan FLC, Wieringa P, Baker MB, Silva JVL, et al. Bioprinting of stem cell spheroids followed by post-printing chondrogenic differentiation for cartilage tissue engineering. Advanced Healthcare Materials. 2023;**12**(19):e2203021

[57] Ayan B, Celik N, Zhang Z, Zhou K, Kim MH, Banerjee D, et al. Aspiration-assisted freeform bioprinting of prefabricated tissue spheroids in a yield-stress gel. Communications on Physics. 2020;**3**. DOI: 10.1038/s42005-020-00449-4. Epub 2020 Oct 16. PMID: 33251340; PMCID: PMC7695349

[58] Daly AC, Davidson MD, Burdick JA. 3D bioprinting of high cell-density heterogeneous tissue models through spheroid fusion within self-healing hydrogels. Nature Communications. 2021;**12**(1):753

[59] Duarte Campos DF, Lindsay CD, Roth JG, LeSavage BL, Seymour AJ, Krajina BA, et al. Bioprinting cell- and spheroid-laden protein-engineered hydrogels as tissue-on-chip platforms. Frontiers in Bioengineering and Biotechnology. 2020;**8**:374

Section 4

Application of Cell Culture in Tissue Engineering

Chapter 8

Autologous Adipose Tissue Stem Cell-Derived Terminally Differentiated Retinal Pigment Epithelium Monolayer on a Biocompatible Scaffold, to Restore Vision in Age-Related Macular Degeneration Patients

Dimitrios Bouzianas

Abstract

In Western countries, age-related macular degeneration (AMD) is the common cause of central visual loss in the elderly leading to gradual blindness. Studies implicate the retinal pigment epithelium (RPE) as an important player in AMD pathogenesis, as progressive loss of RPE cells and photoreceptors lead to poor visual acuity. Several strategies aim to prevent the RPE degeneration by saving the damaged retinal cells or replacing them. Cell rescue provides retinal cells with trophic or immunomodulatory factors, whereas cell replacement aims to repair and regenerate neuroretina providing new cells. Currently, a major limitation is cell loss from subretinal injections of bolus cell suspensions. The most promising studies focus on tissue transplantation or RPE cell patches hosted on implantable scaffolds made of advanced biomaterials. The latter support the development of artificial 3D retinal constructs improving functional integration capacity and increasing the survival of implanted cells into the damaged retina. However, there is no consensus on the optimal RPE source, delivery strategy, cell dose and animal model for testing. This chapter aims to improve the efficacy of RPE grafting suggesting an optimal cell source, an efficient delivery method, and the topography of retina damage as a determining factor to calculate the effective therapeutic dose.

Keywords: adipose-derived mesenchymal stem cells, age-related macular degeneration, advanced therapy medicinal products (ATMPs), biocompatible scaffolds, cell replacement therapy, retinal pigment epithelium cells, photoreceptor cells, cell cultures

1. Introduction

In current industrialized countries, approximately 1.75 million individuals are affected by loss of ocular light-sensing retinal tissue. However, the disease does not always lead to rapid visual loss; depending on the type and severity of AMD, blindness may occur for months to many years [1].

The human RPE epithelium forms early in development and subsequently remains dormant, undergoing minimal proliferation throughout normal life [2]. Many blinding diseases are characterized by the progressive loss of photoreceptor cells (cones and rods) and lack the ability to regenerate in mammals, including humans. Very few existing conventional therapies for retinal diseases slow the progression of the diseases. Basically, there are two potential options to deal with this progressive condition: First, to rescue the damaged tissue by immunomodulation or with retinal support cells that secrete trophic cytokines [3]; second, to repair and regenerate the impaired neuroretinal tissue with cell replacement therapy [4].

RPE cell loss occurs early in age-related macular degeneration (AMD), and cell replacement is the aim of several stem cell therapy programs. Cell replacement therapy is an attractive option for retinal diseases as the eye is an easily accessible and readily monitored human organ; additionally, the risk of cell transplant rejection is limited due to the autonomous and partially privileged immune structure [4]. Therefore, the development of *advanced therapy medicinal products* (ATMPs) has become a fast-growing field and presents an opportunity for preserving or restoring vision even in advanced stages of retinal degeneration, when photoreceptor degeneration has already occurred [5]. As there are currently almost no elective therapies available for AMD and other RPE disorders, RPE replacement, especially in later disease stages, has significantly accelerated progress in the field of RPE biology [2]. The effective integration of the transplanted RPE cells into the damaged area to achieve the reconstruction of the neural circuitry and the long-term survival and functionality of the transplanted cells are currently the major targets for successful transplantation [6, 7]. However, a major limitation is donor cell loss associated with common delivery methods such as subretinal injections of bolus cell suspensions. Advanced biocompatible materials to develop implantable artificial 3D retinal constructs and support their functional integration in the damaged retinal microenvironment may be a potential solution to the main problems of cell transplantation, improving the integration capacity and increasing the survival of implanted cells [8].

Currently, there is no consensus regarding the optimal strategy for RPE replacement therapy, and multiple RPE sources, delivery methods, cell doses, and recipient animal models have been investigated with variable results [9]. This chapter aims to clarify the importance of these factors in successful RPE transplantation and focuses on selecting the optimal RPE source, the most efficient method of cell delivery, and the calculation of therapeutic dose. We suggest that the degree of retina damage should be a determining factor in assessing the therapeutic cell dose. Moreover, this proposal could provide a way out of the existing problem of determining the effective therapeutic dose of cell preparations in the field of regenerative medicine to achieve the best results for the benefit of AMD patients and other people, who resort to regenerative medicine to restore their damaged health.

2. Architecture and function of normal retina

The human retina is the light-sensitive tissue lining the inner surface of the eye. It is a complex neural network composed of multiple cell layers with different features and functions (**Figure 1**). The main objective of this network is light transduction through electrical pulse, encoding, and packing visual input for visual pathways and superior visual elaboration [2].

Retina is a 2-layered cup formed by the retinal pigment epithelium (RPE) monolayer (outer layer) and multilayer neurosensory retina (inner layer), separated by the subretinal space (**Figure 2**). *The RPE* is a cuboidal, pigmented cobblestone monolayer (**Figure 3**) of polarized columnar epithelium of post-mitotic epithelial cells. These cells are located in Bruch's basement membrane, while their apical surface contacts the outer segment of retinal photoreceptors. (**Figure 4**) [14]. The hexagonal network formed by RPE cells does not allow cell overlapping or for the formation of voids, thus permitting higher area coverage with the lower cell surface tension. *The innermost layer or neuroretina* is made up *by a complex circuitry* of six main neuronal cell types, including (from inner to outer): photoreceptors (PRCs) – rods and cones – horizontal, bipolar, amacrine and retinal ganglion cells (RGCs) (**Figures 1** and **4**). PRCs are located in the most external layer of neuroretina, and they have a close anatomical relation with the RPE monolayer. *Bruch's membrane* (BrM, **Figure 5**) is an organized 2–5 mm-thick acellular sheet divided into five layers and mainly composed of elastin, collagen I-V, laminin and fibronectin. Structurally, this multi-laminar and highly specialized structure separates

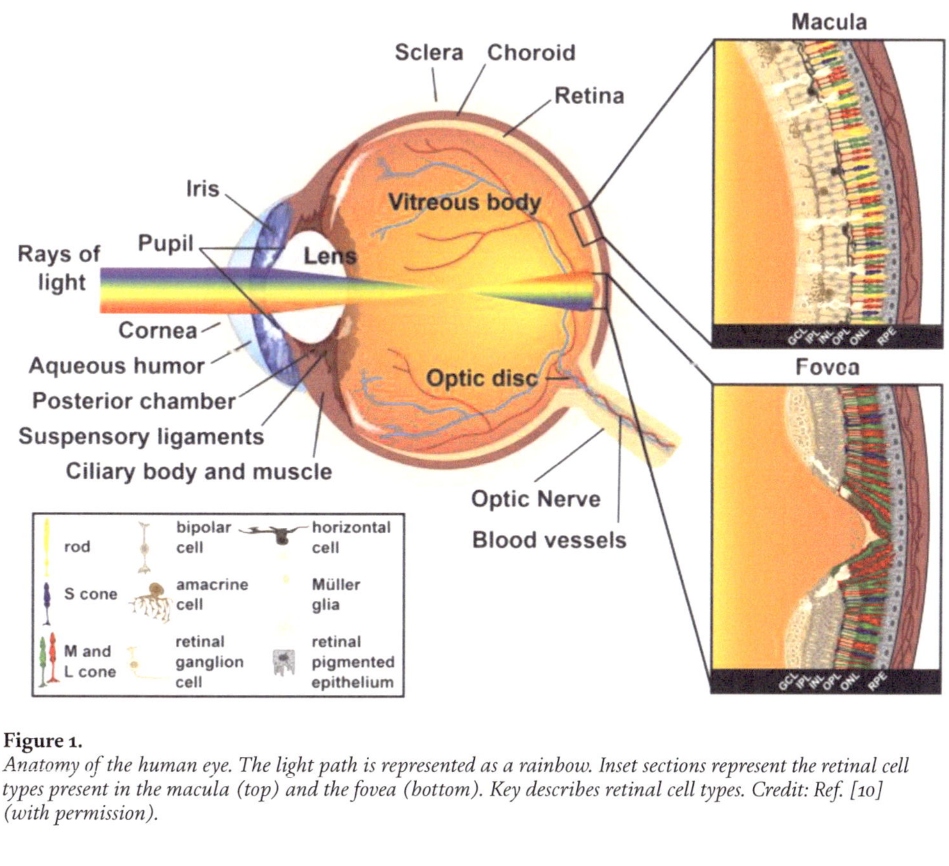

Figure 1.
Anatomy of the human eye. The light path is represented as a rainbow. Inset sections represent the retinal cell types present in the macula (top) and the fovea (bottom). Key describes retinal cell types. Credit: Ref. [10] (with permission).

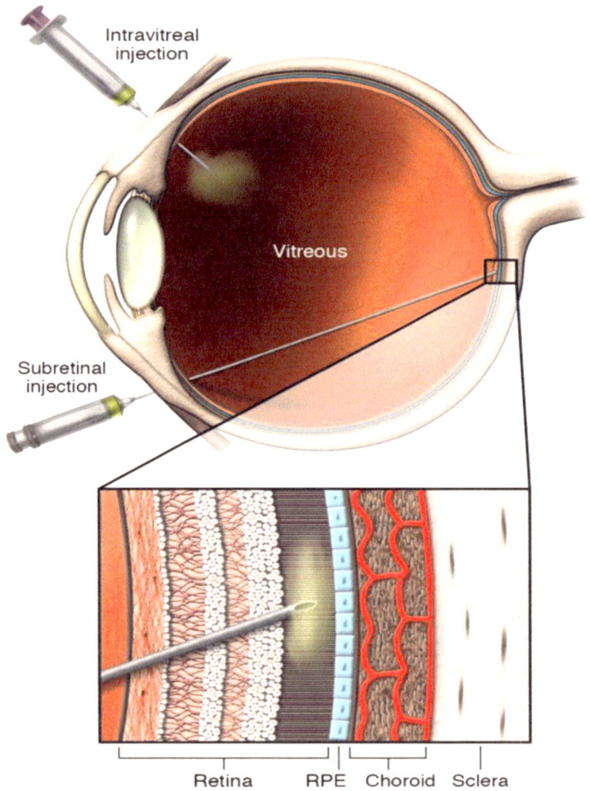

Figure 2.
Intravitreal versus subretinal delivery. The first is less technical with fewer risks associated with structural damage to ocular tissues; however, PRE cell delivery to the disease tissue can be problematic because of the diffusional volume of the vitreous. Subretinal delivery is more technically challenging, however, the cell concentration is much higher. Credit: Ref. [11] (permission copyright clearance Center, Inc).

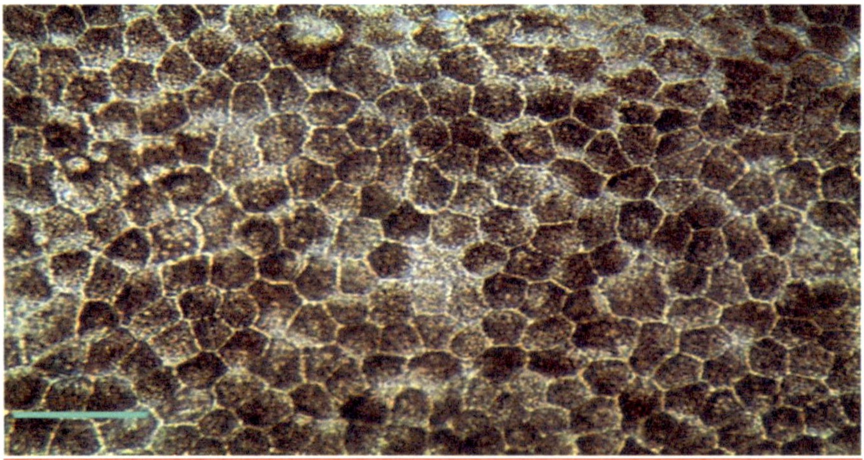

Figure 3.
The RPE layer is a sheet of polygonal mostly hexagonal cells. Cells differentiated from the iPSC of a healthy donor and cultured for 3 months. Credit: Ref. [12] (with permission).

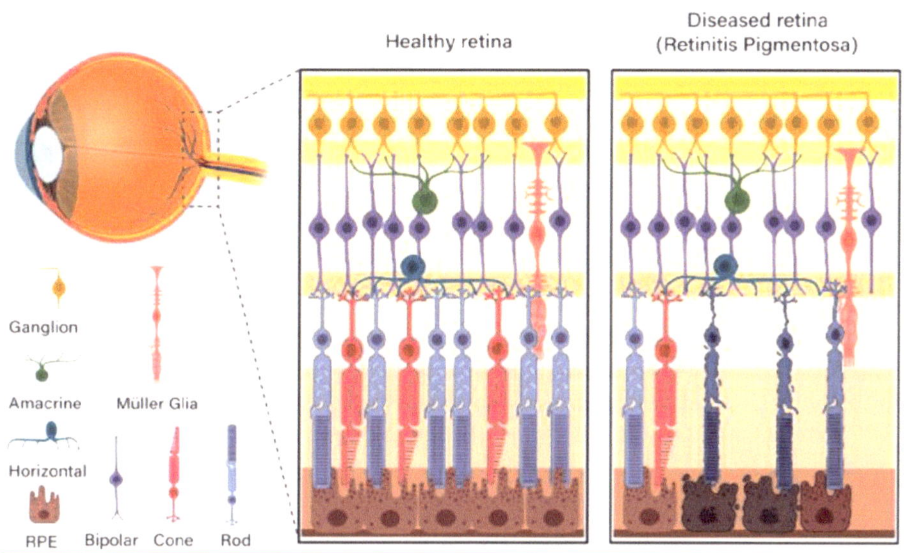

Figure 4.
(a) Retina structure and its main neural cell types. (b) Schematic of a healthy versus diseased retina. Retinal degenerative diseases damage the physiology of the retina firstly affecting the RPE cells and the rod and cone photoreceptors, causing irreversible vision disfunctions in the patient. – Credit: Ref. [13] (with permission).

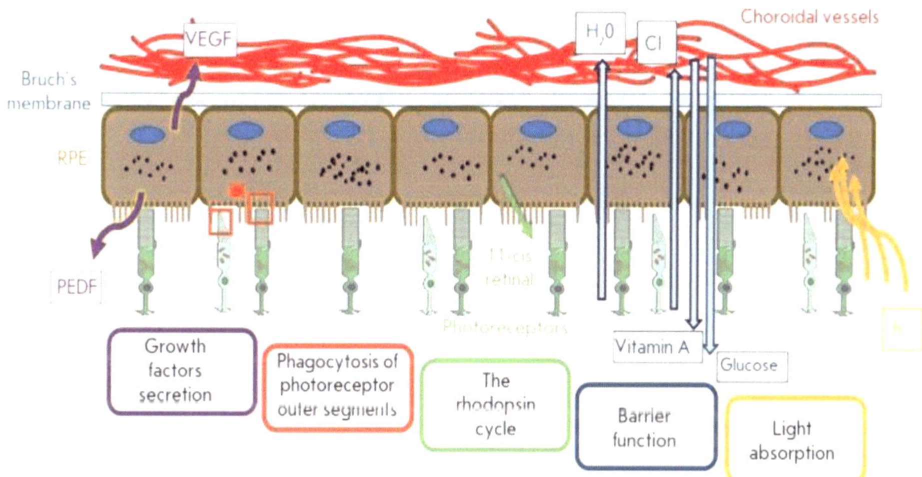

Figure 5.
Main functions of RPE. Implementation of the visual cycle - barrier function - secretion of hormones and growth factors - phagocytosis of used photoreceptor discs - absorption of light. Credit: Ref. [12] (with permission).

the RPE layer from the choroid; functionally, it supports the exchange of nutrients and oxygen supply between the neural retina and the capillaries of the choroid layer. The choroid is a vascular layer of the eye and lies between the sclera and the BrM (**Figure 2**). Its capillaries supply oxygen and nourishment to the outer layer of RPE. Sclera is a tough white sheath around the outside of the eye ball (**Figure 2**). The fenestrated choriocapillaris, BrM, and RPE form the outer blood-retinal barrier [7, 8, 15, 16]. Fovea is a small depression in the middle of the retina and is responsible for high-resolution vision [7, 17]. This retinal region is a relatively small area of 1.8 mm^2 and represents a cone-rich area extremely important for human vision, as it facilitates important tasks such as facial recognition, driving, reading, writing, etc. It is estimated that there are approximately 200,000 cones, a cell number easily produced by organoid technology [8]. Region surrounding the fovea is called the macula (**Figures 1** and **6**), the 6-mm centermost area of the neurosensory retina rich in cones.

RPE represents a pillar to maintain the structural and functional integrity of the entire retina. It supports the PRCs, providing metabolic and transport functions essential for homeostasis of the neural retina (**Figure 5**). The hallmark of the RPE layer is the presence of melanosomes. Stray light passing through the PRC layer is absorbed by the contained melanin granules and is also prevented from reflecting on the PRCs themselves. In the absence of melanin, the visual image would be degraded [7]. RPE also phagocytoses the shed photoreceptor outer segments; when it degenerates, as happens in AMD, eventually, PRCs degenerate with it [7, 16].

Light enters the eye through the cornea, passes through the pupil, lens and strikes the retina (**Figure 1**). Visual information from the retina transmits to the brain through optic nerve fiber. In this visual signal processing, a sequence of events takes place, and every cell plays a definite role. PRCs, the first-order specialized neurons, have the structure and metabolic privilege to convert light into electricity through the visual cycle, a molecularly-driven complex procedure. The PRC-generated electrical impulse is then transmitted to bipolar cells (second-order neurons) and finally to the optic nerve core, which is composed of ganglion cells (third-order neurons) [19].

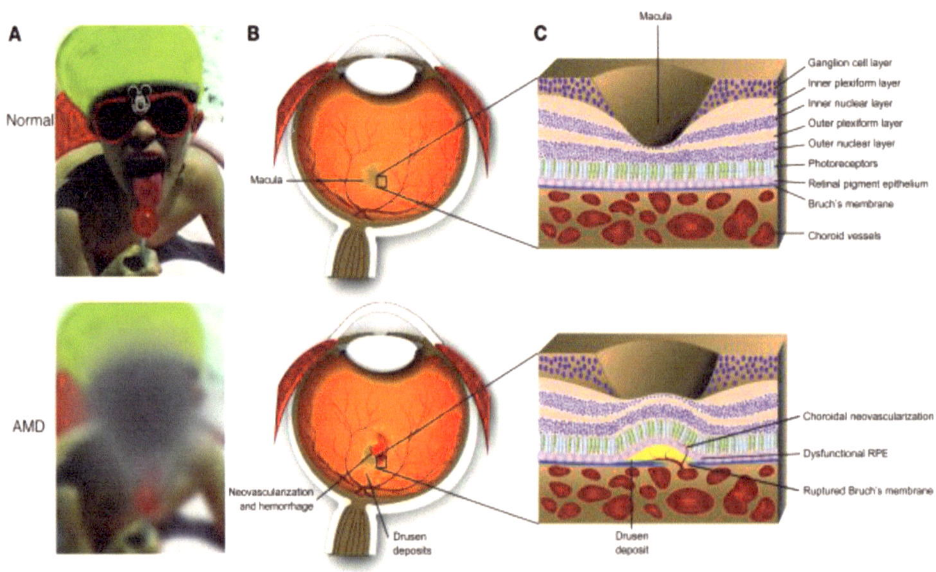

Figure 6.
Loss of central vision in AMD patients. (A) Visual field seen by a person with normal vision and by an AMD patient (loss of central color vision). (B) A normal human eye compared with an AMD eye depicting deposits of drusen (yellow spots) and choroidal neovascularization with subretinal hemorrhage in the macula (red). (C) Cross-section of a normal eye through the macula shows retinal neuronal layers, RPE, Bruch's membrane, and choroid vessels. Credit: Ref. [18] Oct (with permission).

3. Age-related macular degeneration (AMD)

RPE layer is a fragile photosensitive part of the central nervous system prone to degenerative diseases, including AMD. With the adjacent light-sensitive photoreceptors form a functional unit, lining the back of the eye. Its dysfunction plays an important role in degenerative retinal diseases, characterized by non-functional or loss of PRCs leading to deterioration or permanent vision loss [9].

AMD is the common cause of progressive central visual loss in the elderly >60 years old, leading to blindness and affecting 1.75 million individuals in Western countries [14]. Several studies point to the RPE as an important player in AMD pathogenesis. Progressive dysfunction and loss of central – not peripheral retinal pigment epithelium cells and photoreceptors lead to poor visual acuity. Central vision is required for detailed work and for tasks like reading and driving. Early disease may be without appreciable vision loss. Complete blindness is not always the worst outcome for AMD patients. It may occur from months to many years, depending on the type and severity of the disease. Nonetheless, central vision disorder is a debilitating event that affects patients' quality of life, e.g., loss of ability to drive, read, facial recognition. (**Figure 6**). Multiple risk factors drive the pathogenesis of AMD, most prominently age, smoking, dietary modifications, and genetic variation in multiple genes [16]. In the retinal microenvironment, loss of the balance between oxidants and anti-oxidants can cause nitrogen species and reactive oxygen to accumulate, triggering chronic inflammation. RPE becomes dysfunctional and, in severe cases, necrotic enhancing drusen deposits and complement system activation [20]. Visual loss caused by AMD cannot normally be reversed; currently, there is no effective therapy for this highly prevalent disorder to reverse the degenerative conditions.

There are two main types of AMD (**Figure 7**) [21]: the initial non-exudative or dry AMD characterized by pigment alterations in the macular area and accumulation

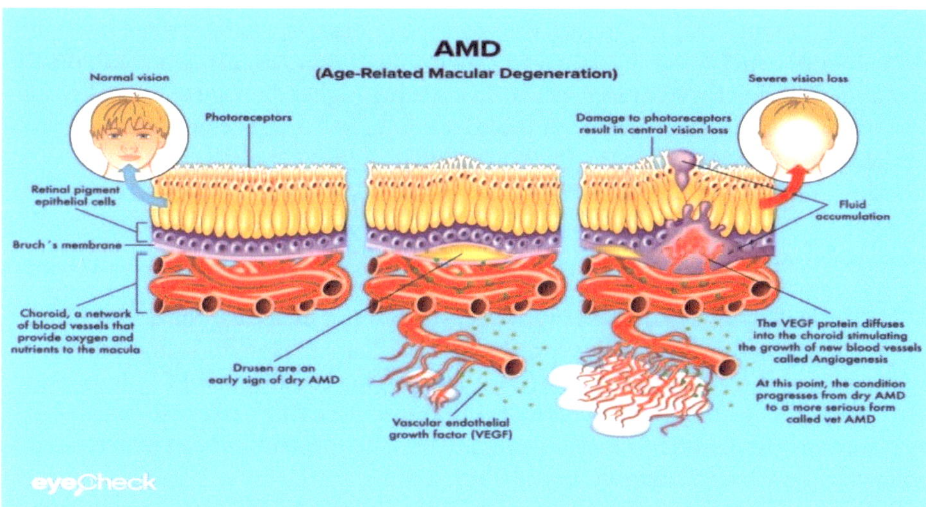

Figure 7.
What is age-related macular degeneration. The intimate relation among photoreceptors, RPE, and choroid is disrupted by drusen (lipid and cellular debris). This separates RPE from Bruch's membrane and the underlying choroidal vessels in association with RPE atrophy and photoreceptor degeneration (hallmarks of dry AMD). In wet AMD, abnormal leaky choroidal neovessels proliferate and penetrate the altered Bruch's membrane protruding into the subretinal space, causing hemorrhage and rapid loss of vision. Credit: https://learn.eyecheck.com/what-is-macular-degeneration (with permission - Mr. J. Anthony Moses CEO Eyecheck (https://eyecheck.com).

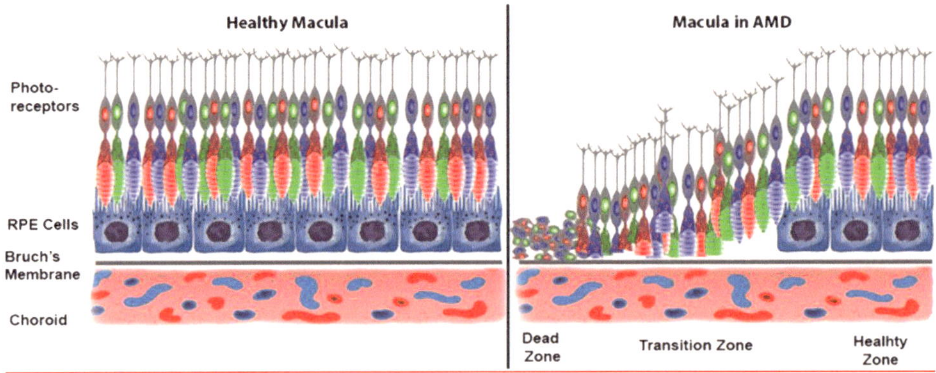

Figure 8.
Healthy macula vs. macula in AMD. The "dead zone" in the macula has minimal viable PRs and RPE cells. The healthy zone is usually toward the periphery of the macula and still has functional PRs with healthy RPEs. In the "transition zone" RPE cells are dead, but the PRs are intact. This is the potential target zone for RPE transplantation. Credit: Ref. [22] (with permission).

of extracellular whitish-yellow deposits of protein, lipid and carbohydrate termed drusen (the pathological hallmark of AMD). Drusens are deposited subretinally at the RPE-Bruch's membrane interface and are accompanied by RPE cell atrophy. Dry AMD constitutes about 90% of AMD cases in the United States. The later exudative stage, or wet AMD, occurs after the ingrowth of the underlying choroidal vasculature into the retina. Drusen formation is accompanied by progressive atrophy of the RPE layer and dysfunction of the overlying PRCs (**Figure 8**). Coalescence of large drusen droplets may cause separation of the RPE layer from Bruch's membrane with subsequent apoptosis of RPE cells; geographic atrophy represents an advanced AMD type with extensive apoptosis leading to gradual loss of central vision. A small ratio of all dry AMD cases (about 10%) progress to the wet type of disease, characterized by choroid neocapillaries penetrating the damaged Bruch's membrane, overgrowing into the RPE layer and overlying the neurosensory retina causing higher damage to central vision [2]. Wet AMD is the most severe but more treatable, and it can be delayed by repeated anti-VEGF medication.

4. Cell replacement therapy

In mammalian organs of the central nervous system, including the ocular PRE, the lack of regenerative capacity represents a significant reason for using stem cells to replace damaged tissues, e.g., in dry AMD, an atrophic condition of PRE. Importantly, RPE cells normally divide and eventually differentiate during early embryogenesis; the resulting post-mitotic RPE cells generally remain dormant but can be activated to proliferate in various circumstances of adult life [2].

RPE loss of function, which occurs early in the course of AMD and precedes photoreceptor death, has been a key therapeutic target for the last three decades [7]. RPE cell transplantation holds promise by replacing dysfunctional RPE cells and lost PRs to slow disease progression and/or restore vision. In clinical trials, RPE cells were injected in suspension and biomaterials [23, 24]. A key factor in increasing a retinal implant's effectiveness is the accomplishment of a substantial integration of the implanted cells [8]. Limitations such as the delivery method, low integration rate, and short-term survival of grafted cells hinder a successful cell-based therapy; moreover, it is unknown if disease pathogenesis due to a genetic mutation may affect a grafted RPE; this would downgrade RPE cell intervention as a temporary treatment and not a cure [7].

Clinical trials have demonstrated safety and modest vision improvements, but their number has been too small. They also do not provide specific data to elucidate if implants act via cell replacement, neuro-protection, immune-modulation or other mechanisms [15]. Due to the absence of predictive or diagnostic markers and the lack of long-term follow-ups to ensure safety, the *in vivo* functionality of RPE grafts has not been fully studied, and their survival remains uncertain even if anatomical integration is achieved [7].

4.1 Cell sources

Cell source is a key component to the success of any cell therapy. Identifying the appropriate cell source and suitable type, as well, represents a major challenge for the successful treatment of human retinal diseases. Current therapy focuses on the direct differentiation of four kinds of somatic stem cell therapy medicinal products (sCTMPs) into RPE and PRC phenotypes: human embryonic stem cells (hESCs), human inducible pluripotent stem cells (hiPSCs), retinal progenitor cells (RPCs) and mesenchymal stem/stromal cells (MSCs) [4, 8, 25]. All of the above stem cell types have been extensively investigated in several preclinical and clinical trials to confirm their effectiveness in repairing damaged retinal tissues. In addition, MSC-derived paracrine factors and exosomes have been tested to understand their efficiency in reversing retinal degeneration.

Embryonic stem cells (ESCs) have unlimited self-renewal capability. These totipotent stem cells, which originate from the inner cell mass of the embryonic blastocyst, can differentiate into cell types from all three embryonic germ layers, namely, ectoderm, mesoderm and endoderm [26]. The generation and establishment of murine and human ESCs and iPSC was a decisive step in the advancement of clinical translation, as their capacity for unlimited expansion provided an inexhaustible source of viable, easy-to-use donor cells. Numerous *in vitro* studies have demonstrated the ability of both human and mouse ESCs to differentiate into different functional retinal types, such as photoreceptors and RPE cells [27]. Treatment of retinal degenerative diseases by hESC-derived photoreceptors with or without RPE cells offers a vast source of transplanted cells for retinal replacement therapy. Using a specific combination of factors, hESCs can be directed to a retinal fate to generate retinal progenitors [28]. The Food and Drug Administration approved the first clinical trial using hESC-derived RPE cells for the treatment of dry AMD and Stargardt disease [29]. Despite encouraging results, the potential of using ESCs in cell replacement therapy for the treatment of retinal diseases is still limited. Ethical issues associated with obtaining hESCs and serious complications, including xeno-transplant immune rejection, have limited their clinical application [30]. Additionally, after several months of naïve ESCs transplantation, teratoma formation containing cells of all three embryonic germ layers from residual undifferentiated stem cells may have been observed [19]. Therefore, stringent release testing strategies are designed to detect cells with tumorigenic potential prior to transplantation.

iPSCs have several advantages as a source of retinal cells for transplantation. They possess unlimited self-renewal capacity and can be obtained from the patients themselves by directly reprogramming adult somatic cells to transit to a pluripotent state [31, 32]. Depending on the combination of soluble agents used, these cells can be induced to proliferate and differentiate into various retinal cell types, including rods, cones, retinal ganglion cells, etc. They can be collected easily and produced at a relatively reasonable price, present no ethical problems, and can be autologous, thus preventing rejection issues. However, they could conserve the epigenetic characteristics of the original cells, harbor the disease genes from the donor (the mutation load of iPS cells passes through several stresses and long culture periods), and lead to tumor formation [33]. A major drawback of iPSCs is their unstable ability to differentiate into the specifically programmed cell type. Additionally, in the case of autologous therapies, it would be advisable for patients with genetic mutations to have their genome edited. When therapies are allogenic, potential rejection may spoil long-term results; therefore, human leukocyte antigens (HLA) matching and possibly immunosuppression would be needed [7].

MSCs are a population of multipotent stromal cells that can be found abundantly in a variety of adult tissues including bone marrow, adipose tissue and dental pulp [34]. They can also be isolated from neonatal tissues and fluids such as umbilical cord, Wharton's jelly, amniotic membrane, amniotic fluid, and placenta. Despite the different tissues from which they are generated, MSCs have some distinct common characteristics [5]. Below we list the specific minimal criteria established in 2006 by the International Society for Cellular Therapy (ISCT), to define a cell as an MSC [35]:

- MSCs are able to adhere to plastic support in standard culture conditions, and this ability is exploited for their isolation.
- Have the potential to differentiate into an array of cell types including osteoblasts, chondrocytes, myocytes, adipocytes, neural cells, etc. (**Figure 9**).
- Do not express hematopoietic and endothelial cell markers such as CD31, CD34 CD45, CD11b, CD11c, CD14, CD19, CD79a, CD86, Stro-1, SSEA-4 and HLA class II molecules.
- Defined by the expression of specific cell surface markers as CD105, CD90, CD44, CD9, CD13, CD29, CD271, CD146 and CD73.
- Capable of avoiding immune cell recognition due to these properties, low levels of expression of class I HLA, zero expression of class II HLA, and low immunogenic capacity; all these properties render MSCs an appropriate source for allogeneic and autologous transplants.
- Wide distribution and ease to harvest.
- Possess minimal susceptibility to malignant transformation.
- Therapeutic efficacy to promote regeneration of multiple cells including central nervous system neurons.

Recent clinical trials to repair retinal degeneration with MSCs have demonstrated their safety profiles and promising potential to delay further retinal dysfunction. Administration of MSCs has revealed significant restoration of the visual system. MSCs have the potential to survive for a long time after injection into the vitreous body and have the capacity to regenerate a crushed optic nerve or protect retinal

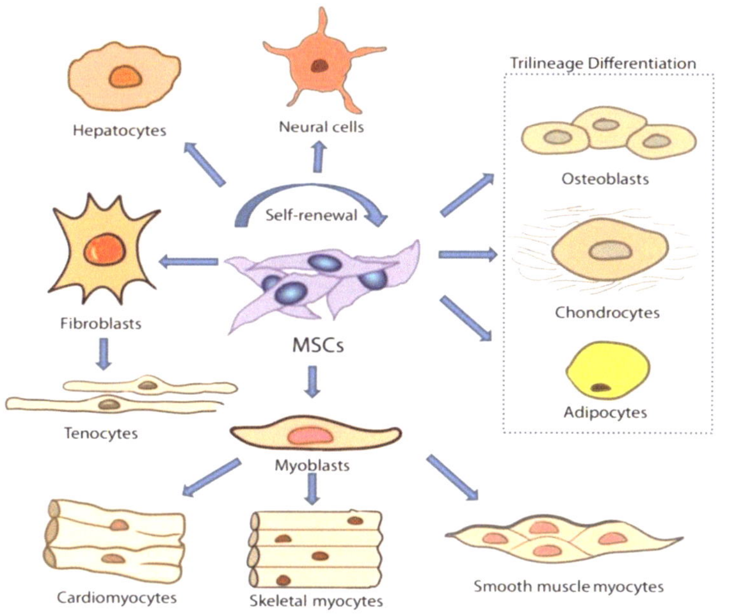

Figure 9.
Mesenchymal stem cell differentiation potential. Credit: Ref. [36] (reproduced under the terms of the CC attribution 4.0 International License).

ganglion cell functions. Neuroprotective effects on degenerated retinal cells could be associated with delaying or even stopping uncontrolled cell death [5]. Due to their neurotrophic properties, pluripotency, hypo-immunogenicity, modulation of immune response to tissue injury, limited ethical concerns, and low risk of tumor formation, MSCs have garnered significant interest in regenerative medicine and emerged as a prospective leading cell source for the treatment of retinal diseases [37]. However, in some clinical trials for the repair of retinal disorders with MSCs, various problems regarding the efficacy and safety profiles were reported, e.g., severe side effects or failure ti achieve the primary end-points of stem cell efficacy. Several clinical trials have shown that some AMD patients – after receiving intra-vitreal transplantation of autologous bone marrow/adipose tissue-derived MSCs experienced devastating outcomes (dense vitreous hemorrhage and rhegmatogenous retinal detachments following severe proliferative vitreoretinopathy and finally lost their vision) [1, 38–40]. Similarly, a high rate of retinal detachment and perforation was observed in AMD patients with geographic atrophy after subretinal injection of MSCs derived from human umbilical cord blood [41]. A probable cause is the trans-differentiation of injected MSCs into myofibroblast-like cells, which can induce fibrosis or proliferative vitreoretinopathy [25]. Complications, which are a significant issue of constant research, have additionally been attributed to the surgical procedure applied, number of administered MSCs, transplantation timing and the intrinsic mechanism of their *in vivo* action [42]. Therefore, for ocular or intraocular delivery of MSCs, it is imperative to follow more careful surgical procedures and more stricter control regarding the quality of MSC preparations [43].

Table 1 contains the most important advantages and disadvantages of the three main stem cell types used for the treatment of human retinal diseases.

	ESCs		iPSC		MSCs	
A/A	**Advantages**	**Disadvantages**	**Advantages**	**Disadvantages**	**Advantages**	**Disadvantages**
1.	Stable genome	Forbidden in several countries due to ethical reasons.	Differentiation capacity in diverse retinal populations	Genomic instability and tendency to carcinogenesis	Desirable differentiation potential	Limited differentiation potential (osteoblasts, chondrocytes, myocytes, adipocytes, etc.) compared to ESCs/iPSCs
2.	Endless self-renewal ability	High costs for production and lower availability	Affordable collection and easy production scaling	Changing ability to differentiate into the desired cell linage	No ethical bans – Low immunogenicity	Susceptibility (minimal) to malignant transformation
3.	Reduced rejection in allogenic transplants	Risk of teratoma formation	Can perform either autologous or allogeneic transplantation	Rejection of allogeneic grafts	Ability to regenerate multiple human-damaged tissues, including CNS neurons	
4.	Totipotent cells, able to differentiate into any type of human somatic cells	Ethical issues arising from collection sources and their use	Without any ethical dilemmas		Wide distribution in human body – easy to isolate, expand and be maintained	

Table 1.
Advantages and disadvantages of ESCs – iPSCs – MSCs [7, 25, 44].

4.2 Cell delivery formulations

In simple cell replacement therapy techniques, such as subretinal delivery of cell suspension boluses, a major challenge is the loss of donor cells in the intraocular microenvironment (**Figure 10**). Currently, cell transplantation with supportive biomaterials is considered as an effective alternative to avoid cell loss, improve survival rate, and increase cell integration with the host retina microenvironment [45].

4.2.1 Cell suspensions

RPE cell suspensions do not appear to be the ideal solution for restoring defective vision in AMD patients. When administered in a suspension formulation, there is an increased risk of cultured or freshly harvested RPE cells not surviving or integrating into the retinal network of aged or damaged Bruch membrane, causing serious complications such as proliferative vitreo-retinopathy [16]. Cell suspension injections are a suboptimal procedure with three major drawbacks: First, loss of retinal cells can be observed due to their dispersion into the host's intraocular tissue (**Figure 11**) [8]. Second, areas of basal lamina can form, alternating with multilayers formed by the uneven distribution of the transplanted RPE cells. Third, cells cannot integrate and function normally for a prolonged period in a defective or diseased basal lamina; additionally, the hostile sub-foveal microenvironment is not conductive to successful retinal engraftment due to glia and host immune insults [13].

4.2.2 Scaffold implants

A transplant-ready RPE monolayer can be generated on certain biomaterials to facilitate cell delivery and *in vivo* engraftment. RPE cells can be grown on bioengineered, nanofibrous scaffolds, and the RPE cell-enriched scaffolds can be transplanted into patients' eyes [13]. An ocular scaffold is a biocompatible construct fabricated to mimic the native retinal microenvironment. It aims to support essential properties of the grafted cells, e.g., adhesion, proliferation, differentiation and long-term survival. In different experimental models, implanted retinal cells on supportive biomaterials slightly improved vision capabilities, achieving a better but limited integration, than suspension cells in the same area [8].

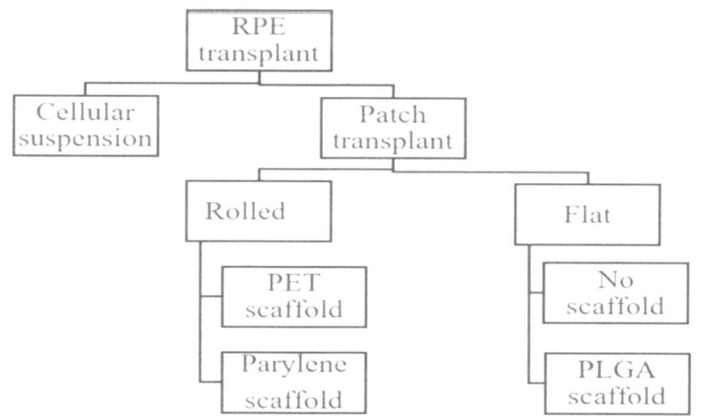

Figure 10.
Scheme of possible RPE transplant delivery system. Credit: Ref. [7] (with permission).

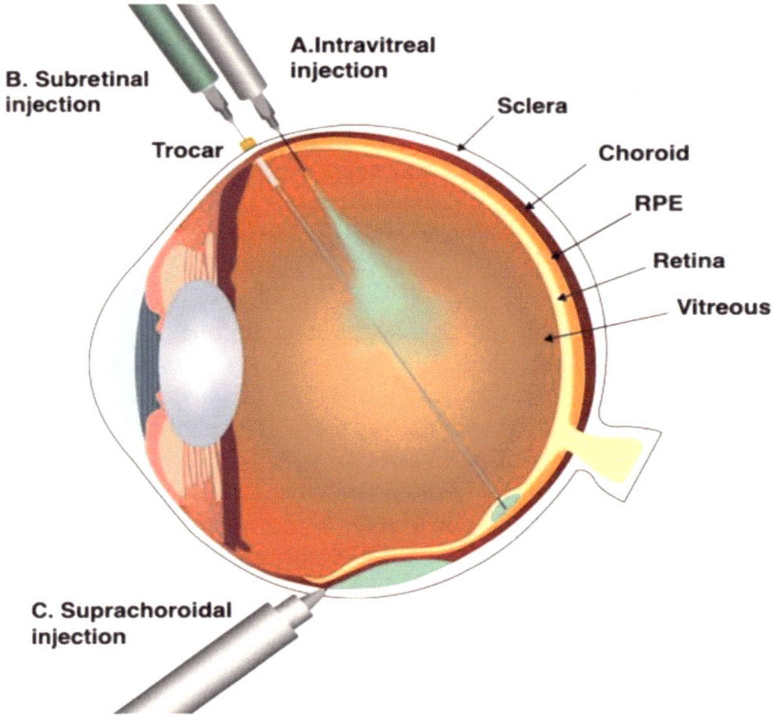

Figure 11.
Surgical approaches for subretinal delivery. (A) Intravitreal injection: Fills the vitreous body with the therapeutic suspension, (B) subretinal injection: Delivers the therapeutic suspension under the sensory retina into the subretinal space (between photoreceptors-RPE), (C) suprachoroidal injection: Delivers cells into to the space suprachoroidal space (between choroid-sclera). Credit: Ref. [46] (with permission).

Recent advances in material science helped to develop strategies to grow cells as intact monolayers or as sheets on biodegradable or non-degradable polymer scaffolds for transplantation into the eyes. Biocompatible polymers are either synthetic substances, e.g., polycaprolactone (PCL), polylactic acid (PLA) and poly lactic-co-glycolic acid (PLGA), or materials of natural origin, e.g., celluloses, zein, alginates, etc. [47]. Such implants are found to be more promising than the bolus injection approach. Obviously, the supportive biomaterial construct along with the extracellular matrix formed can help maintain cellular integrity, hold the cells in place to increase cell survival, encourage proper cellular alignment, and improve integration with the host retina and visual function [16, 45]. A crucial factor to the successful long-term function of a synthetic biomaterial construct is its ability to replace native damaged BrM, in which the grafted RPE cells can grow and integrate into the native retinal circuit. The surface properties of the underlying substrate scaffold directly influence the cells' ability to form a differentiated monolayer. The physical properties of synthetic membranes include biostability, porosity and suitable mechanical strength for surgical handling [16].

Biomaterial scaffolds provide a platform for reconstituting RPE sheets with the neighboring choroid and neurosensory retina. They:

- must eliminate the possibility of injected RPE cell suspensions refluxing into the vitreous cavity, which can cause fibrosis, leading to retinal detachment and blindness [16];

- may ensure the correct orientation of rods and cones and their functional vicinity with underneath RPE cells [48];
- should create a functional microenvironment to favor adhesion, survival and differentiation of transplanted retinal cells;
- must be nontoxic when implanted in the subretinal area;
- must be below 10 μm in thickness, in order not to affect the focal length of the eye, provide mechanical stability and flexibility to resist implantation strain and facilitate retina curvature;
- are positioned between the RPE and choriocapillaris, so they should ideally degrade and be replaced by the natural extracellular matrix that separates these tissues;
- should be biocompatible, biologically inert (not induce local tissue damage) and sterilizable, neither triggering cytotoxic processes nor inflammatory responses. A very promising biomaterial is *silkworm silk fibroin* because it fulfills all these requirements [8];
- could be derived from natural sources (chitosan, gelatin, collagen, cellulose, alginates, etc., are more preferred as they have higher biocompatibility, excellent biodegradability, and minimal toxicity) or made synthetically (polylactide-*co*-glycolide (PLGA), polycaprolactone (PCL), polylactic acid (PLA), fibronectin, polyurethane, etc.) [49].

Researchers have used biomaterials based on 3D printed *Bruch's* membranes, alginate, silk, cellulose, and hyaluronic acid [13] to create scaffolds to:

- support the development of transplanted retinal cells
- act as substitutes for Bruch's membrane or
- act as retinal cell delivery systems.

Comparing biodegradable scaffolds with non-biodegradable counterparts seem to perform better: they keep the focal length of the eye unchanged, they do not leave foreign materials in the subretinal space, they allow better vascularization, the new RPE sheet is smoothly incorporated into the host's RPE network [50]. In addition, transplanted polymers that undergo rapid bulk degradation such as poly lactic-*co*-glycolic acid (PLGA) significantly lower local pH and cause widespread retinal injury when delivered in sufficient quantities.

5. Terms and conditions for a long-term functional retinal cell implant

Precise integration of the implanted cells into the existing host cellular network in the damaged macula area is a prerequisite to increasing the effectiveness of retinal implant therapies. Approaches using cell suspensions involve easer technical maneuvers; however, their functional outcomes seem to be inferior. Indeed, the success of retinal transplantation based on cell suspension injection is hindered by limited cell survival and lack of cellular integration. Currently, the ultimate balance between practicability and clinical outcomes remains unclear. Among the different experimental approaches tested, patch implants showed satisfactory outcomes with a reasonable incidence of adverse events. For the treatment of AMD, we propose to administrate – on appropriate scaffolds – in the form of monolayer as many mature RPE cells as are required to functionally cover a damaged area [7, 16].

5.1 Factors to consider in establishing a long-term functional RPE implant

RPE cell implants are able to survive and make connections to the host retina with visual improvement, if the graft integrates and/or communicates with the host retina [17]. The limitations of suspension cell injections can be overcome by an efficient strategy using 2D or 3D scaffolds. The current major obstacles to successful retinal cell transplantation are the integration of the scaffold into the impaired area of the host retina and the functional reconstruction of the neural circuitry [6]. In contrast to cell suspensions, ex vivo growth of an RPE monolayer has the advantage of performing experimental readouts of its functionality before surgical implantation [23]. Importantly, most biomaterials used for retinal regeneration have not been tested under *in vivo* conditions [8]. For AMD patients, because implanted RPE cells will need to be functional, possibly for decades, properties such as cell attachment to an aged Bruch's membrane and survival of transplanted cells in a degenerative environment present a challenge [17]. Although safety has been demonstrated, the cell replacement mechanism and efficacy remain difficult to validate [15].

5.1.1 The basic principle of cellular implant generation: Replicating the in vivo *healthy RPE layer*

Aged human Bruch's membrane does not support attachment and survival of transplanted RPE. Compared to an RPE monolayer patch, injectable cell suspensions do not reliably form a functional RPE monolayer and exhibit poor long-term survival. Transplantation of an intact RPE sheet onto a suitable substrate mimicking normal Bruch's membrane represents a feasible approach to overcome this hurdle [16]. Scaffolds must be engineered in such a way that they can mimic the conditions of the host retinal microenvironment and promote the structural and functional integration of cells into the existing native neuronal network. A monolayer RPE patch aims to recapitulate the anatomy of a normal RPE layer. Also, it promotes the formation of tight junctions between neighboring retinal cells (**Figure 12**) to establish cell polarity, which is necessary for the RPE cells to support the photoreceptor functions.

5.1.2 Disadvantages of suspension-injected RPE grafts

There have been several investigations into the inoculation of RPE cell suspensions. There is also evidence that suspension-injected RPEs do not always survive long -term and frequently fail to form a functional monolayer. This is a significant challenge for injections of suspended cells because the therapeutic effects of grafted RPE cells depend on their polarization and spatial arrangement in a subretinal monolayer. Unfortunately, the ability of cells to migrate from the transplant site to the damaged host retina is strongly prevented by the hostile environment of the

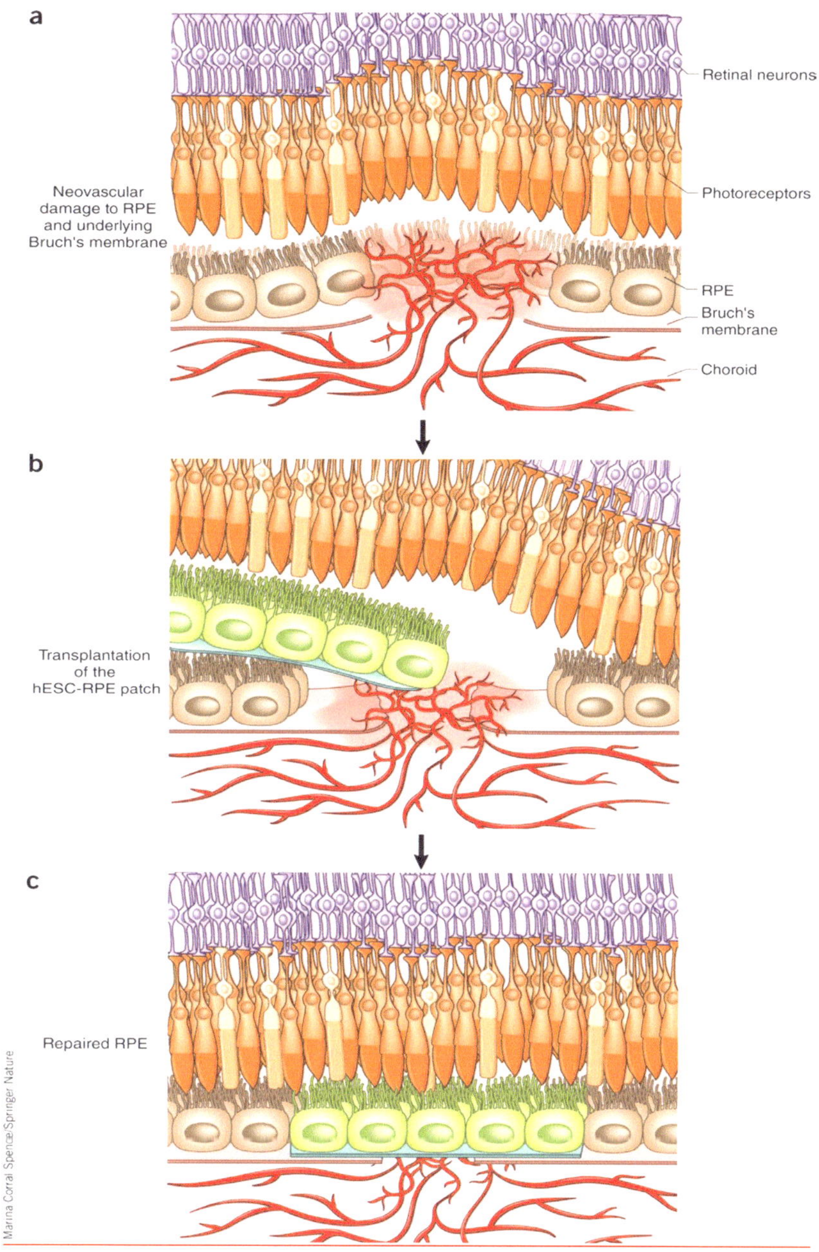

Figure 12.
RPE replacement therapy for wet AMD. (a) Choroidal blood vessels penetrate through the RPE, leaking fluid in the subretinal space. (b) An hESC-RPE monolayer on a plastic substrate (green) is delivered to the areas of vessel leakage and RPE - retina detachment. (c) Patient photoreceptors (orange) recover through interaction with transplanted RPE cells; cytokines from RPE cells cause choroidal vessels to stop proliferating and leaking. Credit: Ref. [50] (with permission).

degenerating retina [19]. Both preclinical and clinical trials have shown that suspension-injected RPE eventually loses most of its morphological and functional properties, resulting in low efficacy of this type of cell replacement therapy [43]. RPE cells have a limited migratory ability, and it is unknown if they can migrate away from the injection site, how far they will travel in a diseased eye, and if the benefit is similar to animal studies. Additional limitations include low survival, low cell integration at the transplant site, and difficulty in maintaining the injected cells in the target areas. Conclusively, the current golden standard technique for cell delivery in retinal therapies, that is, subretinal bolus injection of RPE cells, is governed by limitations inherent to the transplantation process. It frequently results in disorganized and poorly localized grafts with low rates of cell survival due to poor incorporation of donor cells and cell regression into the vitreous cavity [28].

5.1.3 Cell delivery system: Retinal damage should determine the properties of engineered scaffolds

It has been shown that aged or damaged human Bruch's membrane does not support a functional suspension-injected RPE implant in the long-term. Due to the disappointing outcomes, infusion of cell suspension is currently considered not to be a suitable cell replacement approach for retinal regeneration [16]. Presently, cell attachment on a scaffold represents the preferred therapeutic approach. Critical to successful subretinal transplantation is the ability to grow a functioning RPE monolayer on a substrate, having the ability to replace a damaged Bruch's membrane. Such a biocompatible construct may have the potential to preserve or restore vision in patients affected by blinding diseases such as AMD (**Figure 13**). The choice of substrates with proper physical and chemical properties should assist the engineered cell retinal construct be functional even for decades [7]. The type of support should be more biologically relevant and mimic, at some points, the endogenous Bruch's membrane [8]. Scaffolds for retinal cell transplants can be fibrous or cylindrical or made of hydrogels to closely imitate the microstructure of the extracellular matrix or the vertical cell arrangement in the retina layers or the mechanical properties of retina, respectively. The candidate scaffolds on which RPE cells will be placed may be

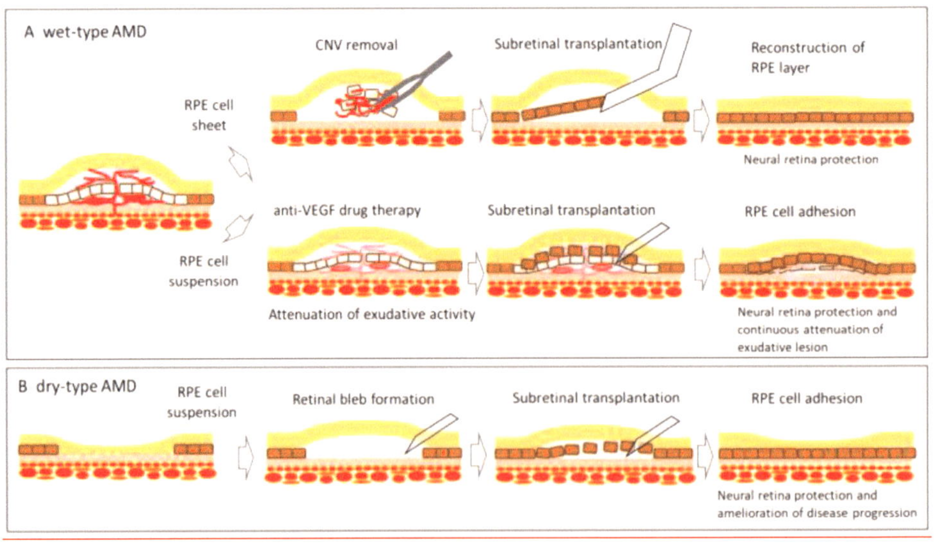

Figure 13.
Wet- and dry- type AMD cell therapy: A conceptual mode of action using pluripotent stem cell-derived RPE cell products. Credit: Ref. [51] (with permission).

designed to act either as supporting devices or as artificial BrMs, and their selection should preferably have the following characteristics [16]:

- for easy handling, to have proper flexibility and mechanical resistance
- for the exchange of nutrients, to be porous
- for subretinal implantation, to have compatible thickness
- for biodegradable substances, not to produce toxic by-products.

Various scaffolds with well-defined characteristics such as physico-mechanical properties, ultrafine structure and biocompatibility have been investigated for retinal tissue engineering and repair. Many synthetic polymers such as poly (e-caprolactone) and poly (ethylene glycol) diacrylate, polylactic acid, poly-glycerol sebacate, polylactic-co-glycolic acid, etc., have also been tested for their ability to deliver RPE cells or photoreceptors to the subretinal area, or as a Bruch's membrane substitute. Studies have shown that polymers present different advantages [13]:

- increased rate of proliferation and differentiation,
- higher cell survival and number of cells delivered to the subretinal area,
- induction of retinal differentiation to generate functional neurons.

The needs of both RPE and photoreceptor replacement therapy are not likely to be met by a single polymer substrate alone. It is known that RPE cells have a limited ability to adhere and function on a damaged or aged Bruch's membrane; therefore, the support provided by a polymeric material must be essential for RPE cell localization, survival, or directed functional differentiation. However, a permanent barrier in the subretinal space would compromise photoreceptor transplant functionality [13]. To overcome this obstacle, we would have to make a porous scaffold with natural materials, e.g., collagen, elastin, etc., in order to mimic healthy Bruch's membrane as closely as possible. Biodegradable polymer scaffolds may increase survival cell rate, and it also appears that the adhesion and orientation of transplanted retinal cells are promoted by the scaffold porosity (dimensional conformation). In addition, cell orientation can be affected by scaffold topography because of its positive influence on cell morphology, proliferation, migration and adhesion [52]. Good results on biomaterials and RPE cells have been obtained from both, *in vitro* and *in vivo* studies; however, how long the transplanted cells could survive has still to be determined [53]. Unfortunately, research in the field is still in the early stages, clinical trials are very few and in very early stages of development, and the lifetime of retinal grafts has not been correlated with any prognostic indices [8].

5.1.4 Interventional timing

In RPE cell suspension replacement therapy, the time point at which retinal cells should be injected is a critical factor in successful transplantation. The remaining dysfunctional RPE cells do not allow the attachment of new cells, as they still occupy Bruch's membrane. An ideal invasive timing would be when the endogenous host-damaged RPE cells have been removed, but photoreceptors are still present and functional. For photoreceptor preservation strategies, early intervention leads to better vision preservation [17].

5.1.5 Site of graft placement

However, in some cases, transplantation may cause a retinal detachment (**Figure 14**) [17]. The point of injecting an RPE implant is also a significant factor. It should be near areas of the macula or geographic atrophy as the AMD macula is affected. However, in some cases, the transplant can cause a retinal detachment, which represents a high risk for macular vision disorder. In patients with geographic atrophy, there are probably several foci of degenerative cells, and inoculation of multiple retinal implants in an equal number of injection sites is recommended [17].

5.1.6 Cell density of RPE layer: An important factor to calculate the effective injected cell number

The RPE sheet is a monolayer in the outer retina composed of polygonal, mostly hexagonal cells and is highly adapted to local physiological requirements (**Figure 3**).

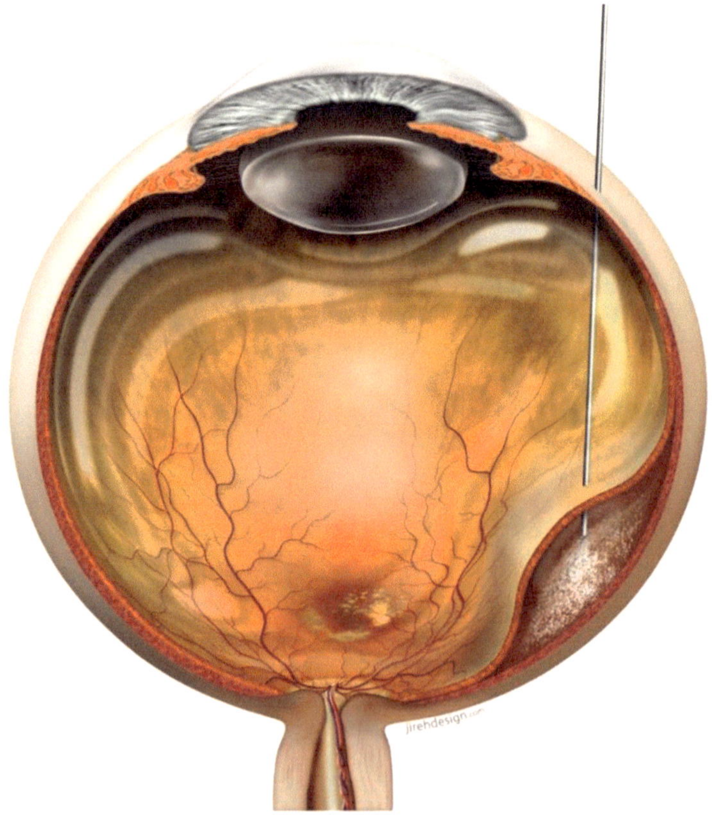

Figure 14.
Depiction of the subretinal injection site. Credit: https://jirehdesign.com/stock-eye-illustrations/eye-surgery/retina-surgery-illustrations/subretinal-injection-suvr0049/ (with permission - Mr. Mark Erickson CEO JirehDesign www.JirehDesign.com).

Increasing RPE cell numbers by transplantation can effectively treat atrophic RPE diseases, including AMD. Currently, there is no standardized guidance for the injection volume and cell concentration, and the exact number of cells required for efficient transplantation remains unclear [17].

Patient age, location of retinal damage, and disease status are three factors that affect RPE cell density and shape. Across the posterior pole, significant spatial variations can be seen in terms of RPE cell morphology, photoreceptor density and distribution, as well. The healthy foveola is composed of strictly mononuclear cells, mostly hexagonal and of uniform size. It exhibits the highest RPE cell density, and only cones are found. In the human eye, there are approximately 137 million photoreceptors, 7 million cones and 130 million rods. The rods are located at a distance of approximately 6 mm from the foveal center and about 200 from it. Their number decreases toward the periphery but more slowly than the cone numbers. Instead, the cones can be seen in the center of the fovea. Their density lowers from the center to the periphery in a radial manner. Regarding apoptotic RPE cells, age significantly increases their proportion in erderly patients. The majority of them are mainly confined to the macula area and their number is more decreased in peripheral regions of Bruch's membrane [54]. In aging, although RPE cell apoptosis occurs in the macula, the overall number remains constant and decreases in the periphery. This observation suggests that peripheral RPE cells migrate from the periphery to the center of macula where they can replace the dead cells of the RPE layer. Indeed, although during aging, the density of RPE cells in the macula region remains unchanged, a more variable number of neighboring cells can be detected in this RPE area of the macula.

Various structural changes can be observed in an aged RPE layer, such as basal deposits, accumulation of drusen, calcification of the elastin layer, diffuse layer thickening, laminar deposits, collagenic cross-linking, etc. Age-dependent extracellular matrix changes within Bruch's membrane may induce apoptosis in cells in the overlying RPE because disruption of epithelium-basement membrane interactions leads to apoptosis. These changes in the extracellular matrix of Bruch's membrane are the hallmark of elderly AMD patients at autopsy and may be related to apoptosis in the overlying RPE [54].

The number of RPE cells in normal human eyes decreases by approximately 0.3% per year (in younger eyes it is estimated to be ~0.07 per 100,000 cells) [55]. In older eyes loss increases up to 1.96% of macular RPE cells per year. This can translate into an annual loss of 1960 cells per 100,000 of RPE cells in the human macular; that is, a maximum of nearly 20% of RPE cells in the macula would be lost per decade, which represents a significant rate of loss. The local variations in normal cell density, combined with variations in disease presentation, make systematic assessment of the clinical status of AMD patients a prerequisite prior to the production of any RPE implant. Accurate topography of the lesion is required as a guide – by making further anatomical and integrity studies, such as optical coherence tomography (OCT) analysis – to calculate the extent of the damaged area and the approximately (exact) number of dead RPE/photoreceptor cells to engineer a potential fully functional implant.

5.1.7 Autologous adipose tissue mesenchymal stem cell-derived terminally differentiated RPE (ADMSC-RPE): a promising cell population for retinal regeneration

ADMSCs are a promising cell therapy with regenerative properties. These multipotent stem cell-like reside in the stromal vascular fraction (SVF) of adipose tissue, and their properties can be compared to those of bone marrow stromal cells (BMSCs) properties comparable to those of bone marrow stromal cells (BMSCs). A major advantage over BMSCs is that much larger amounts of regenerative ADMSCs can be harvested by applying minimally invasive procedures during liposuction. ADMSCs can perform their paracrine action by secreting a plethora of pro-regenerative cytokines and growth factors to enforce the regeneration of damaged tissues [16]. MSCs meet all the specifications recommended for human ADMSCs by the relevant department of the International Society for Cellular Therapy (ISCT) [35]. Hunan ADMSCs under standard culture conditions exhibit adherent properties, proliferate, differentiate, express the surface markers CD13, CD29, CD44, CD73, CD90, CD105, CD146 and CD271 and are negative for the antigens CD31, CD34, CD45, Stro-1, SSEA-4 and HLA-DR [44]. Similar to many other types of stem cells, they express a high proliferative rate and have the ability to differentiate into cells originated from other embryonic germ layers. Under *in vitro* conditions, they have the potency to differentiate into osteoblasts, adipocytes and chondroblasts; they can also be induced to differentiate into myogenic, neurogenic, angiogenic, hepatic and retinal cell lines [56]. By choosing autologous fatty tissue to isolate ADMSCs, we avoid the destructive side effects of a potential RPE transplant rejection. These properties make ADMSCs attractive candidates for cell replacement therapies and a proper cell source for the treatment of several retinopathies [48, 57].

In humans, RPE cells undergo early terminal differentiation starting at 4–6 weeks of gestation. These cells normally remain dormant throughout life, dividing rarely or not at all [2]. They can be activated upon release from the niche that normally maintains a quiescent state. Escape from the niche causes massive RPE proliferation, leading to devastating outcomes, e.g., destructive proliferative vitreoretinopathy (PVR) or disc-shaped scar formation. Although MSCs have emerged as a prospective leading cell source for the treatment of retinal diseases, such destructive clinical results have reduced their credibility as a safe source. The proliferative response of RPE can be reparative in dry AMD to partially counter RPE cell atrophy [2], or after receiving intraocular injection of autologous BM- or AT-derived MSCs [40]. Mature RPE cells derived preferentially from the most readily available AT-MSCs appear to be a safe and functional cell source for retinal replacement therapy. This technique allows for the easy production of a considerable number of MSC-derived RPE cells seeded on an appropriate scaffold and maintains the physiological and functionally crucial RPE polarization [7]. Due to their terminal differentiation state, the scaffold-implanted RPE cells lack proliferative capacity and remain dormant within the retinal microenvironment, mimicking the *in vivo* physiological state. Therefore, the repaired RPE monolayer is not compromised by cell overgrowth which may cause uneven distribution of RPE cells, multilayer formation or even retinal detachment.

5.1.8 Photoreceptor replacement therapy

Photoreceptor damage is one of the main characteristics found in retinal degeneration diseases, such as Retinitis Pigmentosa or AMD. Such degenerative diseases are currently turning into irreversible conditions attributed – on the one hand, to the insufficient intrinsic regeneration rate in the human retina and – on the other hand, to the lack of effective treatments to arrest the progressive loss of photoreceptors. Cell and tissue engineering approaches focus on the preservation of photoreceptors by replacing damaged RPE; this technique has been shown to slow the degeneration of photoreceptors and even restore partial visual functions [21, 58]. To tackle their loss and restore vision in affected AMD patients, photoreceptor cell replacement therapy has been proposed as a potential treatment modality [17, 45]. Although it seems that photoreceptor degeneration may be retarded by RPE replacement, a 'retinal sheet', that is, an organized construct that can be easely incorporated into native tissue will likely be needed to replace impaired photoreceptors to restore vision [6].

5.1.8.1 Photoreceptor transplantation in advanced AMD stages

In advanced stages of AMD, photoreceptor replacement can be an effective therapeutic strategy to substitute this type of retinal cell, which eventually dies due to support inability by RPE cells. This therapeutic option either replaces damaged photoreceptors with new ones or induces the secretion of neuro-protective or immune-modulatory factors that promote the functional recovery of damaged cells [43]. There is a wide time-frame to apply a therapeutic photoreceptor intervention, which increases the probability of successful cell transplantation because of the prolonged periods of survival time of neurons of the inner retinal layers after photoreceptor degeneration. However, there is a very difficult condition, namely, the implantation of functional photoreceptors that will make new synapses with the inner neighboring bipolar cells and the reversal of the synaptic remodeling that occurs after the degeneration of photoreceptors [59]. Although post-mitotic transplanted photoreceptors have been proved to be able to regain relative vision functionality [60], to date, no one has achieved a significant restoration of visual function.

5.1.8.2 Photoreceptor preservation

The development of physical substrates for artificial RPE sheets is the main objective for biomaterial engineers. Preserving photoreceptors is not an easy task. Difficulties include keeping them alive until transplantation time and replacing those already been lost, by successful integration of new into neural retina. Depending on the donor age, cultured photoreceptors can be maintained alive for 24–48 h and after transplantation the survival rate is less than 20% [61]. Importantly, the ratio of functional photoreceptors among alive ones is unclear. It is confirmed that a polarized photoreceptor monolayer construct may successfully get integrated into the neural retina by using high-tech 3D scaffolds [62].

5.1.8.3 Photoreceptor transplantation with biomaterials

Both natural biological and synthetic polymeric materials are used to construct scaffolds for photoreceptor implants. Such biological materials that mimic natural architectures include alginate, collagen, etc. Synthetic polymers, such as poly-caprolactone (PCL), poly (lactic-co-glycolic) acid (PLGA), etc., express higher mechanical strength and controlled degradation rates [52]. We would not recommend the use of scaffolds with the capacity to facilitate the differentiation of photoreceptor progenitors (PRPs). Similar to the destructive behavior of MSCs in the repaired RPE monolayer, uncontrolled overgrowth of PRPs could compromise the functional integrity and survival of the repaired photoreceptor network.

5.1.8.4 Retinal cell transplantation must be done in a timely manner before loss of photoreceptors

Among patients with progressing AMD there is a fear that a risky surgical procedure will further worsen their already impaired vision, so they consider their remaining reduced vision. For this reason, in clinical trials, only severely disabled patients are treated with few healthy native photoreceptors capable of interacting with the implanted patch. However, one should take into consideration the adverse molecular microenvironment that prevails throughout the central nervous system and makes the regeneration of photoreceptors (and all cell repair processes) hardly reversible [8]. The greater the loss of photoreceptors, the more hostile the retinal microenvironment.

5.1.9 About animal models: Rat models are a favorable choice for the study of retinal degeneration

Pig eyes are considered ideal for improving translational surgical techniques due to size similarity to human eyes [63], however, large numbers of pigs are costly to house and difficult to immunosuppress for a long time. These inhibitory factors make it difficult to use pig eyes to assess the tolerability of a human retinal cell implant. There are many mouse models for studying hereditary retinal degeneration in humans. However, the mouse eye is too small to reliably inject into the subretinal space and functionally evaluate a photoreceptor graft on a polymer scaffold. Current research is using rat models to study retinal degeneration diseases and even to implant polymer-supported RPE grafts and full-thickness retinal sheets [64]. Their wide experimental use is attributed to (i) the cheap price that allows housing in large numbers and testing many variables simultaneously; (ii) the size of eyes, which is large enough for subretinal transplantation of a biodegradable cell implant.

5.1.10 Clinical validation is essential for post-transplantation

The retinal transplant can be safely and inexpensively observed with bio-microscopy at frequent intervals following transplantation [45]. In addition, advanced optical systems have been invented for the in *in vivo* imaging of retinal grafts, including spectral domain-optical coherence tomography (SD-OCT) and adaptive optics-scanning laser ophthalmoscopy (AO-SLO). These imaging systems are valuable for studying cellular changes of retinal transplants monitoring treatment responses and disease progression [17].

6. Conclusion: key points for successful retinal transplantation

1. A cell product (suspension or implant) is a drug, "living drug," and as with conventional drugs, we must know its therapeutic dose.

2. MSCs (and stem cells in general) have unlimited proliferative capacity. Once infused, they expand beyond the initial administered dose:

    - making the determination of the therapeutic cell dose a serious problem

    - causing RPE detachment due to overproduction of MSC-derived RPE cells.

3. MSCs are considered the safest stem cells (**Table 1**), and their selection represents the best cell source for RPE transplantation.

4. Adipose tissue-derived mesenchymal stem cells are increasingly used in retinopathies due to their ease of obtaining and preparing. Implantation of autologous ADMSC-derived RPE cells protects from the destructive side effects of possible transplant rejection.

5. There is no risk of rejection after administration of live drugs consisting only of autologous ADMSC-derived terminally differentiated RPE cells.

6. Due to their terminal differentiation state, the *in vivo* proliferative capacity of scaffold-implanted RPE cells is negligible or zero; they will remain dormant within the retinal microenvironment, mimicking the *in vivo* physiological state. The repaired RPE monolayer will not be compromised by cell overgrowth which may cause uneven distribution of RPE cells, multilayer formation or even retinal detachment.

7. In AMD replacement therapy, *in vivo* simulation of a healthy RPE layer or photoreceptor network should be the target of subretinal delivery of adipose-derived and scaffold-engineered human retinal cells.

8. Cell suspension preparations proved to be ineffective in achieving reconstruction of the neural circuitry and long-term survival and functionality of transplanted cells.

9. Cell implants on biocompatible scaffolds are the current trend in AMD treatment because they have been shown to integrate efficiently and functionally into the damaged retina.

10. The selection of a substrate with appropriate physical and chemical properties should be a determining factor for the long-term functional performance of the engineered retinal tissue and mimics at some points the endogenous Bruch's membrane. Depending on the clinical status of patients, the candidate scaffold on which RPE cells will be placed may be designed to act either as a supporting device or as an artificial BrM.

11. Clinical parameters should determine the design of AMD cell therapy, which will be different for each patient, not only for patients belonging to different types (dry or wet) but also for patients of the same type. Damage extent and location will determine the:

    - concentration of implanted cells

    - composition and physicochemical properties of scaffold (biodegradable, or non-biodegradable)

    - interventional timing.

12. Retinal cell transplantation must be done in a timely manner before loss of photoreceptors.

13. Long-term post-transplantation monitoring with advanced *in vivo* imaging systems (AO-SLO, SD-OCT) will be a prerequisite.

14. The proposed cell replacement procedure for AMD patients will be:

    - an indicative personalized therapeutic regimen, where dose of cell preparation and scaffold properties will be determined by scientific criteria and not empirically

    - a possible solution to the existing problem of determining the therapeutic dose of cell preparations in the field of regenerative medicine.

Glossary

Adaptive optics-scanning laser ophthalmoscopy (AO-SLO): It is an imaging method for monitor living retinal cells; it makes use of adaptive optics to provide sharper images of the retina or cornea obtained with scanning laser ophthalmoscopy (SLO), removing any optical aberration. SLO utilizes confocal laser microscopy to examine the human retina or cornea for diagnostic imaging purposes.

Adipose tissue-derived mesenchymal/stromal stem cells (AD-MSCs): A highly promising class of multipotent mesenchymal stem cell-like cells for cellular regenerative therapies. They reside in the stromal vascular fraction (SVF) of adipose tissue and have the ability to secrete a multitude of pro-regenerative growth factors. In general, their properties are comparable to those of stromal cells in the bone marrow (BMSCs). All criteria for MSCs recommended by the relative Committee of the International Society for Cellular Therapy (ISCT) are met by human/mouse ADMSCs. They have the potency to differentiate into osteoblasts, adipocytes, and chondroblasts, as well as neural and retinal cells *in vitro*.

Age-related macular degeneration (AMD): The common cause of severe central visual loss in the elderly in industrialized Western countries, associated with loss of face recognition and ability to read or drive a vehicle. Gradual blindness is attributed to damage to Bruch's membrane or choriocapillaris or photoreceptors or RPE layer. There are two distinct AMD types, neovascular (NV) or dry and non-neovascular (NN) or wet. About 10% of dry AMD patients progress to the wet form of AMD. NV AMD is more severe but it can be delayed with repeated anti-VEGF medication, in contrast, no effective treatment exists for NN-AMD.

Amacrine cells (ACs): Retinal inter-neuronal cells with inhibitory properties. In the inner plexiform layer (IPL), they make synaptic connections with bipolar and ganglion cells. They reach across several bipolar cells to regulate signals directed at retinal ganglion cells. So far, around 30 subtypes have been identified.

Bipolar cells (BCs): A type of retinal interneurons which transmits signals to retinal ganglion cells that have received from photoreceptors.

Bruch's membrane (BrM): It is described as a highly specialized and pentalaminar collagenous structure that separates the RPE cells from the choroid, the underlying vascular layer. It is an organized acellular sheet 2–5 mm-thick, and its five layers are mainly composed of elastin, collagen I-V, laminin and fibronectin. Functionally, it supports the exchange of nutrients and oxygen supply between the neural retina and the capillaries of the choroid layer.

Choriocapillaris: The innermost layer of the choroid (vascular layer of the eye that lies between sclera and BrM) underlying Bruch's membrane. It consists of compacted anastomotic capillaries, which supply oxygen and nourishment to the outer layer of RPE and photoreceptors and removes their metabolic wastes. The fenestrated choriocapillaris, BrM, and RPE form the outer blood-retinal barrier.

Cone photoreceptors: Retinal photoreceptors that mediate color vision and function optimally in bright light. There are three types of cones that mediate trichromatic vision in humans: S-M-L-cones sensitive to short-medium-long wavelength light, respectively. The human retina contains 6 million cones residing in the foveola in a densely packed form. Cones mediate the high-acuity vision necessary for face recognition, writing, reading, driving, and fine tasks such as sewing.

Drusen: They are extracellular whitish-yellow deposits of protein, lipid, and carbohydrate deposits located under the retina. The exact relationship between degenerative macular disease and drusen is not clear. Drusen formation, the pathological hallmark of AMD is accompanied by progressive atrophy of the RPE layer and resulting dysfunction of the overlying photoreceptor cells. Merge of large drusen can cause separation of RPE from Bruch's membrane leading to RPE cell apoptosis. If the latter is extensive, it leads to an an advanced form of AMD called geographic atrophy.

Dry or non-neovascular AMD: The initial non-exudative step or dry AMD, which is characterized by pigment alterations in the macular area and accumulation of drusen. They are deposited subretinally at the RPE-Bruch's membrane interface and accompanied by RPE cell atrophy. Dry AMD constitutes about 90% of AMD cases in the United States, and visual loss caused by AMD cannot normally be reversed, as there is currently no effective therapy for this highly prevalent disorder to reverse the degenerative conditions.

Fovea: In the *middle of the retina*, there is a small depression called fovea (which constitutes an area of 1.8 mm^2), which is responsible for high-resolution vision. Fovea is rich in cone photoreceptors; within the fovea there are approximately 200,000 cones, a number of cells that can easily be produced by the current organoid technology. This relatively small region is extremely important for human vision; it makes tasks such as reading, facial recognition and driving.

Ganglion cell layer: The innermost retina layer is composed of ganglion cells and displaced amacrine cells. Retina ganglion cells transmit the synaptic input – received by bipolar cells and amacrine cells – to the brain thalamus, hypothalamus, and midbrain via long projections that form the optic nerve.

Geographic atrophy (GA): An advanced form of AMD characterized by well-defined areas of dead photoreceptors, dead underlying RPE cells and atrophic subjacent choriocapillaris. If GA occurs in the fovea, the patient loses high-acuity vision.

Human embryonic stem cells (hESCs): RPE cells and/or photoreceptors have been demonstrated by numerous protocols, and ESCs are omnipotent cells with unlimited self-renewal capability. They originate from the inner cell mass of the embryonic blastocyst and can differentiate into cell types from all three embryonic germ layers, namely, ectoderm–mesoderm–endoderm. The generation of RPE and/or photoreceptor cells from *in vitro* differentiation of mouse and human ESCs has been demonstrated by numerous protocols and offers huge potential for cell replacement therapy in treating retinal degenerative diseases. Clinical application has been limited by ethical issues related to obtaining hESCs. In addition, serious complications, including xeno-transplant immune rejection, as well as teratoma formation after several months of transplantation (containing cells of all three embryonic germ layers) have prevented their widespread application in cell replacement therapies.

Human inducible pluripotent stem cells (hiPSCs): They can be obtained from the patients' themselves by directly reprogramming adult somatic cells to transit to a pluripotent state. Several types of retinal cells including rods, cones and ganglion cells have been derived from *in vitro* differentiation of iPSCs using a combination of soluble factors. They can be collected easily and produced at a relatively reasonable price, possess unlimited self-renewal capacity, present no ethical problems, and can be autologous, thus preventing rejection issues. However, they could conserve the epigenetic characteristics of the original cells, harbor the disease genes from the donor (the mutation load of iPS cells passes through several stresses and long culture periods), and lead to tumor formation. When cells are autologous, genome editing would be required in patients with genetic mutations.

Horizontal cells (HCs): Regulate the signal that emerges from several rods and cones.

Inner nuclear layer (INL – also inner synaptic layer)*:* It consists of the cellular bodies of the following retinal neurons: amacrine, horizontal, bipolar; it also contains the bodies of Müller cells (the major type of glial cells in the retina).

Inner plexiform layer: An area of the retina that is made up of a dense reticulum of synaptic connections (fibrils) formed by interlaced dendrites of ganglion cells, bipolar cells, and amacrine cells (cells of the inner nuclear layer). The IPL contains the synapse between the second-order and third-order neurons in the visual pathway.

Inner segment: It is a part of the photoreceptor cell where the Golgi apparatus, endoplasmic reticulum, ribosomes and abundant mitochondria are contained; there is also a cilium connecting it with the corresponding outer segment and a synaptic connection with horizontal cells.

Intravitreal injection: It is used to administer medications or other substance to treat a variety of retinal disorders. It is performed to place medicines inside the eye, near the retina. The drug is delivered into the vitreous humor, the jelly-like substance inside the eye. *Intravitreal* drug delivery has become the gold standard for the treatment of many retinal diseases, including neovascular AMD, diabetic retinopathy and retinal vein occlusion, the most common conditions treated with intravitreal anti-VEGF drugs.

Macula: The macula is an oval-shaped small area in the central retina of humans and other animals. It is the pigmented part of the retina located in the very center of the neurosensory retina (in the center of the macula is the fovea). The diameter of the human macula is approximately 5.5 mm and is subdivided into six smaller areas: fovea, parafovea, perifovea, umbo, foveola, and foveal avascular zone. The human macula contains 6 million cones in total, residing in the foveola in a densely packed form. The macula is responsible for the central, high-resolution color vision mediated by cones. If the macula is damaged, e.g., in macular degeneration disorders, this type of vision is impaired. Clinical monitoring of the macula is done from the pupil, as in ophthalmoscopy or retinal photography.

Mesenchymal stem/stromal cells (MSCs): They are a population of multipotent stromal cells that can be found abundantly in a variety of adult tissues such as adult bone marrow, adipose tissue, and dental pulp. They can also be isolated from neonatal tissues and fluids such as the umbilical cord, Wharton's jelly, amniotic membrane, amniotic fluid, and placenta. Although MSCs are derived from different tissues, they share some common features. In recent clinical trials for the treatment of retinal disorders, MSC has shown encouraging safety profiles. In addition, they have demonstrated promising efficacy in having the ability to delay retinal degeneration.

Müller cells (MCs): The most common and abundant type of glial cells in the vertebrate retina (followed by astroglia and microglia). The MC is the only retinal glial cell that shares a common cell lineage with retinal neurons. Müller cells provide trophic and anti-oxidative support to photoreceptors and retinal neurons and regulate the tightness of the blood-retinal barrier.

Nerve fiber layer (NFL): The retinal nerve fiber layer (RNFL) or nerve fiber layer, comprises the axons of retinal ganglion cells. These ganglion cell axons collect the visual impulses produced by the rods and cones and travel through the bodies of ganglion cells.

Outer segment (OS): A subcellular photoreceptor organelle filled with proteins essential for the photoconversion cascade, including light-sensitive pigments. They undergo renewal on a daily basis by discarding their discs (tips), which are phagocytosed by the RPE cells. These physiological mechanisms enable the OSs to maintain constant length while shedding accumulated toxic photo-oxidative compounds.

Photoreceptors (PRCs): Retinal neurons specialized in photo-transduction. They convert light, electromagnetic radiation into cell membrane potential, whose changes result in altered neurotransmitter release from the synaptic terminals of photoreceptors. The light is converted into electrical signals that travel through other retinal neurons to reach the optic nerve. There are two main types of light-sensitive cell photoreceptors in the human eye: rods and cones. Rods are responsible for vision in poor light and at night (scotopic vision); they do not mediate color vision and have a low spatial acuity. Cones enable vision at higher light levels (photopic vision), are capable of color vision, and have a high spatial acuity.

Retina: A thin light-sensitive layer located at the back of the eye. It is a two-layered cup formed by the multilayered neurosensory retina and RPE monolayer. The subretinal space lies between them. This intricate structure is essential for vision. In the center of the retina is the macula, the part responsible for central and fine-detail vision needed for tasks such as reading.

Retinal pigment epithelium (RPE): A retinal cell monolayer attached to Bruch's membrane and closely associated with the outer segments of the overlying photoreceptors. These segments are phagocytosed by the apical villous processes of RPE cells on a daily basis. RPE cells have a hexagonal morphology and are usually pigmented containing melanin granules. The RPE layer forms a natural barrier to the choroidal capillaries and removes harmful metabolic waste produced by photoreceptors in response to light.

Retinal progenitor cells (RPCs): A population of human RPE stem cells.

Rod photoreceptor: The second type of retinal photoreceptor cells that mediates vision under dim light conditions. One hundred twenty million rods are contained in the human retina, and unlike cones, rods have only one type of photosensitive pigment called rhodopsin. Rods do not mediate color vision, and they have poor visual acuity because many rod cells share one connection to the optic nerve.

Scaffolds: They are natural or synthetic biomaterials that mimic the microenvironment of native retinal tissue. They can increase the survival of transplanted cells, support their integrity, encourage proper cell integration with the host retinal cellular network and organize the orientation of photoreceptor cells. The scaffolds significantly decrease the likelihood that injected suspensions of RPE cells regress into the vitreous cavity, thereby helping to avoid potential fibrosis and retinal detachment.

Spectral domain-optical coherence tomography (SD-OCT): OCT angiography is one of the breakthroughs in the advancing field of ophthalmic imaging, a non-invasive *in vivo* imaging modality combining both quantitative and qualitative assessment of vasculature seen on standard angiography. This imaging modality offers volumetric resolution of the head of the optic nerve, retina, and choroid. Early hemodynamic alterations in the vasculature of choroid and retina layers can be readily detected. Because of its non-invasive nature, standard angiographic techniques that monitor changes in the superficial layers of choroid and retina, e.g., indocyanine green angiography (ICGA) and fundus fluorescein angiography (FFA), respectively, utilize OCT.

Subretinal space: A potential space that lies between RPE monolayer and the neurosensory retina. Subretinal infusion is considered a specialized cell delivery method for retinal implants.

Vitreous cavity: An intraocular cavity that occupies the area between the retina and crystalline lens. It is filled with a clear gel, the vitreous humor, which may regulate the concentration of ions in the retina.

Wet or neovascular AMD: The advanced exudative stage or wet AMD occurs after ingrowth of the underlying choroidal neovascular vessels (CNV) through the damaged Bruch's membrane into the RPE layer and neurosensory retina. An estimated proportion of 10% of dry AMD patients progress to the wet form of AMD, leading to gradual loss of central vision.

"Wet"" AMD is the most severe but more treatable, and it can be delayed by repeated anti-VEGF medication.

Notes

The reason for writing this chapter was a published paper reporting three cases of vision loss after bilateral intravitreal injections of autologous adipose tissue-derived stem cells in patients with non-neovascular AMD, obtained in 2015 at a private clinic in Miami Florida – United States [1]. Unfortunately, stem cell injection caused rhegmatogenous retinal detachment combined with vitreous hemorrhage, ocular hypertension and hemorrhagic retinopathy. One year after cell therapy both patients' clinical conditions worsened finally without any light perception.

The next generation of therapeutics approved for the clinic given to regeneration or repair of damaged tissues or organs is expected to be cell preparations termed "living drugs"; this name is due to their ability to dynamically and temporally respond to changes during their *ex vivo* manipulation and after *in vivo* administration. A cell product (suspension or implant) is a living drug containing living cells that retain the *in vivo* ability to proliferate. Once injected, the cells expand beyond the initial administered dose, making determination of the therapeutic cell dose a serious problem in the modern field of regenerative medicine. In contrast, it is unthinkable to administer a conventional drug whose dose of active ingredients is unknown. Stem cells in general have unlimited proliferative capacity and could be likened to a new unruly horse without a bridle in a meadow (obviously, its rider is in danger at any moment from its unpredictable behavior). Although stem cell biology is very developed, there are still many gaps in basic properties such as nature, structure, proliferation, differentiation, mutagenesis, oncogenesis, mechanism of action, etc. Therefore, the enthusiasm for their therapeutic administration in various diseases is still premature and dangerous.

It should be preferable to avoid the administration of stem cells and their undifferentiated progeny for regenerative purposes; instead, we prefer live drugs containing only stem cell-derived terminally differentiated cells, whose proliferative capacity is negligible or zero. AD-MSCs are increasingly used due to the ease of their isolation and manipulation; subretinal delivery of human RPE cells, photoreceptors, or both – differentiated from AD-MSCs to replace the damaged cells in AMD patients – is currently being investigated in several approved clinical trials worldwide. However, one must take into account the individual clinical parameters that are different for each patient, which inevitably lead to personalized therapeutic regimens. Cell replacement therapy should be considered as an individualized therapy and be planned as such. In the present chapter, the proposed treatment process for AMD is illustrated in the individualized treatment method of each patient who is given cell replacement therapy. It could also serve as a solution to the existing problem of determining the therapeutic dose of cell preparations in regenerative medicine to avoid devastating outcomes as in the aforementioned cases of the three women.

Author details

Dimitrios Bouzianas
Biopharmaceutical and Regenerative Medicine Laboratories, Thessaloniki, Greece

*Address all correspondence to: dimbou55@gmail.com

References

[1] Kuriyan AE, Albini TA, Townsend JH, et al. Vision loss after intravitreal injection of autologous "stem cells" for AMD. The New England Journal of Medicine. 2017;**376**(11):1047-1053

[2] Stern J, Temple S. Retinal pigment epithelial cell proliferation. Experimental Biology and Medicine (Maywood, N.J.). 2015;**240**(8):1079-1086

[3] Kramer J, Chirco KR, Lamba DA. Immunological considerations for retinal stem cell therapy. Advances in Experimental Medicine and Biology. 2019;**1186**:99-119

[4] Bacci GM, Becherucci V, Marziali E, et al. Treatment of inherited retinal dystrophies with somatic cell therapy medicinal product: A review. Life (Basel). 2022;**12**(5):708

[5] Martinez Velazquez LA, Ballios BG. The next generation of molecular and cellular therapeutics for inherited retinal disease. International Journal of Molecular Sciences. 2021;**22**:11542

[6] Gagliardi G, Ben M'Barek K, Goureau O. Photoreceptor cell replacement in macular degeneration and retinitis pigmentosa: A pluripotent stem cell-based approach. Progress in Retinal and Eye Research. 2019;**71**:1-25

[7] Raimondi R, Zollet P, De Rosa FP, et al. Where are we with RPE replacement therapy? A translational review from the ophthalmologist perspective. International Journal of Molecular Sciences. 2022;**23**:682

[8] Jemni-Damer N, Guedan-Duran A, Fuentes-Andion M, et al. Biotechnology and biomaterial-based therapeutic strategies for age-related macular degeneration. Part II: Cell and tissue engineering therapies. Frontiers in Bioengineering and Biotechnology. 2020;**8**:588014

[9] Koster C, Wever K, Wagsta P, et al. A systematic review on transplantation studies of the retinal pigment epithelium in animal models. International Journal of Molecular Sciences. 2020;**21**:2719

[10] Hussey KA et al. Patterning and development of photoreceptors in the human retina. Frontiers in Cell and Developmental Biology. 2022;**10**:878350

[11] DiCarlo JE et al. Gene therapy and genome surgery in the retina. The Journal of Clinical Investigation. Jun 2018;**128**(6):2177-2188

[12] Kharitonov AE et al. Possibilities for using pluripotent stem cells for restoring damaged eye retinal pigment epithelium. Acta Naturae. Jul-Sep 2018;**10**(3):30-39

[13] Marcos LF, Wilson SL, Paul Roach P. Tissue engineering of the retina: From organoids to microfluidic chips. Journal of Tissue Engineering. 2021;**12**:20417314211059876

[14] Sachdeva MM, Eliott D. Stem cell-based therapy for diseases of the retinal pigment epithelium: From bench to bedside. Seminars in Ophthalmology. 2016;**31**:25-29

[15] Rizzolo LJ, Nasonkin IO, Adelman RA. Retinal cell transplantation, biomaterials, and in vitro models for developing next-generation therapies of age-related macular degeneration. Stem Cells Translational Medicine. 2022;**11**(3):269-281

[16] van Dongen JA, Harmsen MC, Stevens HP. Isolation of stromal vascular fraction by fractionation of adipose tissue. Methods in Molecular Biology. 2019;**1993**:91-103

[17] Jonesa MK, Lua B, Girmana S, et al. Cell-based therapeutic strategies for replacement and preservation in retinal degenerative diseases. Progress in Retinal and Eye Research. 2017;**58**:1-27

[18] Chen J et al. Overstaying their welcome: Defective CX3CR1 microglia eyed in macular degeneration. The Journal of Clinical Investigation. Oct 2007;**117**(10):2758-2762

[19] Demb J, Singer JH. Functional circuitry of the retina. Annual Review of Vision Science. 2015;**1**:263-289

[20] Datta S, Cano M, Ebrahimi K, et al. The impact of oxidative stress and inflammation on RPE degeneration in non-neovascular AMD. Progress in Retinal and Eye Research. 2017;**60**:201-218

[21] Zarbin M. Cell-based therapy for retinal disease: The new frontier. Methods in Molecular Biology. 2019;**1834**:367-381

[22] Bharti K. Regenerating retinal pigment epithelial cells to cure blindness: A road towards personalized artificial tissue. Current Stem Cell Reports. Jun 2015;**1**(2):79-91

[23] da Cruz L, Fynes K, Georgiadis O, et al. Phase 1 clinical study of an embryonic stem cell–derived retinal pigment epithelium patch in age-related macular degeneration. Nature Biotechnology. 2018;**36**:328-337

[24] U.S. National Library of Medicine. Study of subretinal implantation of human embryonic stem cell-derived RPE cells in advanced dry AMD – Full text view – ClinicalTrials.gov. 2020. Available from: https://clinicaltrials.gov/ct2/show/NCT02590692 [Accessed: Jul 08, 2020]

[25] Lin Y, Ren X, Chen Y, et al. Interaction between mesenchymal stem cells and retinal degenerative microenvironment. Frontiers in Neuroscience. 2021;**14**:617377

[26] Cowan CA, Klimanskaya I, McMahon J, et al. Derivation of embryonic stem-cell lines from human blastocysts. The New England Journal of Medicine. 2004;**350**:1353-1356

[27] Sakada F, Ikeda H, Sasai Y, et al. Stepwise differentiation of pluripotent stem cells into retinal cells. Nature Protocols. 2009;**4**:811-824

[28] Verbakel SK, van Huet RAC, Boon CJF, et al. Non-syndromic retinitis pigmentosa. Progress in Retinal and Eye Research. 2018;**66**:157-186

[29] Mu Y, Zhao M, Su G. Stem cell-based therapies for age-related macular degeneration: Current status and prospects. International Journal of Clinical and Experimental Medicine. 2014;**7**:3843-3852

[30] Cohen CB. Ethical and policy issues surrounding the donation of cryopreserved and fresh embryos for human embryonic stem cell research. Stem Cell Reviews. 2009;**5**:116-122

[31] Takahashi K, Yamanaka S. Induction of pluripotent stem cells from mouse embryonic and adult fibroblast cultures by defined factors. Cell. 2006;**126**:663-676

[32] Yamanaka S. Induced pluripotent stem cells: Past, present and future. Cell Stem Cell. 2012;**10**:678-684

[33] Bracha P, Moore NA, Ciulla TA. Induced pluripotent stem cell-based

therapy for age-related macular degeneration. Expert Opinion on Biological Therapy. 2017;**17**:1113-1126

[34] Mead B, Logan A, Berry M, et al. Concise review: Dental pulp stem cells: A novel cell therapy for retinal and central nervous system repair. Stem Cells. 2017;**35**:61-67

[35] Dominici M, Le Blanc K, Mueller I, et al. Minimal criteria for defining multipotent mesenchymal stromal cells. The International Society for Cellular Therapy position statement. Cytotherapy. 2006;**8**:315-317

[36] Pountos I. Mesenchymal stem cells. In: Adult Stem Cells: Recent Advances. UK: University of Leeds; 2016. Available from: www.smgebooks.com

[37] Mannino G, Russo C, Longo A, et al. Potential therapeutic applications of mesenchymal stem cells for the treatment of eye diseases. World Journal of Stem Cells. 2021;**13**:632-644

[38] Leung EH, Flynn HW Jr, Albini TA, et al. Retinal detachment after subretinal stem cell transplantation. Ophthalmic Surgery, Lasers & Imaging Retina. 2016;**47**:600-601

[39] Rong AJ, Lam BL, Ansari ZA. Vision loss secondary to autologous adipose stem cell injections: A rising problem. JAMA Ophthalmology. 2018;**136**:97-99

[40] Khine KT, Albini TA, Lee RK. Chronic retinal detachment and neovascular glaucoma after intravitreal stem cell injection for usher syndrome. American Journal of Ophthalmology Case Reports. 2020;**18**:100647

[41] Ho AC, Chang TS, Samuel M, et al. Experience with a subretinal cell-based therapy in patients with geographic atrophy secondary to age-related macular degeneration. American Journal of Ophthalmology. 2017;**179**:67-80

[42] Gounari E, Komnenou A, Evangeli K, et al. Intravitreal administration effect of adipose-derived mesenchymal stromal cells combined with anti-VEGF nanocarriers, in a pharmaceutically induced animal model of retinal vein occlusion. Stem Cells International. 2022;**2022**:2760147

[43] Singh MS, Park SS, Albini TA, et al. Retinal stem cell transplantation: Balancing safety and potential. Progress in Retinal and Eye Research. 2020;**75**:100779

[44] Xiang XN, Zhu SY, He HC, et al. Mesenchymal stromal cellbased therapy for cartilage regeneration in knee osteoarthritis. Stem Cell Research & Therapy. 2022;**13**:14

[45] Han IC, Bohrer LR, Gibson-Corley KN, et al. Biocompatibility of human induced pluripotent stem cell-derived retinal progenitor cell grafts in immune-compromised rats. Cell Transplantation. 2022;**31**:1-25

[46] Ladha R et al. Subretinal therapy: Technological solutions to surgical and immunological challenges. Frontiers in Medicine (Lausanne). Mar 2022;**9**:846782

[47] Alaa EE, Salwa S, Nermeen AE. Biomaterials for tissue engineering applications and current updates in the field: A comprehensive review. AAPS PharmSciTech. 2022;**23**(7):267

[48] Prantl L, Eigenberger A, Brix E, et al. Adipose tissue-derived stem cell yield depends on isolation protocol and cell counting method. Cells. 2021;**10**:1113

[49] Ramírez, Rodríguez GB, Patrício TMF, Delgado López JM. Natural polymers for bone repair. In: Bone Repair

Biomaterials. Woodhead Publishing; 2019. pp. 199-232

[50] Bharti K. Patching the retina with stem cells. Nature Biotechnology. 2018;**36**:311-313

[51] Maeda T et al. Trends of stem cell therapies in age-related macular degeneration. Journal of Clinical Medicine. 20 Apr 2021;**10**(8):1785

[52] Yang TC, Chuang JH, Buddhakosai W, et al. Elongation of axon extension for human iPSC-derived retinal ganglion cells by a nano-imprinted scaffold. International Journal of Molecular Sciences. 2017;**18**:2013

[53] Jemni DN, Guedan DA, Cichy J, et al. First steps for the development of silk fibroin-based 3D biohybrid retina for age-related macular degeneration (AMD). Journal of Neural Engineering. 2020;**17**:055003

[54] Del Priore LV, Kuo Y-H, Tezel TH. Age-related changes in human RPE cell density and apoptosis proportion in situ. Investigative Ophthalmology & Visual Science. 2002;**43**(10):3312-3318

[55] Panda-Jonas S, Jonas JB, Jakobczyk-Zmija M. Retinal pigment epithelial cell count, distribution and correlations in normal human eyes. American Journal of Ophthalmology. 1996;**121**:181-189

[56] Zou J, Wang W, Neffe AT, et al. Adipogenic differentiation of human adipose derived mesenchymal stem cells in 3D architectured gelatin based hydrogels (ArcGel). Clinical Hemorheology and Microcirculation. 2017;**67**:297-307

[57] Zhang Y, Zhang D, Wei W, et al. Effects of RPE-conditioned medium on the differentiation of hADSCs into RPE cells, and their proliferation and migration. Experimental and Therapeutic Medicine. 2017;**14**(4):3699-3707

[58] McGill TJ, Osborne L, Lu B, et al. Subretinal transplantation of human central nervous system stem cells stimulates controlled proliferation of endogenous retinal pigment epithelium. Translational Vision Science & Technology. 2019;**8**:43

[59] Mandai M, Fujii M, Hashiguchi T, et al. iPSC-derived retina transplants improve vision in rd1 end-stage retinal-degeneration mice. Stem Cell Reports. 2017a;**8**(1):69-83

[60] Barber AC, Hippert C, Duran Y, et al. Repair of the degenerate retina by photoreceptor transplantation. Proceedings. National Academy of Sciences. United States of America. 2013;**110**:354-359

[61] Sortwell CE, Pitzer M, Collier TJ, et al. Time course of apoptotic cell death within mesencephalic cell suspension grafts: Implications for improving grafted dopamine neuron survival. Experimental Neurology. 2000;**165**:268-277

[62] Jung YH, Phillips MJ, Lee J, et al. 3D microstructured scaffolds to support photoreceptor polarization and maturation. Advanced Materials. 2018;**30**:e1803550

[63] Thompson JR, Worthington KS, Green BJ, et al. Two-photon polymerized poly (caprolactone) retinal cell delivery scaffolds and their systemic and retinal biocompatibility. Acta Biomaterialia. 2019;**94**:204-218

[64] McLelland BT, Lin B, Mathur A, et al. Transplanted hESC-derived retina organoid sheets differentiate, integrate, and improve visual function in retinal degenerate rats. Investigative Ophthalmology & Visual Science. 2018;**59**(6):2586-2603

Section 5

Application of Cell Culture in Bioprocessing

Chapter 9

Perspective Chapter: Advanced Process Control and Automation with Special Focus on Emerging Continuous Bioprocessing

Letha Chemmalil, Chris Chumsae, Gloria Li and Anthony Leone

Abstract

Legacy batch processing carried out in pharmaceutical and biopharmaceutical sectors is undergoing transformation to adopt the next generation continuous processing to produce safe and effective drugs with better efficiency and consistency at a reduced cost. To facilitate innovative continuous processing, enabled by an end-to-end process with a single uninterrupted production scenario, it is essential to generate real-time or near-real-time data using process analytical technology (PAT), which has been defined by the FDA as a system for designing, analyzing, and controlling manufacturing through timely measurements to ensure final product quality. Based on quality by design (QbD) principles, PAT-enabled data monitoring is essential for the timely control of critical process parameters (CPPs) and critical quality attributes (CQAs) to keep the process in a desired state of control to achieve a predefined product quality. Based on QbD philosophy, quality cannot be tested into products; it should be built-in or should be by design. Deployment of PAT tools for real-time monitoring is integral to align with the guiding principles of QbD-enabled workflow to enhance process and product understanding to administer a control strategy to keep the process within the design space. Aim of this chapter is to highlight the recent advancements in PAT tool-development to monitor and control CPPs and CQAs.

Keywords: cell culture, continuous bioprocessing, process analytical technology, quality by design, critical quality attribute, critical process parameters

1. Introduction

Pharmaceutical and biopharmaceutical medicines are diverse entities of therapeutics with shared common mission of advancing scientific knowledge into the development of breakthrough treatment modalities for complex medical conditions. While pharmaceutical small molecule drugs are leveraging traditional chemical synthesis, the large biopharmaceutical drugs are being produced in living organisms, employing recombinant DNA (rDNA) technology leveraging the deeper understanding of biochemical pathways and human physiology. Widely known rDNA technology, often referred to as genetic engineering, is the method of recombining two or more genetic DNA sequence-fragments from different species to create a novel gene with a unique function. Such genetically engineered expression vector, carrying the gene of interest, is then integrated into a host cell to express a desired protein through gene expression during the cell culture process. Animal cell culture process leveraging rDNA technology is an indispensable tool to produce protein therapeutics such as monoclonal antibodies (mAbs), lymphokines, interleukins, hormones and other drugs for various disease indications including oncology, cardiovascular diseases, and inflammatory diseases [1]. Applications of rDNA technology is spanning over various fields including healthcare, agricultural sector, environmental segment, and others. The contribution of rDNA technology is significant in the healthcare arena for developing vaccines, biopharmaceutical drugs, diagnostics, etc. [2]. In the environmental sector, rDNA technology is applied to generate biological pesticides and microbes to maintain environmental sustainability [3]. In the agricultural sector, genetic tools are employed to introduce foreign gene to achieve higher yield with desired qualities such as disease resistance and nutritional enhancement [4].

For biopharmaceutical drug development, the commonly used host cell is Chinese hamster ovary (CHO) cells over bacterial, fungal and yeast cells due to the specific need of eukaryotic expression systems. Prokaryotic expression systems are not suitable for achieving eukaryotic post-translational modifications such as desired glycosylation [5]. Upstream bioprocess development as well as pilot and production scale manufacturing are continuing to evolve to produce high quality drugs at reduced cost. Although small molecule medicinal chemistry drugs can be synthesized with utmost reproducibility, biopharmaceutical drugs produced in living cells face inherent variability. Therefore, effective control strategy is critical to keep the process in a steady state with the guidance to operate strictly within the well-established boundaries of the design space. One of the ways in which the cost-effective manufacturing of biopharmaceutical drugs with reduced variability can be achieved through the adaptation of QbD enabled integrated continuous bioprocessing. Fully automated connected end-to-end process eliminates human intervention and hence reduces risk and improves drug safety. Despite the significant progress made in the field of PAT-enabled continuous bioprocessing, exemplified by the successful research conducted at Novartis-MIT Center for Continuous Manufacturing, the end-to-end integration is still not advancing to the forefront [1]. With the emergence of digital transformation, the opportunity for a connected continuous bioprocessing with the adaptation of intensification and data integration is not far from reality.

Mammalian cell culture with engineered cell line is a prerequisite for the modern-day biopharmaceutical processing to produce biological therapeutics such as mAbs, fusion proteins, hormones, interferons, tissue plasminogen activator, EPO, colony stimulating factors, clotting factors, enzymes, vaccines, and others. In addition to cell culture, remarkable growth is on the horizon for the tissue culture, leveraging 3D cell culture model, mimicking the physiological microenvironment, to form a fully developed organs, resembling the structure and function of their natural counterparts. The complex 3D cell culture is leveraging multidisciplinary technologies including bioengineering, mechanics, material science and chemistry [6]. Irrespective of the type of cell culture being considered, it is essential to supplement continuous stream of growth medium containing nutrients and growth factors to the culture. Monitoring and controlling of essential media components as well as minimizing the levels of certain metabolites such as lactate is critical to the steady growth and proliferation of cells. Maintenance and controlling of cell's surrounding environmental factors such as pH, temperature, osmolarity, oxygen and carbon dioxide are equally significant. Advent of rDNA technology in conjunction with the adaptation of animal cell culture have landmarked various breakthroughs. Recombinant proteins therapeutics, recombinant vaccines, CAR-T Cell therapy and gene therapy are some of the examples of milestones achieved throughout the bioengineering journey. The birth of biotech began to accelerate subsequently to the 1982 approval of rDNA-based insulin produced in bacteria. 60–70% of recombinant therapeutics produced since then have been in mammalian cells [7], including the first recombinant therapeutic drug tissue plasminogen approved in 1986. Approval of recombinant vaccines against hepatitis A & B also was resulted from the application of rDNA technology [8, 9].

Despite the tremendous progress made in the biopharmaceutical drug development in conjunction with the enhanced understanding of human biology, the surging drug development cost and increased failure rates are major industry-wide concerns. Drug development trajectory is directly opposite of Moore's' law, predicted by Intel CEO Gordon Moore stating that the computing power of a microprocessor would double every two years while the cost would reduce by half. In the pharmaceutical sector, to reduce the overall failure rate requires a transformational shift from the current platform of clinical trial to embrace relatively risk-free approaches such as the utilization of organ-on-a-chips (OoC) platform to screen out non-potential candidates early in the program, without investing too much time and resources into the lengthy and expensive clinical trials. OoC is an emerging technology in which the microfluidic technology is coupled with biology to model human physiology ex-vivo. OoC, leveraging tissue engineering, was awarded as one of the top emerging technologies by the World Economic Forum in 2016 [10]. The ex-vivo model generated using OoC provides an in depth understanding of the drug safety and efficacy of potential drug candidates even before it enters the clinical trial.

From the process perspective, modernization of cell culture process and downstream purification steps can improve process efficiency to cut down the overall cost of manufacturing. QbD driven continuous bioprocessing is the most effective strategy for agile manufacturing to achieve targeted product quality. Integration of intensification of processes along with the establishment of design spaces is critical to the success in concert with continuously monitoring and controlling of the upstream cell culture process to maintain the cell genotype and phenotype. To improve upstream bioprocessing efficiency, selection, and optimization of the specialty media specific for the cell type is critical. Establishing the correlation between process parameters and desired product attributes is being leveraged to establish a design space using DoE. To enable the faster delivery of high-quality biopharmaceuticals at maximum efficiency and enhanced flexibility, modernization of the process is required with a transition from the legacy batch processing to the next generation continuous bioprocessing along with the integration of PAT. Bioinformatics plays an integral role in the effective deployment of integrated continuous bioprocessing. Adaptation of mechanistic modeling to create digital twin to facilitate data driven operation is critical to establish an agile and cost-effective bioprocessing with reduced cost and product variability.

2. Continuous processing

For the last several decades, pharmaceutical and biopharmaceutical companies have been adhering to the traditional batch processing to manufacture therapeutic drugs. Recent technological advancements have encouraged drug manufacturers to divert from batch processing to embrace more efficient continuous processing. The FDA is taking proactive steps to encourage drug makers to implement emerging technologies, including continuous operation to improve product quality [11]. Although continuous operations are widely embraced and routinely practiced in the chemical engineering field, their implementation in the biopharmaceutical and pharmaceutical sectors is requiring additional assessments [12]. Recently published white paper series from the MIT-Strathclyde symposium on continuous manufacturing is highlighting the current state of thinking on the modernization of pharmaceutical and biopharmaceutical processing landscape [12]. Based on the recent shift in paradigm to adopt the guiding principles of engineering concepts and product design, drug companies are embracing risk-based approaches to build quality into the product with an informed process understanding. With the increased emphasis on safety and quality of drugs with simultaneous reduction in cost, there is an increased acceptance of science-based approach for structured process development. To implement a well-established chemical engineering knowledge in the drug manufacturing landscape, FDA and ICH have drafted guidance for PAT and QbD, respectively [13, 14]. While several industries have undergone a manufacturing evolution using continuous technologies, the transition is still in the early stage for the pharmaceutical industry. The setback is even greater for biopharmaceutical sector due to the lack of innovative technologies to modernize such complex operations. The biopharmaceutical platform is getting more dynamic with the introduction of new modalities such as bi & tri-specific molecules, antibody-drug conjugates, cell therapy, mRNA technology, gene therapy, aptamer technology, aptamer-drug conjugate, and other continuously evolving drug development platforms. The recent development in continuous bioprocessing with the incorporation of PAT tools in conjunction with digital technology is leading the way to build a continuous manufacturing with digital transformation to support the current paradigm shift.

3. Process analytical technology (PAT)

The scientific and risk-based framework of PAT is not only intended to modernize pharmaceutical and biopharmaceutical processes through innovation, but also facilitate the enhancement of process and product understanding [13]. FDA is encouraging manufacturers to use the PAT framework to develop and deploy efficient innovative approaches in pharmaceutical development and operation. With an essentially important role in health care, drug manufactures need to employ innovation and apply cutting edge scientific and engineering technologies [13]. Current practice of generating offline analytical data is not only inefficient, but also can exhibit various logistical challenges. Because of longer analytical turnaround time associated with the offline testing, timely process control is not achievable. Also, due to the prolonged sample storage and multiple sample transfers associated with offline testing, issues such as sample stability and data integrity are legitimate concerns. With real-time or near-real-time testing, PAT-tools provide the opportunity for timely control of the process to build desired product qualities into the product. This type of control strategy is critical for supporting continuous processing in which there is no option to send samples for offline testing and waiting for results. Therefore, PAT is referred to as the analytics of future to support the next generation of continuous processing.

PAT has been designed to analyze and control CPPs, CQAs and performance attributes of raw materials and in-process samples with the goal of ensuring desired final product quality. With the utilization of PAT as an enabling technology for assessing the quality of intermediates during in-process steps, the need for final product testing can be eliminated, resulting in a significant reduction in lead time. Various PAT technologies including 1D- & 2D- LC, Raman, Mid-IR, Near-IR, Flow VPX, RI sensors, sequential injection analysis, Mass spec-based on-line multi-attribute methods and other emerging technologies can be deployed for real-time or near-real-time monitoring and control of product titer, product quality, glycan profiling, peptide mapping, metabolite profiling and others. Implementation of PAT tools in bioprocessing can support the supply chain to achieve cost reduction via enabling Real-Time-Release Testing (RTRT) of therapeutic products, which is aligned with the framework of innovative drug development guiding principles. Therapeutic drug development continues to evolve with the increased emphasis on science and engineering principles to improve the efficiencies of both manufacturing and regulatory landscapes. The goal of PAT is to enhance the process and product understanding as well as to control the manufacturing process to build quality into the products through innovation. The emphasis of building quality into the product requires increased understanding of the process and product knowledge as well as the multi-factorial relationships between materials, process variables, environmental factors, and their effects on product quality.

4. QbD-driven process development

QbD provides a robust framework for the design and deployment of science-based processes to achieve a pre-defined product quality. Data driven risk analysis is carried out to understand how the process design affects quality target product profile (QTPP) that identifies the CQAs and critical material attributes (CMAs). Implementation of risk management strategy, formalized design of experiments (DoE), and advanced data analysis techniques as well as process modeling and control are critical to maintain a desired product quality. Testing products at the end of the manufacturing process limits the options to employ control strategies to remediate the process. By combining PAT with QbD, companies can move away from traditional quality approaches, and employ data-driven strategies to deliver high quality drugs at a reduced cost. Regulatory agencies are the active proponents of QbD implementation, inspired by its power to produce high quality drugs with reduced variability. ICH has provided guidance to define terms such as CQA, CPP, QTTP and design space [14]. PAT- enabled QbD requires upfront investment in money, time, and resources. However, the return on investment is significant in terms of reduced product variability, risk minimization, cost reduction and improved regulatory compliance. Once the design space is established, adjustment within the design space is free of new registration with the regulatory agencies. QbD approaches also facilitates RTRT, which enables the process engineers to make informed decisions to provide timely intervention. QbD offers a better understanding of the manufacturing process, making the process scale-up relatively easier. As continuous process-performance verification is done in real-time with QbD, formal process validation is not a requirement. Drug manufacturers are enthusiastic about the PAT/QbD approach, attributed to the ability to manufacture high-quality drugs at a reduced cost. Finally, the patients are benefited from receiving safe and efficacious drugs.

QbD enabled-knowledge gained from raw materials and in-process samples along with additional insights gained from extended characterization performed using biophysical, biochemical, and microbiological techniques offers opportunity to implement better control strategy. To establish a statistically enabled design space for the safe operation, real-time or near-real-time monitoring of process parameters and product quality attributes are better served than relying upon the offline data. The flexibility to operate within the established design space helps to build desired quality into the products. Implementation of PAT systems in pharmaceutical industries have been resulted from the regulatory authorities' initiative to improve and modernize the pharmaceutical industry to enhance product quality with the adaptation of QbD/PAT concepts. The common theme of QbD philosophy is to build quality into the products instead of testing the product to ensure quality. QbD/PAT enabled control strategies ensure the development of robust and efficient processes to deliver high quality drugs with desired product quality at reduced variability. QbD/PAT principles will be the norm rather than an exception for the QbD driven next generation continuous bioprocessing.

5. QbD enabled continuous bioprocessing

To facilitate next-generation continuous bioprocessing, deployment of PAT tools for real-time or near-real-time monitoring is integral for process understanding and timely process control to keep all CPPs within the boundaries of the design space to align with the guiding principles of QbD to achieve predefined CQAs. While conventional offline testing is done in labs outside of the process operation area, PAT enabled inline, at line and online analyses are carried out within the immediate vicinity of the process operations. To reduce the cost while improving product quality, continuous operation is a highly recommended platform for pharma and biopharma companies to improve efficiency and reduce cost. There is an alignment between the major stakeholders to provide a strong foundation for building and facilitating the implementation of continuous processing. There are four pillars for providing foundational stability to continuous processing, which is based on the interplay between the government, regulatory agencies, the vendors and the drug companies. Government is in support of continuous processing, attributed to the benefit of reducing the cost of drugs. Regulatory agencies are in concert with the strategy, owing to the benefit of high quality drugs with improved safety and regulatory compliance. Vendors are excited about the opportunity to design and develop new enabling technologies to support the new platform. Pharma and biopharm companies are in full alignment due to the results of various economical benefits including, increased speed, enhanced flexibility, reduced manufacturing footprint, improved robustness, minimized variability, enhanced product quality, diminished product failure, etc. The next generation continuous bioprocessing resides on four pillars consisting of process intensification, single-use technology, real-time data monitoring and RTRT as elucidated in the following sections. Connectivity between these four pillars is essential to develop and deploy a fully integrated continues bioprocessing.

5.1 Process intensification

The paradigm shift in the adoption of next generation bioprocessing is impacting all areas of biopharmaceutical operations. The process intensifications (PI) of upstream and downstream processes require adaptation of perfusion technology and the deployment of multi-column chromatography, respectively. PI involves design, development and deployment of efficient technologies and novel devices to bring dramatic changes to the process operation with spectacular improvements to the process plants, without sacrificing product safety and product quality. Design of single unit facility can lead to substantial reduction in footprint, significant decrease in equipment size, attrition in energy consumption and minimization of waste generation. Ramshaw, one of the pioneers in the field, defined process intensification as a strategy for making significant reductions in the footprint of chemical plant, either by minimizing the size of individual equipment or reducing the number of unit operations by consolidating multiple operational units into fewer units [15]. PI applies to the common scientific areas of chemical process engineering, mathematics, physics, quantum chemical approaches, classical molecular simulations, thermodynamics, classical mechanics, transport phenomena, numerical mathematics, electrodynamics, chemical kinetics, etc. [16]. PI allows biomanufacturers to produce more product using less raw materials and smaller equipment in a reduced space. With the innovative principles applied in processes and equipment design, PI brings significant benefits in terms of process efficiency at a lower capital and operating expenses to produce higher quality products. While the implementation of PI improves productivity and product quality, the single use scale-down technologies help to reduce the size of the facility to bring the overall cost down. Without having appropriate PAT tools to generate real-time or near-real-time analytical data, the upstream and downstream continuous bioprocessing is inconceivable.

5.2 Single use systems (SUS)

Single use bioprocessing systems are being utilized predominantly for pre-clinical and clinical manufacturing than in commercial setting. SUS provides tremendous economic benefits for small scale manufacturing, attributable to the significant reduction in facility footprint and other operational simplicity. For large scale commercial manufacturing, stainless-steel bioreactors are preferred as it is economically more beneficial. In addition to the miniaturization of facility footprint, the disposable single use bioreactors do not have to undergo cleaning, sterilization, and validation as they come in ready-to-use plug-and-play format. Adoption of upstream process intensification strategies has added pressure to the downstream to adopt similar intensification processes to handle higher titers evolved from the upstream intensification process. Adaptation of disposable chromatographic columns had faced resistance as chromatographic resins are quite expensive and they are intended to be re-used for several cycles without having major issues. Multi-column continuous chromatography systems are now getting traction as this technique eliminates the inherent limitations of traditional large columns to be utilized fully to its maximum capacity during the capture affinity step. In multi-column chromatography, the large column is split into several smaller columns with flexible valve configuration allowing the breakthrough of the product stream from the 1st column to enter the 2nd column, which enables the utilization of the full capacity of the column resin through overloading. Modular platforms such as BioSC from Sartorius could connect and orchestrate multiple downstream modules with a single platform and unified software to increase efficiency in operation of the connected platform.

5.3 Tools for real-time data monitoring

Measurement tools such as sensors and probes that are immerced in sample sources such as a bioreactors or directly into the sampling interfaces are labeled as in-situ in which the sample is not removed from the process stream and can be invasive or noninvasive. Inline measurement is continuous without removing the probe or samples from the process. Online measurement is very similar to inline with a key difference in which the sample is diverted from the manufacturing process and may be returned to the process stream. Online analysis usually involves diverting a portion of the product from the main process line to perform measurements on the diverted portion of the product through a sampling loop, flow cell or sampling interface. The diverted sample can be either re-introduced into the process stream or to the waste, depending on the application. Both inline and online measurements offer the ability to continuously carryout measurements. At-line and offline measurements usually is requiring manual collection of samples to perform analysis separately from the process. For at-line measurements, the sample is being removed and analyzed in close-proximity to the process stream, while offline analysis is carried out apart from the process. When it comes to real-time and near-real-time data monitoring, a fast response from an inline sensor (in-situ) is considered real-time. On the other hand, in-situ with sampling bypass and ex-situ through sampling modules are considered as near-real-time. For example, data generated using in-situ spectroscopic measurements in which probes are immersed in the source is real-time. Same measurements taken ex-situ using a loop or flow-cell is categorized as near-real-time. UPLC systems interfaced with a sampling interface utilized for the measurements of titer, nutrients, metabolites and CQAs from upstream and downstream unit operations also falls in the category of near-real-time. Other near-real-time analyses include µSIA-based glycan analyis, amino acid analysis, peptide mapping, etc.

5.4 Real-time release testing (RTRT)

RTRT recognizes that an appropriate combination of CPP with pre-defined material attributes may provide greater assurance of product quality than end-product testing [17]. RTRT is based on monitoring and controlling of the process using PAT tools to ensure desired product quality of in-process samples and final end-products. The PAT component of RTRT includes a valid combination of measured material attributes and process controls [17]. In-process testing of process parameters and attributes to enhance process understanding along with the deployment of control strategy can be used to justify the replacement of routine end-product testing. RTRT is a system for product release that gives assurance that the product meets the desired quality, based on the process understanding and product knowledge acquired during the process steps. RTRT recognizes that an appropriate combination of process control and pre-defined material attributes to achieve a desired CQA may provide greater assurance of product quality than traditionally performed end-product testing. Release of a product can be a combination of RTRT approach for certain critical quality attributes (CQAs) and a more conventional evaluation for certain other quality attributes. The application of RTRT may offer advantages from a manufacturer's perspective as well as from a regulatory point of view to gain enhanced knowledge throughout the process. Other potential benefits include real time monitoring and the opportunity to provide feed-back or feed-forward controls. RTRT strategy should be based on a thorough understanding of the process and of the relationship between process parameters, in-process material attributes and product attributes. RTRT comprises a combination of process controls utilizing PAT in combination with relevant process control and material attributes. PAT-based data monitoring along with timely process control could enable RTRT to replace traditional end-product testing.

6. Existing and emerging PAT tools

To support next-generation continuous bioprocessing to fulfill QbD-driven PAT initiatives, various emerging PAT technologies are now available for online monitoring and controlling of pharmaceutical and biopharmaceutical processing. Some of the PAT techniques unique to the biopharmaceutical sector consists of online monitoring of product titer, amino acid analysis, peptide mapping, glycan profiling and CQA assessments using chromatography-based hybrid techniques. Other product attributes can be monitored and controlled using techniques such as NIR, FT-IR, UV/Vis, Flow VPX, Fluorescence spectroscopy, LC/MS, and other techniques. At-line fluorescence detection and chemometric modeling is an alternative to UPLC-PSM-based approach for purity based peak pooling. IR sensors and Solo-VPE are useful for the measurements of real-time protein concentration of downstream samples. FTIR, Raman and NIR are useful for product quality assessments based on chemometric modeling. LC/MS interfaced with online autosampler is suitable for online monitoring of multi-attributes simultaneously. Process and product understanding accomplished on real-time and near-real-time monitoring can provide opportunities to keep the process in a desired state via potential feedback control. Compared to offline or at-line testing, inline and online measurements can augment speed, sample integrity, and convenience. There are many tools available to facilitate process understanding, continuous improvement, and development of scientific risk-mitigation strategies [13]. Some prominent tools applicable for biopharmaceutical applications are presented in subsequent subsections.

6.1 HPLC/UPLC based PAT tool

For online monitoring of upstream titer and nutrients as well as assessing upstream and downstream product quality of biopharmaceutical drugs with the option to perform timely process control, UPLC system interfaced with a process sample manager (PSM) can be utilized. A 1D- & 2D-LC system interfaced with an appropriate online sampling device is useful for various applications including online titer measurements, online product quality assessments and online quantitation of amino acid contents during upstream cell culture process [18]. In all cases, samples are withdrawn automatically from bioreactors through a FISP probe or equivalent device with appropriate filter discs to remove all debris from the bioreactor culture samples before introducing it into the HPLC/UPLC. Since the upstream samples require purification prior to product quality assessment, the 1st dimension of 2D-LC is leveraged for protein purification while the 2nd dimension is utilized for product quality assessment [19]. A UPLC system with PSM autosampler is better suited for upstream titer measurement and downstream product quality measurements [20]. The same UPLC system with PSM autosampler can facilitate online purity assessment during peak pooling [21]. Near-real-time data generated using online testing using chromatographic, spectroscopic, and other techniques help to enhance process and product understanding and provide an opportunity to control the process to achieve a pre-defined final product quality. Also providing speed and efficiency to improve sample integrity with the elimination of protein degradation and undesired post translational modification associated with prolonged sample storage prior to the offline testing.

Figure 1 and **Table 1** exihibit acceptable comparability of a representative batch bioreactor titer results generated using online UPLC with PSM autosampler vs. at-line and offline results of the same sample set. Statistical analysis confirmed that there are no statistical differences between the three data sets. **Figure 2** illustrates acceptable comparability of online titer data generated using UPLC/PSM vs. at-line UPLC data of a perfusion bioreactor sample set. Online amino acid profile of a typical cell culture media overlaid with amino acid standard is shown in **Figure 3**. As shown in **Figure 4**, the spike and recovery of online OPA method is equivalent to the offline AccQ-Tag method. Tabulated recovery data is shown in **Table 2. Figure 5** is illustrating the power of 2D-LC interfaced with a sampling device such as segFlow for online measurement of titer and product quality of bioreactor samples from a single analysis work stream.

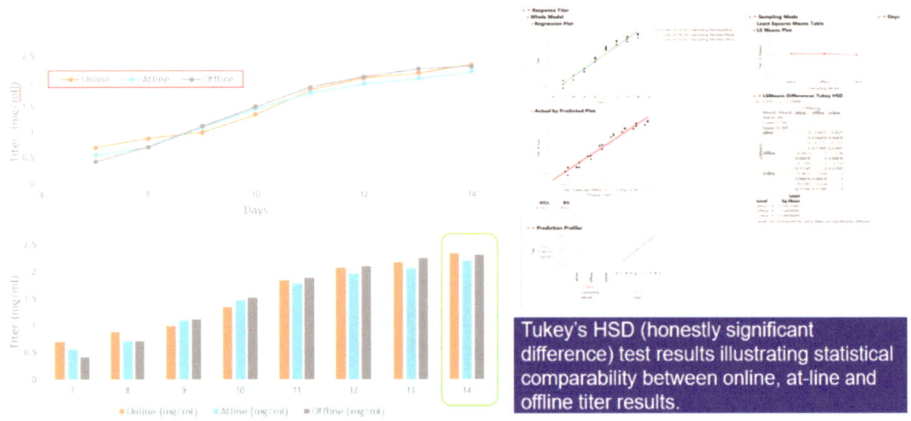

Figure 1.
Online at-line and offline mAb titer data from batch bioreactor using1D-LC with PSM.

Days	At-line (g/L)	Online (g/L)	Offline (g/L)
7	0.7	0.56	0.426
8	0.88	0.71	0.713
9	1	1.1	1.122
10	1.35	1.47	1.511
11	1.83	1.78	1.891
12	2.07	1.96	2.099
13	2.17	2.06	2.253
14	2.34	2.2	2.305

Table 1.
Titer data for mAb-1 in three sampling modes (At-line, Online, and Offline).

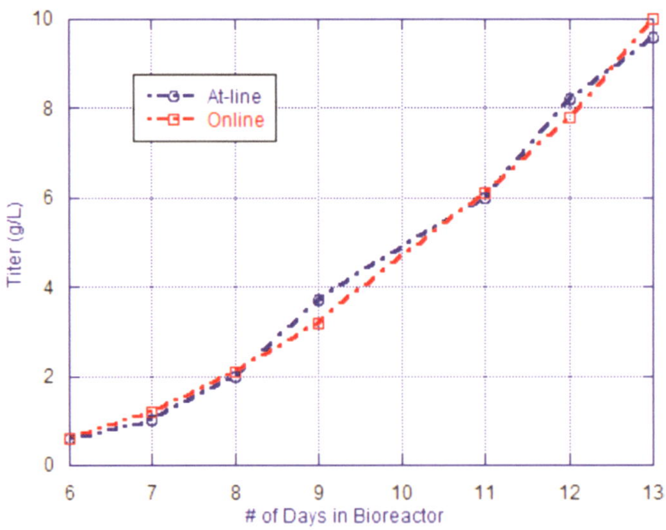

Figure 2.
Online and at-line mAb titer data from an ATF bioreactor using 1D-LC with PSM.

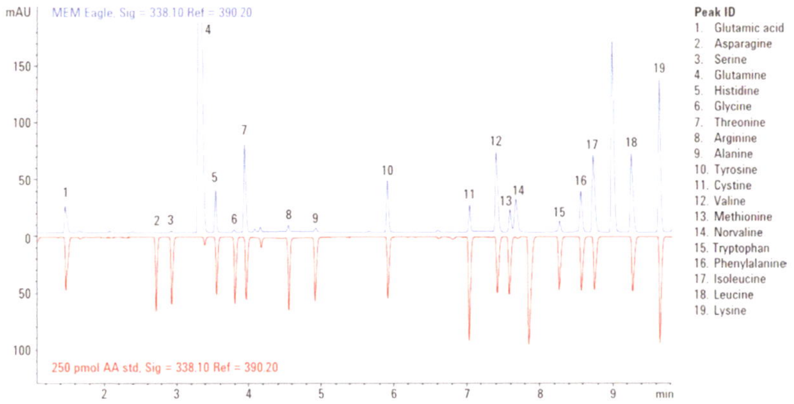

Figure 3.
Representative online amino acid profile of MEM media (blue) mirrored against amino acid standard (red) generated using 1D-LC with OPA derivatization.

Figure 4.
Spike & Recovery study results of online OPA method vs. conventional offline AccQ-Tag method demonstrating acceptable recovery (80–120%) of 6 essential amino acids for both methods.

Amino acid	AccQ-Tag method (500 mM spike) (%)	OPA method (125 mM spike) (%)	AccQ-Tag method (500 mM spike) (%)	OPA method (125 mM spike) (%)
Tyrosine	95.7	98.2	102.6	114.3
Valine	101.9	102.5	108.2	118.2
Isoleucine	95.3	101.2	101.0	116.6
Leucine	101.3	109.2	107.7	125.5
Phenylalanine	90.3	93.4	97.4	109.4
Tryptophan	90.7	92.8	96.6	107.6

Table 2.
Spike & Recovery study results of online OPA method vs. conventional offline AccQ-Tag method demonstrating acceptable recovery (80–120%) of 6 essential amino acids for both methods.

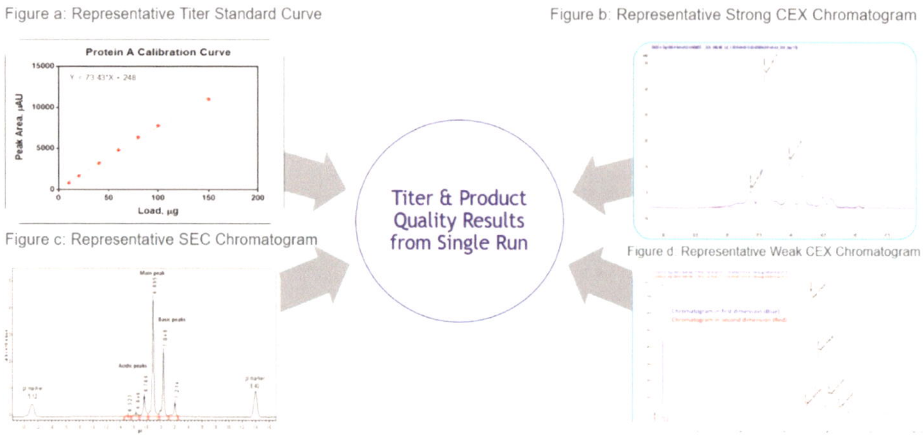

Figure 5.
Online chromatograms of 2D-LC system-3 with ^{1}D Pro-A and ^{2}D product quality.

6.2 Sequential injection analysis (µSIA system)

For complex online analytical techniques requiring laborious sample preparation workflow, a contemporary approach such as µSIA would be ideal. This novel system featuring a fully automated sample preparation modules can be programmed to execute various commands within the workflow using python scripting. The system can be interfaced with online sampling devices such as SegFlow or FIA lab's new built-in architecture (proSIAmpler) to withdraw samples from bioreactors. For N-linked glycan analysis, the script can be written to withdraw samples followed by protein-A based affinity purification within the system's architecture. The purified protein is then be subjected to PNGase-F digestion to release the glycan from the protein. The released glycan subsequently can undergo derivatization and clean-up step before transferring to an integrated online HPLC/UPLC equipped with a fluorescence detector for N-linked glycan mapping. Similarly, for online peptide mapping, the samples withdrawn from the bioreactors are subjected to protein-A purification followed by denaturation, reduction, alkylation, and a subsequent clean-up step to remove the excess reagents. Then desired protease such as trypsin is added to digest the protein to generate assorted peptides. The peptides are then analyzed on an integrated online HPLC/UPLC interfaced with Mass Spectrometry for peak identification and characterization. This architecture provides an ideal platform for the online Multi-Attribute Method (MAM) to monitor multiple critical quality attributes (CQAs) simultaneously. In addition, this platform is suitable for online amino acid analysis using AccQ-Tag, complementary to the online OPA derivatization presented above using the Agilent system.

6.3 LC/MS/MS for online metabolite analysis

LC/MS with integrated inline samplers such as SegFlow is suitable for online metabolite analysis of samples directly from bioreactors. The identification and quantification of the metabolites during cell culture process provides an insight into the homeostasis of the growing cells as well as to gain understanding of how the levels of metabolites influence the phenotypic nature of the living cells to enhance insight into the mechanisms of cellular functionality [22]. In contrast to proteomic analysis data, the metabolite analysis data is more dynamic and challenging [23]. It was identified that molecules related to carbohydrate and amino acid metabolic processes provide insight regarding the dynamic changes happening during the cell culture process [24]. The complexity of metabolome; consisting of varying concentrations of carbohydrates, ketones, organic acids, amino acids, lipids, and other assorted natural products along with their shorter half-life makes it difficult to analyze the entire metabolome. Various statistical and mathematical modeling as well as computational modeling and simulations have been developed for metabolome analysis. The most powerful experimental approaches are LC/MS and NMR-based techniques for structural elucidations. In addition to the multivariate analysis and chemometric modeling of the data generated from LC/MS and NMR, the data can be applied to conventional statistical and AI-based machine-learning techniques to gain understanding of the fate of metabolites through simulation. It has been shown that this deep learning algorithm is reliant on large numbers of training data to significantly enhance the accuracy of the model [25]. Additionally, mass spectral libraries like the Golm metabolome database, linking mass spectrum and chromatographic retention time to specific compounds, have been developed [26]. Various machine-learning and deep learning software packages for different tasks in metabolomics analysis are available. Ion-pair chromatography coupled with mass spectrometry is gaining popularity in virtue of its power to analyze both hydrophobic and hydrophilic compounds in a single analysis in contrast to performing RP-HPLC and HILIC runs separately for hydrophobic and hydrophilic compounds [27].

6.4 Spectroscopic techniques with chemometric modeling

Inline spectroscopic measurements utilizing Raman, FT-IR and NIR provide the platform for real-time monitoring of various nutrients, metabolites, and excipients to keep the process in a state of control. Multivariate analyses are often necessary to extract critical process knowledge for real-time control and quality assurance [13]. Real-time and near-real-time testing of critical product quality provides timely control of process parameters to achieve a desired product quality. With the convenience of integrating these PAT tools to upstream and downstream unit operations through fiber optics provides tremendous flexibility to interface with different unit operations for QbD-driven continuous processing. The process performance can be evaluated rapidly with the utilization of multivariate analysis, chemometric modeling and design of experiments using real-time PAT data. A PLS model can be built using the spectroscopic data for the real-time monitoring. AI-based machine learning techniques can transform the spectroscopic data into structural prediction of therapeutics. Inline monitoring features of spectroscopic techniques in conjunction with chemometric modeling provides an excellent platform for testing certain attributes throughout the product life cycle including in-process measurements. An alternative to the use of sophisticated predictive modeling techniques using FTIR with expensive ATR crystals, single use FT-IR is evolving with the use of silicon-based inexpensive ATR in combination with proprietary algorithms. These innovative systems can be connected directly to the bioreactors for continuous monitoring and control of nutrients and metabolites like glucose and lactate using a single-point standard calibration, reducing the effort of time-consuming chemometric modeling.

6.5 Online multi-attribute method (MAM)

Recent advances in process analyzers are making real time monitoring and control of multiple CQAs using a single analytical work stream during manufacturing is becoming a reality. Online MAM, targeted for measuring multiple product quality attributes simultaneously using a single workflow has received tremendous traction due to the benefit of saving time and resources. Various attributes including protein truncation, amino acid sequence coverage, post translational modifications, glycan profiling, glycation measurement, host cell protein monitoring, charge distribution, product related impurity assessment and more can be accomplished from a single MAM analysis. Simultaneous monitoring of multiple CQAs has significant advantage for QbD driven drug development paradigm in which the goal is to build desired quality into the product to achieve a quality target product profile (QTPP), revolutionizing the traditional approach of analytical testing. With the acquisition of multiple CQAs near-real time, prompt control of process parameters is feasible. Post translational modifications including asparagine and glutamine deamidation, succinimide and pyroglutamate formation, asp-isomerization, methionine oxidation, tryptophan oxidation, glycosylation, glycation, etc. can be monitored near-real time to provide feed-back or feed-forward control to modulate the process promptly to produce therapeutics with a desired product quality attributes. As FDA has been encouraging drug companies to adapt PAT as an enabling platform to modernize the manufacturing via QbD driven continuous processing, adaptation of MAM in an online setting to measure multiple product qualities straight from the bioreactors and other in-process steps is a revolutionary pathway. This approach will be in perfect alignment with the proposed paradigm shift from end-product testing to in-process monitoring and control. With the deployment of MAM in an online setting, the information-rich product quality data acquired at near-real-time can help to guide timely process control to establish a fully integrated automated online PAT tool.

MAM can overcome the challenges of obscured measurements of CQAs in conventional methods due to the co-elution of multiple components in each peak [28]. Furthermore, conventional analytical techniques are not capable of monitoring site-specific CQAs. In addition to providing detailed characterization of complex proteins, MAM can be implemented as a tool for process characterization (PC) during QbD-driven bioprocessing. Model fitting and simulations could be performed to evaluate the impact of process parameters on measured product quality attributes. A central composite design can be formulated to model the effects of these process parameters on product quality. Through model fitting and simulation, acceptance criteria of product quality attributes can be established and monitored during PC. Statistical models can be established based on the correlation between the process parameters and product quality attributes. Once the reduced model exhibits statistically acceptable correlation between the process parameters and quality attributes, a tolerance interval can be established through simulation at 95% confidence interval to predict the product quality in reference to the variation of process parameters [28]. In addition to providing comprehensive characterization of therapeutic drugs using QbD principles, MAM helps to reduce the number of assays required in a traditional release testing panel to a single workstream. The online MAM streamline the process further to generate CQAs at near-real-time to support QbD driven continuous bioprocessing. MAM consists of multiple modular components including automated sampling device, sample preparation platform such as μSIA, sampling device such as SegFlow, HPLC/UPLC and high-resolution mass spectrometry with MS/MS capability.

7. Timely process control

Control strategy is derived from product and process understanding that ensures optimal process performance and desired product quality [14]. A control strategy is designed to ensure that CPPs remain in a constant state of control to achieve a desired product quality during manufacturing [29]. The control ranges are determined based on statistical models in which the limits of the input parameters are controlled such that the response variables meet predefined specifications. With the establishment of design space and the operations within the boundaries of the established design space enable to attain pre-defined product quality. For the timely control of the process as well as the synchronization of the operation between the process equipment and PAT analyzers, two-way data communication across systems is essential. The legacy systems mandate the conversion of analog to digital data as well as 4–20 mA conversion of analog data before sending the signal to the distributed control system (DCS). For transmitting analog-based process information, the 4–20 mA loop is the dominant industry standard. The sensor measures a process variable, the transmitter translates the measurements into current signal, the signal travels through a wire loop to a receiver, and the receiver displays or performs an action with the received signal. Continued evolution of new technologies in conjunction with digital transformation is revolutionizing the way the data being transitioned across and between different types of instruments to build smart factories. For example, analytical process interface (API) serves as a juncture promoting communication between two different software packages to virtually connect with each other and serves as an intermediary between the two applications. Established enterprise scientific platform (ESP) workflow with bi-directional capability can push data between process systems and PAT instruments via data connectivity through a unified platform. Digitalization dealing with information processing can be used to improve workflows through the automation of existing processes to serve as the path towards digital transformation.

8. Building lab of the future through AI, ML and digitalization

In addition to the implementation of PAT tools to generate real-time or near-real-time data, an automatic control strategy is essential to maintain the process in a steady state. FDA's Process Validation Guidance [30] states that process knowledge and understanding are dynamic, and it is critical to establish a process control strategy for each unit operation as well as for the entire process. FDA's 2019 guidance on continuous manufacturing recommends input material control, process monitoring and control, RTRT deployment as well as system integration and data management [31]. Drug companies have been investing in the integration of PAT and data systems to assimilate data for controlling the process effectively and providing real-time predictions of the product quality [32]. Establishment of bi-directional feed-loop control for the synchronization between the process systems and PAT devices is inevitable. To this end, establishing a unified digitalized platform orchestrating the data flow back and forth from a centralized data depository is a way to overcome the challenge of data connectivity between incompatible systems. Harnessing the emerging technologies is essential for companies to stay competitive to foster innovation. Pharmaceutical and biopharmaceutical companies, engaged in leading-edge technological endeavors, are interdependent on un-related technologies such as informatics, statistical modeling, multi-variate analysis, artificial intelligence (AI), machine learning (ML) and other emerging technologies to promote continuous growth to stay ahead of the curve through innovation and modernization. Convergence of the enterprise's core technologies with enabling technologies such as informatics, IT and automations is essential to establish a successfully integrated architecture. To facilitate bidirectional data synchronization between the process equipment and analytical instruments, closed loop communication via feedback control is a prerequisite.

For monitoring and controlling of the processes to achieve a desired product quality, orchestrated digitalization of PAT data is the right course of action. Digital technologies are revolutionizing the biopharmaceutical industry, eliminating data communication silos to create more proficient processes through enhanced communication between various data systems. As part of the enterprise modernization and infrastructure connectivity, digital transformation is leading the way to build a smart factory with enhanced digital communication. High-tech companies such as telecommunication and aerospace industries have successfully implemented digital plants, embracing artificial intelligence and digitalization. In response to their success, drug companies are following the suit to join the digital revolution to modernize their manufacturing process. Such digital transformation is an example of a technology convergence in which unrelated technologies are intersecting at a juncture, resulting in a powerful technological transformation. With the utilization of intelligent systems such as DCS (distributed control systems) or HMI/SCADA, real-time monitoring, visualization, and control of both process parameters and product quality attributes is becoming a reality. Informed insight through data integration and feedback communication as well as timely control of inter and intra lab systems have made enormous progress through digital transformation. With digitalization and centralized visualization, informed decision can be made in real-time such that timely control strategies can be facilitated to achieve better plant performance and improved product quality. Legacy systems lacking opensource software are difficult to institute communication with other systems and require custom approaches to establish connectivity through digitalization. With the development of data analytics and visualization tools, monitoring the status of plants across the globe can be achieved virtually from anywhere in the world.

9. Building digital twin

The initial scientific conceptualization of digital twins was originated from NASA as a computerized simulation model to improve physical-model simulation of spacecraft [33], and now being extended to the pharmaceutical and biopharmaceutical companies. Digital twin, the digital equivalent of the physical system/process serves as the digital counterpart for process monitoring and computer modeling to predict the fate of a product through simulation. Building a virtual replica of a given process, leveraging information and communication technologies in the form of digitalization, enables cross-functional communication and synchronization of multiple activities [34]. Number of technologies such as machine learning, artificial intelligence and advanced robotics can influence the way the digital twins can be crafted and deployed. Digital twin technology is based on the connectivity between analytical instruments integrated with the process equipment. The data is transmitted and communicated to the digital twin through digitalization and various integration technologies to synchronize between the virtual and real system. The concept of digital twin, generating the best model using the input and output data enables the prediction of performance of the process. In addition to predict end-product quality, process bottlenecks can be identified and eliminated through simulation and timely process intervention. Using Big Data analytics, it is possible to access the data for a rapid decision making. With the use of algorithms based on artificial intelligence, digital twin acts as an efficient and intelligent alternative to test, predict, and solve problems virtually.

10. Summary

Pharmaceutical and biopharmaceutical companies have imprinted their signatures on successful development of diverse drug modalities for various disease indications. Despite having many successes, the drug companies are facing enormous challenges on producing cost effective drugs with highest product quality and adequate reproducibility. While the synthetic version of pharmaceutical drugs can be produced with minimal variability, the biopharmaceutical drugs produced in living cells tend to suffer significant challenges of inconsistency. Therefore, control strategy plays an integral role in manufacturing of high-quality biopharmaceutical drugs with minimal variability. Contrary to the pharmaceutical operations, the biopharmaceutical landscape is intensely more complex with convoluted multi-factorial processes as well as its inherent variability associated with its production in living organism. In addition to the complexity of in-process matrix components, the larger size of the biopharmaceutical drugs contributes additional challenges to perform reliable product characterizations to ensure safety and efficacy. As a result, many of the spectroscopic, chromatographic, and mass spectrometric techniques developed and deployed in the pharmaceutical sector for real-time data monitoring cannot be extended to the biopharma application without making significant modifications to their architectural design. Various novel and emerging techniques specific for biopharmaceutical applications are in the process of being developed and deployed.

With the ongoing efforts to make changes to biopharmaceutical landscape to transition from batch to continuous bioprocessing, there is an enhanced emphasis on PAT implementation. While the legacy batch process provides the flexibility to wait for offline data for decision making before advancing to the subsequent unit operation, continuous bioprocessing requires real-time or near-real time data to make timely decision to execute the control strategy. In terms of choosing the most appropriate PAT techniques for upstream and downstream testing, the ideal scenario would be to use an existing offline method either in an inline, at-line or online mode. With the emergence of contemporary sampling devices that can be integrated into the existing and emerging analytical tools, the scope of the PAT implementation and timely control of the processes are steadily advancing. UPLC-PSM with modular technology tailored to draw samples from bioreactors and downstream unit operations can be deployed as an ideal PAT platform for analyzing and controlling upstream titer and downstream product quality. 1&2-dimensional chromatography interfaced with an appropriate sampling device is suitable for titer measurement, amino acid analysis, and product quality assessments of upstream samples. Fully automated μSIA system is suitable for complex analysis with complicated workflow such as peptide mapping, amino acid analysis and glycan profiling. Additionally, the adaptation of emerging information technology infrastructure can facilitate timely process control.

Integrations of PAT tools to the process equipment is involved with interconnectivity between various parts of the integrated system with a flow of information channeling through the network with close alignment between systems. Sharing data between different platforms is not always straight forward with off-the-shelf data sharing tools, and often requires customized solutions, leveraging innovative technologies. Harmonized data communication platforms can be harnessed for systems with open-source software architecture. However, instruments with restricted access requires special licensing to customize the network architecture with a closed-loop communication feedback. For example, the synchronization of the operation between the process systems and PAT analyzers requires conversion of voltage to current followed by analog to digital conversion using the converter located in the analog input module of PLC or DCS systems. Alternatively, an ESP data automation workflow with standard data connectors can be implemented to establish an integrated online PAT solution for two-way data communication between the process systems and PAT analyzers. Tremendous amount of fundamental scientific, engineering and informatics knowledge as well as the technical tools are available for developing and implementing fully integrated innovative engineering principles into an integrated continuous bioprocessing architecture. A more collaborative approach between scientists, engineers and software teams is required at the interface of these multidisciplinary juncture to successfully integrate PAT tools and bioinformatics to deploy effective engineering controls. A continued effort in integrating multiple skillsets including chemistry, engineering and informatics is required to bridge the existing gaps and establish a fully integrated automatic analytical solution such as PAT to support fully integrated continuous bioprocessing.

11. Conclusion

Recombinant DNA (rDNA) technology and animal cell culture processes are the foundational pillars for the rapidly growing life science sector with potential applications including biopharmaceutical drug development, cell therapy, gene therapy, organ development, and other exploratory scientific approaches targeted for treating life-threatening diseases. Considering that the growth and maintenance of engineered cells in a bioreactor is heavily dependent upon its surrounding environment, it is essential to monitor, and control environmental factors and process parameters. Real-time or near-real-time monitoring and timely control of nutrients, metabolites, pH, CO_2 level, and temperature, impacting product titer and product quality attributes, are critical to achieve a desired target product profile (TPP). Quality by design philosophy along with PAT-enabled control strategy can help to achieve a TPP with the adaptation of integrated continuous manufacturing (ICM). ICM provides various benefits including the alleviation of variability encountered in the widely practiced batch processes, facilitating improved safety and enhanced product quality. Compared to Quality by testing (QbT) approach, QbD pathway provides the flexibility to produce high-quality drugs faster at a reduced cost with improved robustness and efficiency. To establish a state-of-the-art fully integrated end-to-end process platform, it is essential to merge various technological frontiers including process operation units, PAT tools and bioinformatics techniques. Fully automated ICM, enabled by the QbD-driven PAT initiatives, provides enhanced process and product understanding and facilitate a closed loop feedback control to keep the process in a desired state of control.

Acknowledgements

Tanushree Prabhakar[1]; Dhanuka P. Wasalathanthri[1]; Xin Zhang[1]; Mathura Raman[1]; Tanmay Kulkarni[1]; Xia Xu[1]; Priya Singh[1]; Sohil Bhavsar[1]; Li Zhang[1]; Helen Shao[1];

Vivekh Ehamparanathan[1]; June Kuang[2]; Jay West[2]; Zhijun Tan[2]; Yuanli Song[2]; Julia Ding[2]; Chun Shao[2]; Robin Barbour[3]; Ryan Knihtila[2]; Neha Puri[2]; Kyle McHugh[2]; Matthew S. Rehmann[2]; Qin He[2]; Yueming Qian[2]; Jianlin Xu[2]; Michael C. Borys[1]; Julia Ding[2]; Zhengjian Li[2]

Current BMS Staff[1]

Former BMS Staff[2]

Summer Interm[3].

Conflict of interest

The authors declare no conflict of interest.

Notes/thanks/other declarations

The authors would like to express their special thanks to Waters Corporation, Agilent, FIA Labs and Flownamics for their technical assistance and healthy collaboration to establish online capabilities for connecting PAT instruments with upstream bioreactors and downstream AKTA systems.

Abbreviations

AI	Artificial Intelligence
API	Analytical Process Interface
ATR	Attenuated Total Reflectance
CHO	Chinese Hamster Ovaries
CMAs	Critical Material Attributes
CPP	Critical Process Parameters
CQA	Critical Quality Attributes
DCS	Distributed Control System
DoE	Design of Experiments
FDA	Food and Drug Administration
ICH	International Council for Harmonization
mAb	Monoclonal Antibody
MAM	Multi Attribute Method
ML	Machine Learning
OoC	Organ on a Chip
PAT	Process Analytical Technology
PC	Process Characterization
PI	Process Intensification
QbD	Quality by Design
QbT	Quality by Testing
QTPP	Quality Target Product Profile
rDNA	Recombinant DNA Technology
RTRT	Real Time Release Testing
SUS	Single Use System

Author details

Letha Chemmalil[1*], Chris Chumsae[1], Gloria Li[1] and Anthony Leone[2]

1 Bristol Myers Squibb, Devens, Massachusetts, USA

2 Bristol Myers Squibb, New Brunswick, New Jersey, USA

*Address all correspondence to: letha.chemmalil@bms.com

References

[1] Domokos A, Nagy B, Szilágyi B, Marosi G, Zsombor Kristóf Nagy ZK. Integrated continuous pharmaceutical technologies—A review. Organic Process Research & Development. 2021;**25**(4):721-739. DOI: 10.1021/acs.oprd.0c00504

[2] Vipra M, Patil N, Sivaram A. Applications of recombinant DNA technology. In: A Complete Guide to Gene Cloning: From Basic to Advanced Techniques in Life Science and Biomedicine for the Non-Expert. Cham: Springer; 2022. pp. 143-157. DOI: 10.1007/978-3-030-96851-9_9

[3] Aziz MA, Brini F, Rouached H, Masmoudi K. Genetically engineered crops for sustainably enhanced food production systems. Frontiers in Plant Science. 2022;**13**:1027828. DOI: 10.3389/fpls.2022.1027828

[4] Rosano GL, Ceccarelli EA. Recombinant protein expression in Escherichia coli: Advances and challenges. Frontiers in Microbiology. 2014;**5**:1-17. DOI: https://www.frontiersin.org/articles/10.3389/fmicb.2014.00172, 10.3389/fmicb.2014.00172

[5] Leung CM, de Haan P, Ronaldson-Bouchard K, et al. A guide to the organ-on-a-chip. Nature Reviews in Methods Primers. 2022;**2**(33):1-29. DOI: 10.1038/s43586-022-00118-6

[6] Sun M, Liu A, Yang X, Gong J, Yu M, Yao X, et al. 3D cell culture: Can it be as popular as 2D cell culture. Advance Nanobiomedical Research. 2021;**1**(5):1-18. DOI: 10.1002/anbr.202000066

[7] Verma A, Verma M, Anchal Singh A. Animal tissue culture principles and applications. Animal Biotechnology. 2020;**2020**:269-293. DOI: 10.1016/B978-0-12-811710-1.00012-4

[8] Horta D, Forné M, Agustí A, Raga A, Martín-Cardona A, Hernández-Soto JM, et al. Efficacy of Hepatitis B virus vaccines HBVaxpro40© and Fendrix© in patients with chronic liver disease in clinical practice. Vaccines (Basel). 2022;**10**(8):1323. DOI: 10.3390/vaccines10081323

[9] Saber MA, Hatem AS, Fouad YM, Elsayed M, Hassnine AA. Overview on Hepatitis B vaccination. The Egyptian Journal of Hospital Medicine. 2022;**86**:696-699

[10] Hock SC, Siang TK, Chan Lai Wah CL. "Continuous manufacturing versus batch manufacturing: benefits, opportunities and challenges for manufacturers and regulators." Generics and Biosimilars Initiative Journal, 2021. 10, 1. Gale Academic OneFile, link.gale.com/apps/doc/A692876993/AONE?u=anon~e98fa4eb&sid=googleScholar&xid=bfbbffa7 [Accessed 31 May 2023]

[11] Sau (Larry) Lee. Deputy Director of the Office of Testing and Research, and Chair of the Emerging Technology Team, Office of Pharmaceutical Quality, CDER. Modernizing the Way Drugs Are Made: A Transition to Continuous Manufacturing Spotlight on CDER Science. 2017

[12] Rantanen J, Khinast J. The future of pharmaceutical manufacturing sciences. Journal of Pharmaceutical Sciences. 2015;**104**(11):3612-3638. DOI: 10.1002/jps.24594

[13] Guidance for Industry PAT – A Framework for Innovative Pharmaceutical Development, Manufacturing, and

Quality Assurance, U.S. Department of Health and Human Services Food and Drug Administration Center for Drug Evaluation and Research (CDER) Center for Veterinary Medicine (CVM) Office of Regulatory Affairs (ORA) Pharmaceutical CGMPs September 2004

[14] International Conference on Harmonisation of Technical Requirements for Registration of Pharmaceuticals for Human Use considerations (ICH) guideline Q8 (R2)

[15] Ramshaw C. "Higee distillation": An example of process intensification. Chemical Engineer – London. 1983;**389**:13-14

[16] Keil F. Process intensification. Reviews in Chemical Engineering. 2018;**34**(2):135-200. DOI: 10.1515/revce-2017-0085

[17] EMA/CHMP/QWP/811210/2009-Rev1 Committee for Medicinal Products for Human Use (CHMP)

[18] Chemmalil L, Wasalathanthri D, Zhang X, Kuang J, Shao C, Barbour R, et al. Online monitoring and control of upstream cell culture process using 1D & 2D-LC with SegFlow Interface. Biotechnology and Biotechnology Bioengineering. 2021;**118**(9):2593-3603. DOI: 10.1002/bit.27873

[19] Chemmalil L, Naegele E, Kutscher D. An automated sampling system for online monitoring as a process analytical technology (PAT) tool. Biopharma/Pharma. 2021;**2021**:1-8

[20] Chemmalil L, Prabhakar T, Kuang J, West J, Tan Z, Ehamparanathan V, et al. Online/at-line measurement, analysis and control of product titer and critical product quality attributes (CQAs) during process development. Biotechnology and Bioengineering. 2020;**117**(12):1-9. DOI: 10.1002/bit.27531

[21] Chemmalil L, Chumsae C, Ding J. Chromatography based near-real-time measurements & control of product titer, product quality and cell culture nutrients. Biotechnology and Biotechnology Bioengineering. 2022;**118**(9):2593-3603. DOI: 10.1002/bit.27873

[22] De Sousa CS, Hassan SS, Pinto AC, Silva WM, De Almeida SS, Soares SDC, et al. Chapter 1 – Microbial omics: Applications in biotechnology. Omics Technologies and Bioengineering. 2018;**2018**:3-20. DOI: 10.1016/B978-0-12-815870-8.00001-2

[23] Mapelli V, Olsson L, Nielsen J. Metabolic footprinting in microbiology: Methods and applications in functional genomics and biotechnology. Trends in Biotechnology. 2008;**26**(9):490-497. DOI: 10.1016/j.tibtech.2008.05.008

[24] Marcobal A, Yusufaly T, Higginbottom S, Snyder M, Sonnenburg JL, Mias GI. Metabolome progression during early gut microbial colonization of gnotobiotic mice. Scientific Reports. 2015;**5**(1):11589. DOI: 10.1038/srep11589

[25] Alakwaa FM, Chaudhary K, Garmire LX. Deep learning accurately predicts estrogen receptor status in breast cancer metabolomics data. Journal of Proteome Research. 2018;**17**(1):337-347. DOI: 10.1021/acs.jproteome.7b0059

[26] Hummel T, Kobal G, Gudziol H, Mackay-Sim AJEA. Normative data for the "Sniffin'Sticks" including tests of odor identification, odor discrimination, and olfactory thresholds: An upgrade based on a group of more than 3,000 subjects. European Archives of Oto-Rhino-Laryngology. 2007;**264**:237-243. DOI: 10.1007/s00405-006-0173-05

[27] Qiu J, Chan PK, Bondarenko PV. Monitoring utilizations of amino

acids and vitamins in culture media and Chinese hamster ovary cells by liquid chromatography tandem mass spectrometry. Journal of Pharmaceutical and Biomedical Analysis. 2016;**117**:163-172. DOI: 10.1016/j.jpba.2015.08.036

[28] Guan X, Eris T, Zhang L, Ren D, Ricci MS, Thiel T, et al. A high-resolution multi-attribute method for product characterization, process characterization, and quality control of therapeutic proteins. Analytical Biochemistry. 2022;**643**:114575. DOI: 10.1016/j.ab.2022.114575

[29] Taylor C, Marschall L, Kunzelmann M, Richter M, Rudolph F, Vajda J, et al. Integrated process model applications linking bioprocess development to quality by design milestones. Bioengineering. 2016;**8**(11):156. DOI: 10.3390/bioengineering8110156

[30] FDA Guidance for Industry. Process Validation: General Principles and Practices. Rockville, MD: FDA; 2011

[31] FDA Guidance for Industry (Draft). Quality Considerations for Continuous Manufacturing. Rockville, MD: FDA; 2019

[32] Wu H, Office of Pharmaceutical Manufacturing Assessment (OPMA) Office of Pharmaceutical Quality, FDA. Integration of PAT and data systems for manufacturing process control and quality improvement: A brief review of past developments, status, and future considerations. American Pharmaceutical Review. 2020;**11**:939-948

[33] Negri E, Fumagalli L, Macchi M. A review of the roles of digital twin in cps-based production systems. Procedia Manufacturing. 2017;**11**:939-948. DOI: 10.1016/j.promfg.2017.07.198

[34] Lipsmeier A, Kühna IR, Joppena R, Dumitrescub IR. 13th CIRP Conference on Intelligent Computation in Manufacturing Engineering, CIRP ICME '19 Process for the development of a digital strategy. Science Direct Procedia CIRP. 2020;**88**:173-178

Chapter 10

Bioprocess Development and Bioreactor Scale-Up for the Production of Recombinant Lentiviral Viral Vectors in HEK293 Suspension Cell Culture

Julien Robitaille, Aziza Manceur, Anja Rodenbrock and Martin Loignon

Abstract

Therapeutic applications of viral vectors that initially targeted rare monogenic diseases have now grown to a broader set of indications including cell and gene therapy applications and vaccines. This has prompted the need to increase biomanufacturing capacities, which will require adjustments in the biomanufacturing space to increase yield and lower cost of goods of large-scale productions. HEK293 cells have been widely used for the production of viral vectors because they can grow rapidly in suspension and allow for different modes of production: batch, fed-batch and perfusion. Here we review methods and platforms for producing lentiviral vectors in HEK293 cells grown in serum-free media and the principles and challenges of optimizing and scaling up of bioprocesses in various bioreactors. Lentiviral vectors are particularly difficult to manufacture due to their labile nature. These challenges will be considered in view of current processes and future trends emerging to resolve bottlenecks and existing limitations.

Keywords: recombinant lentiviral vectors, HEK293 cells, suspension cell culture, transfection, packaging cells, stirred tank bioreactors, bioprocess, production modes, cell culture scale-up

1. Introduction

Through evolution, viruses have become highly efficient natural vehicles for the transfer of genetic material into living cells. This property has been exploited to develop recombinant viral vectors for R&D and therapeutic applications. Modern viral recombinant vectors are based on non-enveloped adenoviruses (Ads) and adeno-associated viruses (AAVs), or enveloped retroviruses such as lentiviruses (LVs), each bearing unique properties and requirements for engineering and production.

Viral vectors are designed for efficiency and safety. Viral elements that permit carrying a genetic cargo encoding one or many factors for efficient delivery in targeted cells are preserved, while elements necessary for the virus' replication are removed to improve their safety profile and increase payload size. Each class of vectors has limitations and advantages that, when judiciously selected, best serve targeted applications from vaccine to cell and gene therapy. Globally, there are currently more than 1000 ongoing clinical trials making use of viral vectors [1].

The demand for development of novel and improved viral vectors has reached a critical point where demand for efficient and cost-effective manufacturing is increasing faster than current technology is improving. Across vector design, production, purification and characterization, several bottlenecks need to be solved to improve targeting and infectivity; increase titers, batch to batch reproducibility, yields and purity; as well as expand in-process and post-process analytics tools for release and stability monitoring. The development of a viral vector production platform begins upstream with the selection of host cell, also defined as, the manufacturing unit. Once the host cell is selected, the process development is initiated at small scale. Viral vectors can be produced transiently by transfecting host cells with plasmids encoding viral vector elements or with engineered producer cells that stably express all viral vector components including the transgene(s). For most stable expression systems, producer cells are derived from packaging cells, which express all viral vector genes but the transgene(s).

Once a mammalian cell is selected, its productivity will also depend on the process developed at scale. The production process can be initiated using pre-established parameters and off-the-shelf consumables or by customizing an entire solution including expression system, culture media, supplements, etc. In both cases, a process intensification phase followed by scale-up will be needed to improve viral titers. The production process development and scale-up will significantly differ between expression systems and from one vector to another. Different modes of production such as batch, fed-batch and perfusion exhibit different levels of complexity to set up, and have been adapted for transient or stable production. With suspension cells maintained under constant agitation, batch and fed-batch production can be tested at small scale in shake flask, whereas perfusion usually requires a bioreactor. As production is scaled up, a stepwise progression going from smaller to larger bioreactors allows one to set and adjust control parameters with the goal of replicating original conditions and expected titers. Each mode will require its own bioreactor-controlled parameters and configuration. While pH, dissolved oxygen (DO) and agitation rate qualify as bioreactor-controlled parameters, other parameters will impact the process including bioreactor geometry, impeller type and tip speed, head space and mixing capacities affecting mass transfer. This chapter will focus on the production and scale-up of recombinant lentiviral vectors (rLVVs) using serum-free suspension-adapted HEK293 cells.

2. Characteristics and applications of rLVVs

In addition to the natural features borrowed from viruses, molecular engineering of viral vectors has allowed the design of fit for purpose vectors. For example, mutagenesis and pseudotyping is used to change the tropism of a virus for narrowing [2, 3] or increasing the host range [4]. The range of host cells targeted by a lentiviral viral vector can be broadened by incorporating protein G from the vesicular stomatitis virus (VSV-G) in the envelope [5]. The VSV-G pseudotyped vectors will enter the cells *via* an interaction with the widely distributed low density lipoprotein receptor (LDL-R) [6, 7]. The most used rLVV in gene therapy is derived from human immunodeficiency virus 1 (HIV-1) although several non-primate alternatives have been developed such as feline immunodeficiency virus (FIV) or equine infectious anemia virus (EIAV) [8]. Different promoters can be introduced to regulate tissue-specific expression of the payload [9]. In theory, the modifications that can be done on viral vectors are limited by their intrinsic properties including capsid/envelope composition and assembly, cargo size, targeted cells, etc. In practice, viral vector engineering not only has an impact on the vector's properties but also may significantly alter expression system productivity, purification and quality attributes.

Viral vectors are used to efficiently transfer a nucleic acid cargo to their natural target, usually a mammalian cell. These nucleic acid cargos encode information for the expression of one or more transgenes. The transgene(s) can function as a reporter, as a therapeutic, for gene editing or to up/downregulate gene expression. They can be used *in vitro*, *ex vivo* or *in vivo* on a large array of human and animal cells and tissues.

Recombinant LVs (rLVV) are derived from HIV-1, and thus share several common physico-chemical properties. They are enveloped viral vectors with a spherical shape of 80–120 nm in diameter. They are sensitive to several factors such as pH, osmolarity, shear stress, freeze-thaw cycles and temperature [10] and their half-life is in the range of only 3–18 h at 37°C [11]. Because of all the above, rLVVs are considered to be labile and bring a unique set of challenges in terms of bioprocessing. rLVVs have been made replication defective to increase their safety profile, and overall, rLVV safety and efficacy have been improved since their first developments in the early 1990s [12]. This was made possible by transferring only the essential rLVV elements on four separate plasmids used in co-transfection for production [13].

rLVVs are used in a multitude of applications. They can deliver genetic material of up to 11 kilobases (kb) in size with high efficacy in both dividing and non-dividing cells, including difficult to transfect and transduce cells such as human neuron, primary and stem cells. rLVVs were employed in more than 100 clinical trials in 2018 [10]. Lentiviral-based gene therapy has successfully treated multiple genetic blood cell diseases including Wiskott-Aldrich syndrome, X-linked severe combined immunodeficiency (XSCID), X-linked adrenoleukodystrophy, b-thalassemia and metachromatic leukodystrophy [8].

Another key use and success with rLVVs in therapies tested in the clinic is in the field of cancer treatment. Lentiviral vectors are used to deliver Chimeric Antigen Receptors (CAR) into patients' T cells *ex vivo*. The cells expressing the modified receptor are re-introduced into patients where they can recognize the antigen of interest, e.g. CD19 or BCMA and attack tumor cells. Since 2017, six CAR-T cell therapies have been approved for hematological cancers by the Food and Drug Administration (FDA) [14].

3. Cultivation of cells for viral vector production

The most common mammalian cells for the production of viral vectors are human embryonic kidney cells (HEK293), human lung adenocarcinoma cells (A549) and kidney epithelial cells from African green monkey (Vero). These cells can be grown as adherent or in suspension in a bioreactor. Adapting cells from adherent to suspension is challenging, and while the adaptation of HEK293 cells has been achieved by many groups, only a few laboratories have succeeded in adapting Vero cells in suspension [15, 16]. Suspension grown HEK293 cells are the most popular cells for the production of viral vectors in stirred tank bioreactors. They are easy to grow; they can be engineered to develop packaging and producer cell lines and are permissive to the production of many types of viral vectors. In addition, culture media, supplements and vector systems are commercially available.

3.1 Cultivation of cells in two-dimensional (2D) systems

The cells used to make viral vectors can be grown in a two-dimensional (2D) system to provide a surface for the cells to adhere. Several types of vessels can be used in such static systems including tissue culture flasks (T-flasks), cell factories or cell stack. The main advantage of 2D systems is that they are relatively easy to implement. The productions are scaled up by multiplying the number of vessels. However, these systems have a significant footprint and the overall process is labor-intensive [17, 18].

Adherent cells can also be cultivated in stirred tank bioreactors on microcarrier beads made of glass, plastic or other material that provide a surface for the cells to grow. This combination significantly reduces the footprint but the process development is not straightforward and requires substantial expertise. A single-use fixed-bed bioreactor is the most recent device developed to scale up adherent cell expression platforms. It reduces physical space requirements; it simplifies process development and is offered by several manufacturers. While these systems can support the growth of 30–200 M cells per mL of fixed bed, the cells cannot be counted when transfection or infection is performed. The cells are concentrated in a proportionally smaller bioreactor volume, which increases the demand for nutrients. The nutrients can be supplied throughout the production phase either by perfusion or by a medium re-circulation strategy. Despite these challenges, some robust processes have been developed using fixed-bed bioreactors [19]. The largest fixed-bed bioreactors, such as the iCellis 500 from Pall or the Scale-X from Univercells [20], offer surface areas for 2D culture to up to 500 m^2 and 600 m^2 respectively. To obtain a similar growth surface using cell factories, approximatively 950 units would be required [19] while similar cell mass could be grown in 100–1000 L stirred tank bioreactor, depending on the achieved cell density [21]. iCEllis bioreactors have been successfully used to produce lentiviral vectors in chemically defined media [22].

3.2 Cultivation of suspension-adapted HEK293 cells in serum-free medium

The adherent human cell line HEK293T remains the host of choice for the production of viral vectors because it is well characterized and safe for clinical use. However, most adherent cell systems rely on serum supplemented media, generally fetal bovine serum (FBS). This increases the production costs, risk for contamination with adventitious agents and production lot-to-lot variability. While fixed-bed bioreactors have improved production with 2D systems, investments and efforts are being made to replace traditional 2D systems with large-scale productions that offer more practical options for high-titer viral vector–based productions in adherent mode. Preferred host cells for the development of expression systems for viral vector production are robust, have short doubling times and grow in suspension at high cell density in chemically defined media absent of serum or supplements derived from an animal source. Cells that grow in suspension can be cultivated in stirred-tank bioreactors to reduce the footprint and process labor.

The most common systems for viral vector production use mammalian cells that are permissive to vector production while limiting the production of replication competent particles. The human embryonic kidney cell HEK293 is one commonly used cell line for production of viral vectors [23]. Of note, HEK293 cells stably express the adenoviral E1A and E1B-55 k genes that support a helper function necessary for the propagation of AAV and Ads vectors [24].

An example of suspension HEK293 cells is the HEK293SF-3F6 cell line. It was adapted to suspension in 1998 in serum-free media [25] through a series of adaptation steps while reducing calcium concentration. The final clone with demonstrated monoclonality, namely HEK293SF-3F6, was selected on the basis of its doubling time of approximately 24 hours. Since its establishment, the cell line has been used for bioreactor productions of adenovirus [26], AAV [27], lentivirus [28], an Ebola vaccine [29] and influenza virus and viral-like particles [30].

4. Production of recombinant viral vectors by plasmid transfection

When optimizing a transfection process, a Design-of-Experiment (DoE) approach is preferred because of the large number of parameters that could possibly interact and affect the result. DoE is now well established for improving the production of recombinant proteins, especially with the growing implementation of miniaturized parallel automated bioreactors such as the Ambr®15 and Ambr®250 from Sartorius [31, 32]. In general, parameters targeted when optimizing transfection of rLVV in serum-free medium include relative ratio of each plasmid, total quantity of DNA, PEI to DNA ratio, cell culture medium composition and cell density at transfection [33]. As for other biologics applications, we foresee a broader use of the DoE approach, combined with the use of high-throughput parallel bioreactors for rapid and efficient optimization. Improving viral vector titers is multi-factorial and the transfection method alone cannot adequately explain differences in production yields. For rLVV, size and nature of the gene of interest (GOI), the mode of production and the expression system all influence vector quantities and quality. In general, transfection methods are efficient at small scale but costly and difficult to scale up, especially with GMP quality plasmids.

4.1 Transfection reagents

The production of recombinant viral vectors by plasmid transfection is relatively simple and accessible at small scale. Several non-viral transfection methods using cationic reagents have been developed to deliver the plasmids to the host cells. The most common methods include calcium phosphate precipitation, polyethyleneimine (PEI) or cationic lipid such as lipofectamine.

The most affordable transfection method using calcium phosphate co-precipitation with plasmid DNA has been used for decades in HEK293 adherent cells, but also with suspension cells in stirred tank bioreactor [34, 35]. The main drawback however is the sensitivity of the precipitation process to pH or agitation force [36]. Furthermore, calcium phosphate co-precipitation usually requires the presence of animal-derived components in the medium such as serum or albumin [37, 38] that is frowned upon by regulators. In comparison, the transfection efficacy of cationic lipids has been reported to be similar to or even better than calcium precipitation [39] or polyethylenimine (PEI) [40, 41]. However, their use is limited due to their cytotoxicity [10] and higher costs [10, 36, 42]. Non-chemical non-viral methods, such as flow electroporation, are also available for the transfection of HEK293 cells in suspension for the production of viral vectors such as lentiviral vectors [43]. However, the scalability of this method is limited by the need to exchange medium and concentrate the cells.

4.2 Production of rLVVs by plasmid transfection

rLVVs have been produced by transfection in HEK293 cells adapted to grow in suspension in serum-free media using four plasmids. One plasmid contains the *Gag/Pol* genes that encode key enzymes: a reverse transcriptase, integrase and protease. These genes also mediate crucial steps in vector assembly including the recruitment of Viral Env protein, packaging of the RNA payload and binding to the plasma membrane. A second plasmid encodes the Rev gene essential for transporting the unspliced RNA into the cytoplasm by interacting with the cis-acting Rev response element (RRE). For pseudotyping the vector particles, a third plasmid contains VSV-G gene. The fourth plasmid encodes the GOI with its specific promotor and regulatory elements [8, 10, 42, 44]. The development of lentiviral vectors is now at its third generation, which presents additional safety features over previous generations [45].

4.3 Production of rLVVs by plasmid transfection: challenges and opportunities for improvement

The reproducibility of transfection methods can be variable, especially at large scale. PEI and DNA are mixed together in specific ratios usually ranging between 2:1 and 3:1 [11, 46, 47]. The relative ratios of the different viral vector plasmids used for transfection might result in poor packaging and the accumulation of empty vectors and batch-to-batch variability [18]. Because PEI alone is relatively cytotoxic to the cells [36, 48], the PEI:DNA ratios must be carefully optimized to maximize transfection efficacy and minimize cytotoxicity. Once mixed, the polyplexes (complexes formed between PEI and DNA) grow in size over time [49–51]. Efficient polyplexes for transfection have a net positive charge and a time-sensitive size distribution. Polyplexes that are too small or too large will not transfect cells efficiently [52]. In most optimized protocols for HEK293, PEI and DNA are mixed together, and incubated (**Figure 1**) for a maximum of 30 minutes (usually 10–15 minutes) in a total volume accounting for 5–10% of the cell culture volume to be transfected [46, 47, 53, 54]. Other process-related factors, such as the medium and ion concentration used in the plasmid DNA mix, will have an impact on particle formation and transfection efficacy [49, 50, 55–57]. Therefore, the optimal incubation time may vary between processes and products. Given the relatively short mixing time and size of transfection mix, it becomes technically challenging to complete transfection within these parameters at increasing scales. Indeed, the time for mixing large volumes of DNA and PEI and transferring the complex to the bioreactor at a given transfer rate significantly increases with the scale. This also narrows the time window of operation, depending on whether the mix is transferred by gravity or pumped in the bioreactor. When adding DNA to PEI mix or vice versa, addition rates and shear stress introduced by mixing velocity are also important parameters to be considered [55, 58].

The optimal HEK293 cell density for transfecting plasmids for producing rLLVs usually ranges between 1.0 and 2.0E6 cells/mL, when cells are still in the exponential growth phase. While higher densities can easily be achieved in batch cultivation, the transfection process becomes suboptimal. For HEK293 cells, the transfection efficacy and the specific productivity, that is, the number of viral genome copies produced

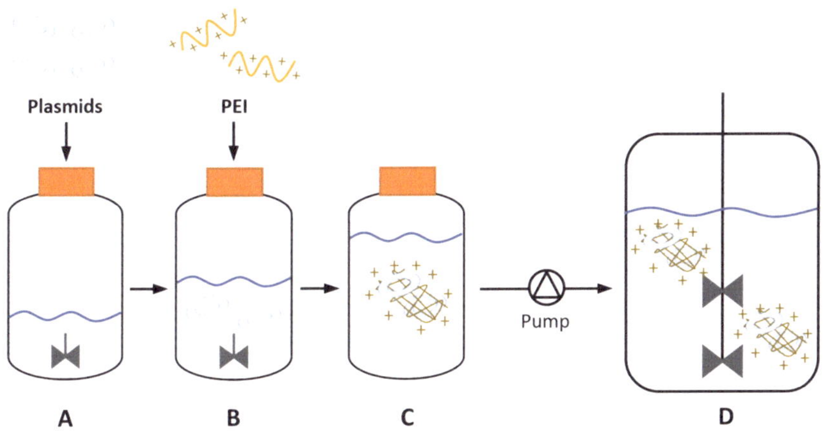

Figure 1.
Example of a polyethyleneimine (PEI)-mediated LVV transfection process for large-scale production. (A) The four plasmids i.e., the gene of interest (GOI), Gag/Pol, Rev and the glycoprotein G of vesicular stomatitis virus (VSV-G) encoding plasmid, are diluted in buffer (Fresh PBS or cell culture medium) and mixed together. (B) PEI, diluted in buffer (Fresh PBS or cell culture medium), is added to the mix via pumping, while agitation allows for constant mixing of the solution. (C) Agitation is stopped, and the mixture is incubated for 10–15 minutes to allow for the PEI-DNA complex to form. (D) The PEI-DNA complex mixture is added to the bioreactor via pumping, with special attention to the pumping rate and tubing size to minimize the shear stress applied to the complex

per cells, decreases significantly when cells density is above 1.0E6 cells/mL [27]. Even when the quantities of plasmid and PEI are proportionally increased at higher cell densities, the titers do not significantly increase and cost-effectiveness is reduced. The mechanism for this is not well understood, but it may be a result of the accumulation of an inhibitory metabolite or secreted protein (shedding of HSPG), or to a change in the physiological state of cells grown at higher densities [59].

Alternative production modes such as perfusion applied to a process intensification approach may improve the physiological state of the cells at high densities and allow for higher volumetric titers and cost-effective transfection. Current literature on production of viral vectors by transfection indicate that cell engineering, vector and capsid design as well as novel and low-cost reagents producing stable complexes and using less plasmid DNA could contribute to increase vector quality, batch-to-batch consistency and cost-effectiveness [60].

5. Development and production of rLVVs with packaging and producer cell lines

Although the production of viral vectors by transfection is relatively simple and accessible, it scales poorly, it is susceptible to batch-to-batch variability and the need for GMP-grade plasmids and PEI to produce clinical material is cost prohibitive. One attractive approach to offset some of these cost- and scale-related challenges is to develop packaging cell lines that stably express the genes essential for viral particle formation. The production of rLVVs in packaging cells necessitates only one plasmid that carries the GOI. The GOI is stably integrated in the genome of producer cells, where rLVV production is controlled by an inducible system [23].

Stable packaging cell lines for lentiviral vectors are challenging to engineer because some viral vector proteins are toxic to producer cells, most notably VSV-G and the protease encoded by the *Gag/Pol* gene. Inducible systems have been used to control the expression of toxic viral proteins with some success. However, many of the tested inducible systems were derived from microbial operons including tetracycline, cumate and coumermycin that can be leaky [8]. Broussau et al. have developed a dual control system by combining tetracycline and cumate operons to regulate the expression of cytotoxic vector elements to produce a stable packaging HEK293 cell line [61] and later used a similar strategy by combining coumermycin and cumate to generate a producer cell line [62].

In producer cell lines, the GOI is also under the control of a molecular switch as part of the inducible system. The production of the rLVV is initiated either directly by inducers in producer cell lines or after transfection of the plasmid with the GOI

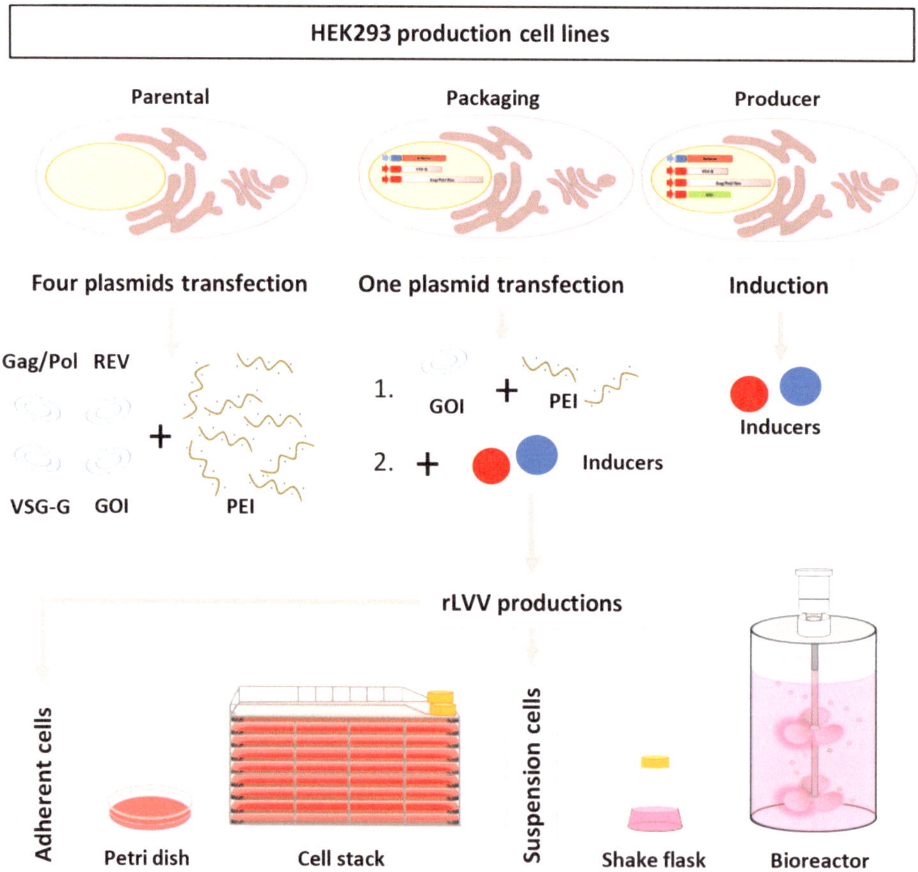

Figure 2.
Evolution of the rLVV expression systems towards the reduction of cost of goods (COGS). Recombinant lentiviral vectors (rLVV) can be produced with adherent or suspension-adapted cell lines. Although fully developed, packaging and producer cell lines represent a significant investment they simplify the production and reduce the costs of plasmids; only the gene of interest (GOI) plasmid is needed for the production with packaging cells. Despite that process development and scale-up for suspension cells can be less strait forward than adherent cells it will in the long term significantly reduce the costs of production. The least cost effective method most labour intensive to produce rLVV is by transfecting adherent cells with 4 plasmids respectively encoding, Gag/Pol, Rev, the glycoprotein G of vesicular stomatitis virus (VSV-G) and the GOI. The most cost effective platform for the production of large quantities of rLVV combine a suspension-adapted producer cell line and a production process in a stirred-tank bioreactor.

combined with inducers in packaging cell lines (**Figure 2**). The lowest overall manufacturing costs are obtained with suspension producers [63] where the product is not contaminated by residual plasmids and transfection reagents. This simplifies the subsequent purification steps. However, generating such a cell line requires 6–12 months for an experienced team, which can be difficult to accommodate with aggressive timelines in competitive commercial landscapes and to address the needs of patients for therapies of last resort. The packaging cell line offers the versatility and speed of execution of the production with the four-plasmid transfection method. In addition, optimizing the process is easier with packaging cell lines.

Generating stable packaging and producer cell lines requires a high level of expertise. A recent review reports a total of 10 stable producer cell lines for the same viral vector (Green Fluorescent Protein) yielding titers varying from 10^5 to 10^7 TU/mL [10]. Of those, only one grew in suspension in serum-free condition, which is the most desirable option for large-scale production [61].

The development of stable packaging or producer cell lines is a major step toward the development of cost-effective and scalable production platforms. The development of molecular switches that are tighter in mammalian expression systems would allow for better control of cytotoxic viral genes and in turn increase the stability of packaging and producer cell lines. High-throughput automated cloning systems and improvements in methods and strategies for selecting high producing clones should allow, in a near future, the development of rLVV producer cell lines capable of producing high titers on a large scale for clinical application.

6. Modes of production: batch, fed-batch, perfusion

rLVV can be produced using transient and or stable systems with HEK293 suspension cells. The production processes supporting each system can be developed with different cultivation modes, batch, fed-batch or perfusion and adapted for each system to consolidate a production platform (**Figure 3**). The decision to produce using one mode or another rests with numerous factors linked to the viral vector characteristics and its corresponding expression system, the targeted titers and batch size needed, downstream processing needs and the availability of appropriate consumables and equipment. Scaling up a process using equipment from different manufacturers with specific characteristics and configuration is not straightforward and will impact the process development and timelines.

6.1 Batch mode

The batch mode is a simple and rapid production process initiated with bioreactor seeding. Depending on the seed train strategy, the bioreactor can be seeded at full bioreactor working volume, or at 25–50% of the final process volume. The seed train strategy is defined by the amplification of cells to obtain a defined volume and cell density to inoculate a bioreactor that is aligned with the targeted final cell density, working volume and production schedule. Under some processes, cells are diluted to the final processing volume after seeding. This simulates cell growth up to the density where the production is initiated either by transfection for HEK293 and packaging cells or by induction of stable producers. The addition of histone deacetylase inhibitors, sodium butyrate or valproic acid on the following day has shown to significantly increase

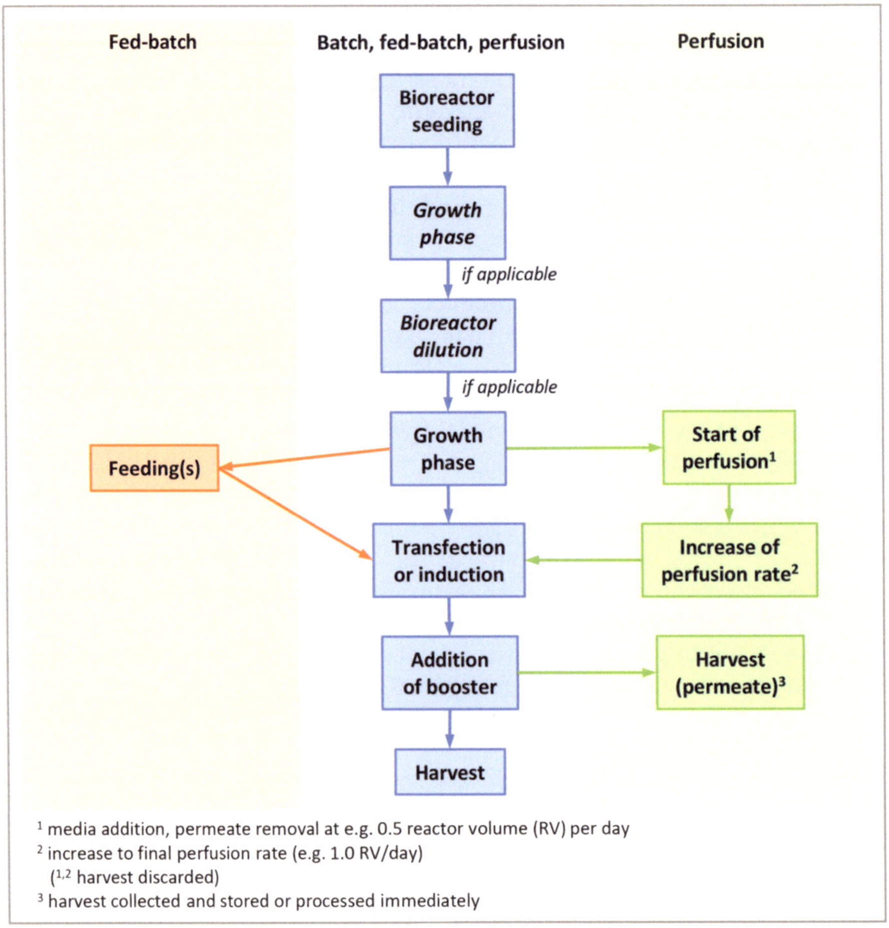

Figure 3.
Process steps for batch, fed-batch and perfusion processes for the production of viral vectors.

titers [64]. Typically, the harvest takes place when maximum titers are obtained, while at the same time aiming for high cell viability to minimize cell debris and other contaminants that reduce yield at subsequent clarification and purification steps.

6.2 Fed-batch mode

A process developed in a fed-batch mode will, for the most part, follow the same steps as a batch mode and be complemented with a feed regimen. The feed regimen consists of one or more additions of feed at a given time and intervals. Feeds are concentrated cocktails of nutrients, and commercially available feeds have been specifically designed for most producer cell lines, including HEK293 cells. The feeds support cell growth and production and help prevent rapid nutrient depletion. For each cell line, several feeds are available and the selection of the best feed and regimen for each process is determined in small-scale studies. Best results usually occur when feeding during the growth phase, but can be done any time during the process to maximize titers. Fed-batch processes typically result in higher cell densities and may also improve the cell's specific productivity and overall volumetric titers. However, the impact of the presence of feed in the harvest on downstream processing and final yield must be evaluated before scaling up the process. Both batch and fed-batch modes are well adapted for the production of viral vectors by plasmid transfection as well as stable producers.

6.3 Perfusion mode

In a production using perfusion mode, a flow system with an in-line separation device is interconnected between the bioreactor and a reservoir for harvesting cell-free spent media pumped out of the bioreactor. The working volume is maintained throughout the process by pumping in equal volumes of fresh media. The separation device is designed to keep the cells in the bioreactor and let through the smaller viral vector particles. The perfusion or media exchange is initiated during the growth phase of the cells, when the final process volume is attained and potentially after dilution. The renewal of media during a perfusion process allows for a constant supply of fresh nutrients, feed and inducers if needed and the removal of metabolites that may become cytotoxic when accumulating. The media exchange rate can vary from 1 to 3 vessel volumes per day (vvd), depending on the expression system.

A perfusion process will allow to increase the cell density by several fold in comparison to a batch or fed-batch process. Depending on the perfusion device and the presence or absence of cell-free media, the viral vectors in the harvest can be purified in batches. This is desirable for unstable vectors such as rLVVs. Stable producer cell-lines are the preferred expression systems to couple to a perfusion mode. In addition, continuous bioprocessing can be implemented by directly feeding clarified permeate into the DSP flow path in a setting where refrigerated storage capacity for the collected permeate may be reduced or even eliminated. Especially in the later scenario, perfusion systems enable the reduction of the production scale and equipment footprint.

There are several types of perfusion systems and retention devices commercially available with different functional designs, including alternating tangential flow filtration (ATF), tangential flow filtration (TFF) and acoustic devices such as the BioSep. ATF and TFF are filter-based devices composed of different types of fibers and variable pore sizes that need to be carefully selected to ensure the membrane is permeable to the viral vector produced [65].

The ATF system (**Figure 4, C-1**) (XCell® ATF System, Repligen) consists of a dip tube, a hollow-fiber module and a diaphragm pump. Culture flows from the bioreactor through the dip tube and hollow-fiber filter when a vacuum is applied to the diaphragm pump. To reverse the direction of the flow, a pressure is applied. The alternating diaphragm movement creates a low shear flow for the cell suspension as well as a backflushing effect across the filter surface of the hollow-fiber filter, in contrast to traditional TFF methods. Filter fouling and retention of the product of interest are thereby reduced, increasing the lifespan of the filter and allowing for longer perfusion processes. The filtered media or permeate is collected in a refrigerated container using a peristaltic pump. Hollow-fiber modules made from different materials and with a range of pore sizes are available from several manufacturers. In general, permeates from ATF systems are cell-free and contain fewer cell debris. However, some devices are subject to fouling depending on the bioprocess parameters (cell density, viability, DNA, feed and protein content, etc.) and process parameters (backflush, permeate flow rates). ATF devices and filters are currently available for bioreactor sizes of up

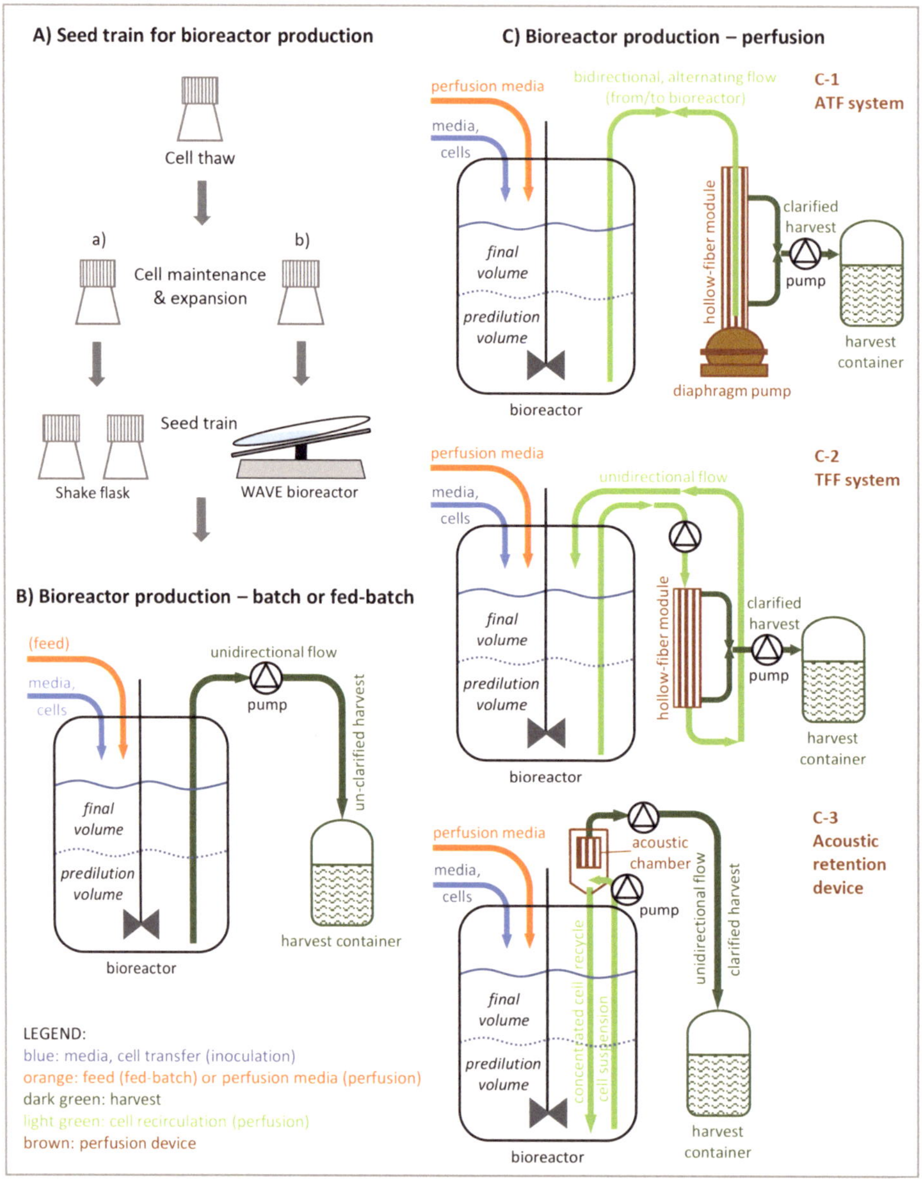

Figure 4.
Seed train (A) and process set-ups (B and C) for viral vector bioreactor production. (A) Cells are thawed and maintained in shake flask, amplified in (a) shake flasks for small-scale bioreactor production (e.g., up to 10 L production volume) or in (b) wave bioreactor to constitute the seed train. (B) Bioreactor set-up for batch or fed-batch production ("feed" is added during the fed-batch process). (C) Bioreactor set-up for perfusion production (C-1: ATF, alternating tangential flow; C-2: TFF, tangential flow filtration; C-3: acoustic retention device). A higher media exchange or perfusion rate in general results in higher sustained cell density and faster removal of product, while a lower vessels volume per day (vvd) causes less strain on filter-based retention devices such as (C-1) or (C2).

to 1000 L (**Table 1**). ATF perfusion technology has been successfully used for viral vectors such as adenovirus, rAAV and rLVVs [71–73].

TFF perfusion systems (**Figure 4, C-2**) are available from different suppliers. With this perfusion technology, the cell culture is pumped through a tubular filter in tangential flow mode, driven by a low-shear peristaltic pump. Most of the feed,

Technology	Example	Bioreactor scale	Pore size/MWCO[1]	References
ATF[2]	XCell® ATF System, Repligen	0.5–1000 L	0.2–0.5 μm/50 kDa	[66]
TFDF[3]	KrosFlo® TFDF® System, Repligen	1–2000 L	2–5 μm	[67]
TFF[4]	Xcellerex™ APS, Cytiva	50–500 L	0.1–0.45 μm 10–750 kDa	[68]
TFF[4]	Cellicon® cell retention solution, Millipore Sigma	3–2000 L	5–15 μm	[69]
ATF[2] or TFF[4]	VHU® System, Artemis BIOSYSTEMS	0.5–1000 L	5–15 μm	[70]

[1]*MWCO: molecular weight cut-off.*
[2]*ATF: alternate tangential flow filtration.*
[3]*TFDF: tangential flow depth filtration.*
[4]*TFF: tangential flow filtration.*

Table 1.
ATF and TFF perfusion technologies.

including the cells, is redirected back to the bioreactor. Part of the clarified culture and components smaller than the pore size of the filter pass through as permeate and are pumped into a refrigerated harvest container. For viral vectors with 20–100 nm diameters, TFF filters with an effective pore size rating of 2–5 μm are recommended for filtering out cells and most cell debris. Repligen has developed a TFDF perfusion device that combines the principles of tangential flow and depth filtration while following the same technical principles as a standard TFF perfusion process. TFDF perfusion has been successfully used for AAV and LV processes [65, 72, 74]. In general, the higher pressure in TFF systems is more likely to cause shear stress than ATF perfusion and is subject to filter fouling [72]. The TFF perfusion technology is currently available at up to 2000 L bioreactor scale (**Table 1**).

An acoustic device (**Figure 4, C-3**), such as the Applikon BioSep (Applikon) uses ultrasonic waves to separate cells from culture medium in a resonator chamber. The chamber is composed of two opposed glass surfaces. When specific frequencies are selected, an acoustic standing field is generated between the glass walls. Culture that is pumped from the bioreactor into the chamber forms loose cell aggregates in the acoustic field, which then sediment out of the field, disaggregate and return to the bioreactor through a dip tube. A second pump removes harvest from the exit port of the resonator chamber. The combined flow of harvest and culture returned to the bioreactor is equivalent to the culture circulated from the bioreactor to the resonator chamber. While the harvest contains some dead cells and debris, an acoustic device is gentle and non-fouling and can operate continuously for up to thousands of hours. Acoustic perfusion technology has been described in the literature for different viral vectors [11, 28, 75, 76]. The Applikon BioSep is currently scalable from 0.1 to 1000 L/day) [77].

7. Scaling up the production of viral vector

For suspension cell platforms such as the HEK293 cells, many options are available for large-scale manufacturing. The stirred tank reactor (STR) remains the gold standard for manufacturing when large quantities of material for clinical trial and commercial production are needed. Other alternatives are available for suspension cell lines, such as rocking motion bioreactors [47] and orbitally shaken bioreactors [78] for viral vector production. Most suppliers only provide these alternatives in a volume range from 25 L to 100 L [79]; however their use at a volume of 200 L and beyond has been tested [80]. The production of both rLVVs and rAAVs by transfection is challenged by the limited scalability of transfection processes and COGS, especially for GMP quality plasmids.

7.1 Principles of cell culture scale-up

While stirred tank bioreactors are the preferred model for large-scale manufacturing, they do present unique scale-up challenges to maintain favorable oxygenation and mixing conditions that can support high viable cell densities. One key objective of scale-up strategies is to limit shear stress linked to hydrodynamic conditions to which HEK293 and other mammalian cells are sensitive. Several factors may influence the hydrodynamic conditions in a given stirred tank bioreactor: the dimension and geometry of the bioreactor, the operating cell culture volume, the gas flow rates (air and O_2) delivered by different types of spargers mixed in the liquid phase and the presence of baffles or other obstacles. Finally, the moving force of the impeller that is dependent on impeller size, type and setup is crucial. A bioreactor can be configured with one or more impellers positioned at variable distances from the bottom of the bioreactor and from each other. The impeller shaft can be positioned within the central axis or off-centered, or parallel to the side wall. Whether the impeller setup creates an upward or downward flow pattern will also affect the hydrodynamic conditions.

Formulas to calculate the agitation rate of the impeller have been developed and are used in scale-up processes (**Table 2**). A typical scale-up strategy consists in selecting one physical property that is kept constant across the different scales. For example, to maintain the volumetric power input and volume of gas per volume of liquid per minute (vvm) constant, the agitation and sparging rates are calculated from their respective equations using equipment dimensions from each scale. The drawback of this approach is that it is physically impossible to keep all the impact parameters constant during scale-up, and some must be prioritized. Bioreactor scale-down models are useful tools to generate large datasets to explore trends that link different

Parameter	Equation	
Volumetric power input P/V (W/m^3)	$\frac{N_p{}^* \rho^* N^{3*} D^5}{V}$	Eq. (1)
Tip speed (m/s)	$\pi^*\ N^*\ D$	Eq. (2)
Reynold number (–)	$\frac{\rho^* N^* D^2}{\mu}$	Eq. (3)
Volumetric oxygen mass transfer coefficient (k_La, h^{-1})	$A^* \left(\frac{P}{V}\right)^{\alpha *} (v_s)^{\beta}$	Eq. (4)

*N refers to the agitation rate (s^{-1}), D to the impeller diameter (m), ρ the density of the liquid (kg/m^3), V to the volume (m^3), μ to the viscosity of the liquid (Pa*s) and v_s to the superficial velocity of the gas into the vessel (m/s). N_p, the dimensionless power number of the impeller often referred as Newton number [81], is an impeller-specific and vessel-specific characteristic of the mixing system. A, α and β are coefficient that might be either estimated from values in the literature [82] or estimated experimentally for each vessel for more accurate results.*

Table 2.
Scale-up criteria equations for agitation rate in reusable bioreactors.

operation conditions and physical constraints to the cell culture performance and productivity [83, 84]. Based on the data generated at smaller scale, a design space for large-scale manufacturing can be established, where all the relevant physical parameters are within acceptable boundaries for cell growth and viral vector production.

The power input per unit of volume (P/V) (W/m^3) is a very common parameter to scale-up agitation rate, as it allows keeping the quantity of energy transferred to the cell culture *via* the impeller constant (per unit of volume) during the scale-up. HEK293 suspension cell culture requires volumetric power input values between 13 and 44 W/m^3 [85–89]. HEK293 cells in small-scale benchtop vessels have been shown to tolerate specific power input of much greater intensity, up to 451 W/m^3 in [90]. Using specific power input for scale-up also correlates well to the oxygenation rate. It has been demonstrated that the coefficient of volumetric oxygen transfer is strongly correlated to the power input volume *via* the Van't Riet correlation [91]. Indeed, increasing agitation tends to break larger bubbles into smaller ones, thus increasing the surface available for oxygen transfer.

The tip speed (m/s, **Table 2**, Eq. (2)) is a parameter linking shear stress applied to the cells and agitation rate. Tip speed of ~1–2 m/s is generally acceptable, although there is no consensus on how to interpret this parameter [92]. Reynold number (Re, **Table 2**, Eq. (3)) is a parameter linked to the agitation rate that represents the ratio of the momentum force of the flow compared to the viscous shear forces. While not used directly for agitation scale-up calculations, the evaluation of the Reynold number allows to establish the flow regime of the bioreactor [82]. A Re value greater than 10,000 corresponds to a turbulent flow regimen, which is what is typically observed in a large-scale bioreactor.

The agitation in the bioreactor has for a long time been considered as the main contributor to shear stress leading to cell damage. However, bubble fragmentation in bioreactors was more recently identified as a highly probable contributor to shear stress [93]. Bubble fragmentation is induced by a combination of impeller velocity and sparging rate. Importantly, very high specific energy dissipation rates are associated with every step of the gas sparging, from bubble formation to bubble fragmentation at the surface. While some bubble fragmentation seems to be well tolerated by robust CHO cells, HEK293 cells could be more susceptible to this phenomenon. The sensitivity of HEK293 and CHO cells to bubble fragmentation was compared using a small-scale perfusion process. The study concluded that the agitation and sparging conditions were harmful to HEK293 but not CHO cells [94].

Pitch blade/elephant ear impellers [92] are commonly used for mammalian cell culture applications. This type of impeller allows for a good compromise between radial and axial pumping, thus allowing for both sufficient oxygen transfer by effective bubble breaking and dispersion, and bulk mixing to maintain a homogeneous mixture. In addition to shear stress, the agitation rate may influence the tendency of suspension adapted HEK293 cells to aggregate [90, 95]. The cell aggregation can affect the growth properties and productivity of the cells and is very difficult to control at larger scales. Sparging is critical to monitor and control during process scale-up. Over- or under-sparging will induce either foaming and cell damage, or oxygen limitations and CO_2 accumulation respectively. One way to fine-tune the bioreactor's oxygenation consists of using a combination of gases (nitrogen, air, oxygen). Air sparging supplemented by oxygen on-demand is the preferred approach used as a scale-up strategy.

The oxygenation rate for a specific bioreactor is linked to fraction of oxygen in the gas composition and the volumetric mass transfer coefficient k_La (**Table 2**, Eq. (4)). The k_La value depends on cell culture medium properties and the surface area available for oxygen transfer [96]. Oxygen transfer from the bubble to the liquid is a limiting factor because the solubility of oxygen in aqueous solutions is low, particularly at 37°C. In bioreactor operation, the most practical way to increase oxygenation is to increase the surface area of the bubbles by increasing the sparging rate and/or reducing the size of the bubbles. Different sparger types will have different pore sizes as well, resulting in variable bubble sizes that will also influence the oxygenation rate greatly. The characterization of the stirred tank bioreactor oxygen transfer coefficient k_La is an important criterion for selecting and operating a bioreactor. The gassing-out method based on removing dissolved oxygen by gassing nitrogen is recommended for k_La determination [81]. Bioreactor suppliers will often provide specific values and guidelines for their equipment. A k_La value for oxygen in the order of 1–10 h^{-1} should cover most standard cell culture applications for viral vector production [97].

Lastly, CO_2 is also as a potential limiting factor for high cell density culture when scaling up. Larger bioreactors tend to accumulate CO_2 at a higher rate than smaller bioreactors because of the difference in mass transfer mechanism between oxygen supplementation and CO_2 removal [98–100]. It has rarely been reported as an issue in scaling up viral vector bioprocesses, likely because viral vectors are still produced at low cell densities. Still, the type of sparger and the sparging rate scale-up strategy must be tested and selected with care for a specific cell line to obtain a good balance between oxygenation, CO_2 removal and shear stress applied to the cells.

7.2 Large-scale production in reusable bioreactors

Reusable bioreactors made of glass and stainless steel such as clean in place (CIP) steam in place (SIP) stirred tank bioreactors have a strong track record for the production of biologics at various manufacturing scales. These bioreactors have been used for decades for monoclonal antibody production up to the 25,000 L scale [101]. Their scalability, while still challenging, is very well documented and studied. They can be used from process intensification to scale-up development, from small-scale screening (0.25 L) to the commercial scale. Mixing of the bulk liquid phase is done by one or more impellers, and gassing is done *via* surface aeration and sparging at the bottom of the vessel. Gas sparging with oxygen-enriched air ensures appropriate dissolved oxygen (DO) levels. CO_2 sparging is used to lower pH value to physiological levels. Both dissolved oxygen and pH are monitored continuously during the production. Electrochemical glass sensors for pH monitoring and amperometric Clark-type sensors for dissolved oxygen levels are commonly used [102]. Both types are available in steam-in-place, clean-in-place versions, suitable for repeated use in bioreactors.

7.3 Single-use bioreactors: configuration and costs of goods matters

One of the heavy trends in the cell culture industry over the last decade is the adoption of single-use bioreactors (SUBs) to replace reusable stainless-steel ones. Benchtop bioreactors up to 50 L are typically made of rigid polymers while pilot scale bioreactors of 50 L–6000 L are inflatable bags that can be inserted in temperature-controlled jackets. Both bench top and pilot scale bioreactors are fitted with their own pre-sterilized impeller, sparger, inlets, outlets and probes [103]. SUBs do not pose any risks of carry-over contamination from one batch to another, or from one product to another between different batches or campaigns, respectively [104].

When the rLVV productivity is similar in suspension *vs*. adherent HEK293 cells, large scale production in STR is usually more cost-effective. Comisel et al. [105]

compared the cost of lentivirus production in packed-bed versus stirred tank bioreactors and concluded that the single-use bioreactor was the most cost-effective means of culturing cells for rLVV production at a scale of 200 L with suspension cells, compared with cell factories or packed-bed bioreactors and adherent cells. Using a suspension cell line for rLVV in a stirred tank bioreactor reduced 95% of the COGS when comparing STR and cell factory, or 17% when comparing STR and packed bed reactor. While the difference is less significant in earlier clinical trials, it does become more and more significant according to their finding as the scale of production increases during the product development process cycle.

Because of these characteristics, SUBs are broadly used for clinical trials, since they do not pose any carry-over risks between campaigns. In addition, their size (up to 2000–5000 L depending on the manufacturer) is perfectly suitable for most current and future viral vector applications. While an attractive solution for new viral vector products, single-use bioreactors do have some limitations. There are some concerns over the potential risks with leachable and extractable compounds from the single-use plastic surfaces [104]. These compounds can both affect the growth of the cell lines and contaminate the final product. Another concern is the lack of uniformity between the different single-use product manufacturers and the reliance of the customer toward its supplier for single-use products [106]. For bioreactors, each supplier has its own exclusive single-use bioreactor design that fits with the equipment, with its geometric characteristics often unique to the system. Unlike traditional stainless-steel vessels with well-established designs, single-use bioreactors come in a variety of designs affecting impeller position and angle, as well as the shape

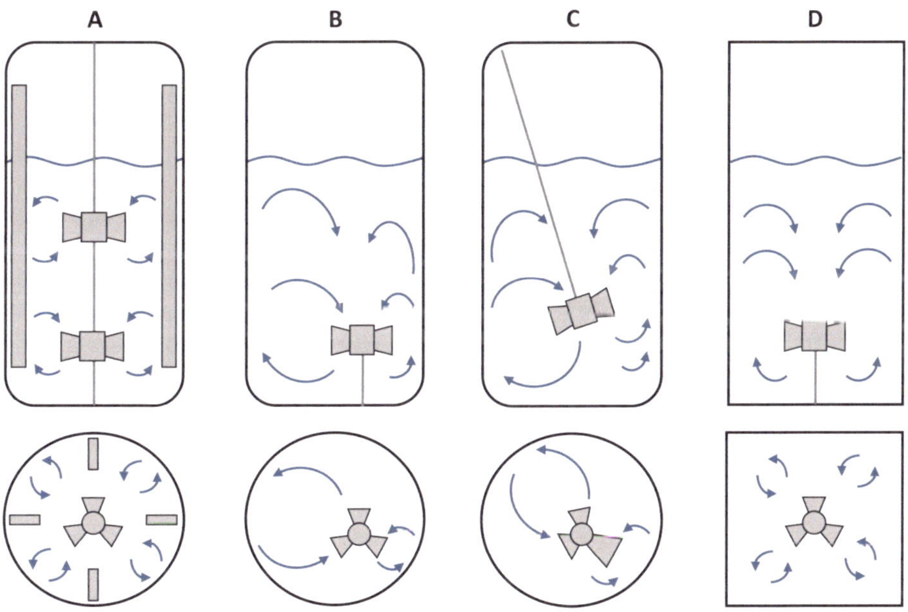

Figure 5.
Bioreactor design for single use bioreactors (SUBs). Comparison of bioreactor design for single-use bioreactors compared with traditional, multi-impeller baffled STR (System A). First row represents the profile view and second row the top view. In order to compensate for the lack of baffles and to prevent vortex formation, asymmetric bioreactor design have been developed for single-use bioreactors, such as with an eccentric position (B), an offset with agitation shaft (C), or by adopting a cuboid design for the vessel instead of a cylindrical one (D). Arrows are an approximation of the movement of the impeller on the flow pattern (axial) for each design [107].

of the vessel itself. These designs have been adopted to increase mixing efficacy in the absence of baffles and multi-impeller agitation shaft (**Figure 5**). These systems increase complexity of scale-up and technical transfer due to geometric dissimilarities across different scale and limited availability of small-scale bioreactor models for scaling-down processes.

8. Conclusions

Decades of research and progress in the development of safer viral vector and biomanufacturing processes has allowed the commercialization of cell and gene therapies. Despite these major accomplishments, there are still numerous obstacles and remaining challenges to produce and commercialize cost-effective cell and gene therapies. There will always be a need for new and improved reagents and analytical methods that reduce the COGs of viral vector biomanufacturing, such as transfection reagents and media that support high viable cell densities. The increase in manufacturer's offering of SUBs with various configurations, cell culture media, and feeds with new formulations has increased complexity in the selection of equipment and consumables to develop optimized expression systems and processes. Further, the COVID pandemic has highlighted the unresolved challenges for low-cost, consistent, traceable and well-characterized raw material and consumables and emphasized the vulnerability of globalized supply chains.

The reduction of the dependency of rLVVs biomanufacturing to high COGs, GMP-grade plasmids and transfection reagents in particular is being addressed by the development of packaging and producer cell lines. One of the main engineering challenges is to tighten the expression systems that regulate on demand the expression of toxic proteins such as VSV-G used in rLVVs pseudotyped particles. Ideally, a packaging or producer cell line should be robust, stable and highly scalable and produce high titers to facilitate GMP compliant processes. While titers are often used to decide on the best clones early in cell line development, all the above criteria should be considered in the evaluation. As such, it is worth considering engineering or subcloning parental and packaging cells to select host cells for their robustness and scalability into which a packing or producer cell line would be developed.

Notwithstanding that rLVVs are from design to production at scale under continuous improvement, the relatively limited information available in the public domain for the biomanufacturing and scale-up of rLVVs, and the lack of reference material and platforms that can be used to benchmark process development, is slowing down progress. The biomanufacturing of rLVVs is further challenged by the fact that it is a labile vector that needs to be processed with relative care. The nature of lentiviral vectors makes them compatible with the development of stable producers, perfusion processes and continuous manufacturing. However, the need for stable producers and complex biomanufacturing processes can only be justified by the need for large quantities of rLVVs. The emergence of rLVVs as powerful vectors for vaccination could be the driver that will accelerate the development of large-scale perfusion processes and continuous manufacturing.

Batch and fed-batch production are more appropriate for cell and gene therapy applications for rare diseases, personalized medicine and production in SUBs where pilot scale productions may be needed. A better understanding of SUBs design space and hydrodynamic environment would benefit their use in various bioprocesses. This knowledge combined with the current scale-up principles should help define optimal operating conditions and configurations in bench top bioreactors and develop reliable and predictive scale-down models for rLVVs production that are representative of manufacturing scale bioreactors. The potential of rLVVs to address unmet medical needs is very promising. To support this, access to more robust cell lines for transient and stable production, improved transfection methods and better scale-up tools and models will contribute to accelerate the development of rLVVs and reduce the COGs for biomanufacturing.

Acknowledgements

The authors would like to thank the human health therapeutic research center of the national research council for supporting the publication of this book chapter. We would also like to thank Dr. Mauro Acchione for reviewing the text.

Conflict of interest

The authors declare no conflict of interest.

Author details

Julien Robitaille, Aziza Manceur, Anja Rodenbrock and Martin Loignon*
Bioprocess Engineering Department of the Human Health Therapeutic Research Center, National Research Council Canada, Montreal, Quebec, Canada

*Address all correspondence to: martin.loignon@cnrc-nrc.gc.ca

References

[1] Bulcha JT, Wang Y, Ma H, Tai PWL, Gao G. Viral vector platforms within the gene therapy landscape. Signal Transduction and Targeted Therapy. 2021;**6**(1):53

[2] Von Seggern DJ, Huang S, Fleck SK, Stevenson SC, Nemerow GR. Adenovirus vector pseudotyping in fiber-expressing cell lines: Improved transduction of Epstein-Barr virus-transformed B cells. Journal of Virology. 2000;**74**(1):354-362

[3] Shi W, Arnold GS, Bartlett JS. Insertional mutagenesis of the adeno-associated virus type 2 (AAV2) capsid gene and generation of AAV2 vectors targeted to alternative cell-surface receptors. Human Gene Therapy. 2001;**12**(14):1697-1711

[4] Cronin J, Zhang XY, Reiser J. Altering the tropism of lentiviral vectors through pseudotyping. Current Gene Therapy. 2005;**5**(4):387-398

[5] Gutierrez-Guerrero A, Cosset FL, Verhoeyen E. Lentiviral vector pseudotypes: Precious tools to improve gene modification of hematopoietic cells for research and gene therapy. Viruses. 2020;**12**(9):1016

[6] Finkelshtein D, Werman A, Novick D, Barak S, Rubinstein M. LDL receptor and its family members serve as the cellular receptors for vesicular stomatitis virus. Proceedings of the National Academy of Sciences of the United States of America. 2013;**110**(18):7306-7311

[7] Nikolic J, Belot L, Raux H, Legrand P, Gaudin Y, A AA. Structural basis for the recognition of LDL-receptor family members by VSV glycoprotein. Nature Communications. 2018;**9**(1):1029

[8] Ferreira MV, Cabral ET, Coroadinha AS. Progress and perspectives in the development of lentiviral vector producer cells. Biotechnology Journal. 2021;**16**(1):e2000017

[9] Li M, Husic N, Lin Y, Christensen H, Malik I, McIver S, et al. Optimal promoter usage for lentiviral vector-mediated transduction of cultured central nervous system cells. Journal of Neuroscience Methods. 2010;**189**(1):56-64

[10] Perry C, Rayat ACME. Lentiviral vector bioprocessing. Viruses. 2021;**13**(2):268

[11] Ansorge S, Lanthier S, Transfiguracion J, Durocher Y, Henry O, Kamen A. Development of a scalable process for high-yield lentiviral vector production by transient transfection of HEK293 suspension cultures. The Journal of Gene Medicine: A Cross-disciplinary Journal for Research on the Science of Gene Transfer and Its Clinical Applications. 2009;**11**(10):868-876

[12] Tomás HA, Rodrigues AF, Carrondo MJT, Coroadinha AS. LentiPro26: Novel stable cell lines for constitutive lentiviral vector production. Scientific Reports. 2018;**8**(1):5271

[13] Sanber KS, Knight SB, Stephen SL, Bailey R, Escors D, Minshull J, et al. Construction of stable packaging cell lines for clinical lentiviral vector production. Scientific Reports. 2015;**5**:9021

[14] Chen YJ, Abila B, Mostafa KY. CAR-T: What is next? Cancers. 2023;**15**(3):663

[15] Logan M, Rinas K, McConkey B, Aucoin MG. Vero cells gain renal tubule markers in low-calcium and magnesium chemically defined media. Scientific Reports. 2022;**12**(1):6180

[16] Shen CF, Guilbault C, Li X, Elahi SM, Ansorge S, Kamen A, et al. Development of suspension adapted Vero cell culture process technology for production of viral vaccines. Vaccine. 2019;**37**(47):6996-7002

[17] Cameau E, Pedregal A, Glover C. Cost modelling comparison of adherent multi-trays with suspension and fixed-bed bioreactors for the manufacturing of gene therapy products. Cell and Gene Therapy Insights. 2019;**5**(11):1663-1674

[18] Wang D, Tai PW, Gao G. Adeno-associated virus vector as a platform for gene therapy delivery. Nature Reviews Drug Discovery. 2019;**18**(5):358-378

[19] Lesch HP, Valonen P, Karhinen M. Evaluation of the single-use fixed-bed bioreactors in scalable virus production. Biotechnology Journal. 2021;**16**(1):e2000020

[20] Leinonen HM, Lepola S, Lipponen EM, Heikura T, Koponen T, Parker N, et al. Benchmarking of scale-X bioreactor system in lentiviral and adenoviral vector production. Human Gene Therapy. 2020;**31**(5-6):376-384

[21] Masri F, Cheeseman E, Ansorge S. Viral vector manufacturing: How to address current and future demands. Cell and Gene Therapy Insights. 2019;**5**:949-970

[22] Pezoa S. Process Optimization for Lentivirus Production in the iCELLis® Nano Using a Chemically Defined, Animal Component Free HEK293 Media. 2022. Available from: https://invitria.com/resources/process-optimization-for-lentivirus-production-in-the-icellis-nano-using-a-chemically-defined-animal-component-free-hek293-media/

[23] Tan E, Chin CSH, Lim ZFS, Ng SK. HEK293 cell line as a platform to produce recombinant proteins and viral vectors. Frontiers in Bioengineering and Biotechnology. 2021;**9**:796991

[24] Imler JL, Chartier C, Dreyer D, Dieterle A, Sainte-Marie M, Faure T, et al. Novel complementation cell lines derived from human lung carcinoma A549 cells support the growth of E1-deleted adenovirus vectors. Gene Therapy. 1996;**3**(1):75-84

[25] Côté J, Garnier A, Massie B, Kamen A. Serum-free production of recombinant proteins and adenoviral vectors by 293SF-3F6 cells. Biotechnology and Bioengineering. 1998;**59**(5):567-575

[26] Kamen A, Henry O. Development and optimization of an adenovirus production process. The Journal of Gene Medicine. 2004;**6**(S1):S184-SS92

[27] Chahal PS, Schulze E, Tran R, Montes J, Kamen AA. Production of adeno-associated virus (AAV) serotypes by transient transfection of HEK293 cell suspension cultures for gene delivery. Journal of Virological Methods. 2014;**196**:163-173

[28] Manceur AP, Kim H, Misic V, Andreev N, Dorion-Thibaudeau J, Lanthier S, et al. Scalable lentiviral vector production using stable HEK293SF producer cell lines. Human Gene Therapy Methods. 2017;**28**(6):330-339

[29] Gélinas J-F, Azizi H, Kiesslich S, Lanthier S, Perdersen J, Chahal PS, et al. Production of rVSV-ZEBOV in serum-free suspension culture of HEK 293SF cells. Vaccine. 2019;**37**(44):6624-6632

[30] Petiot E, Jacob D, Lanthier S, Lohr V, Ansorge S, Kamen AA. Metabolic and kinetic analyses of influenza production in perfusion HEK293 cell culture. BMC Biotechnology. 2011;**11**:84

[31] Xu P, Clark C, Ryder T, Sparks C, Zhou J, Wang M, et al. Characterization of TAP Ambr 250 disposable bioreactors, as a reliable scale-down model for biologics process development. Biotechnology Progress. 2017;**33**(2):478-489

[32] Janakiraman V, Kwiatkowski C, Kshirsagar R, Ryll T, Huang YM. Application of high-throughput mini-bioreactor system for systematic scale-down modeling, process characterization, and control strategy development. Biotechnology Progress. 2015;**31**(6):1623-1632

[33] Riethmüller D. Scalable upstream process development for the suspension-based production of lentiviral vectors for CAR T cell therapies with multiparallel & benchtop bioreactor systems & DoE methodology. Cell and Gene Therapy Insights. 2021;**7**(6):689-700

[34] Jordan M, Köhne C, Wurm FM. Calcium-phosphate mediated DNA transfer into HEK-293 cells in suspension: Control of physicochemical parameters allows transfection in stirred media. Transfection and protein expression in mammalian cells. Cytotechnology. 1998;**26**(1):39-47

[35] Feng L, Guo M, Zhang S, Chu J, Zhuang Y, Zhang S. Optimization of transfection mediated by calcium phosphate for plasmid rAAV-LacZ (recombinant adeno-associated virus–β-galactosidase reporter gene) production in suspension-cultured HEK-293 (human embryonic kidney 293) cells. Biotechnology and Applied Biochemistry. 2007;**46**(2):127-135

[36] Wright JF. Transient transfection methods for clinical adeno-associated viral vector production. Human Gene Therapy. 2009;**20**(7):698-706

[37] Girard P, Porte L, Berta T, Jordan M, Wurm FM. Calcium phosphate transfection optimization for serum-free suspension culture. Cytotechnology. 2001;**35**(3):175-180

[38] Martínez-Monge I, Martínez C, Decker M, Udugama IA, Marín de Mas I, Gernaey KV, et al. Soft-sensors application for automated feeding control in high-throughput mammalian cell cultures. Biotechnology and Bioengineering. 2022;**119**(4):1077-1090

[39] Reed SE, Staley EM, Mayginnes JP, Pintel DJ, Tullis GE. Transfection of mammalian cells using linear polyethylenimine is a simple and effective means of producing recombinant adeno-associated virus vectors. Journal of Virological Methods. 2006;**138**(1):85-98

[40] Guan J-S, Chen K, Si Y, Kim T, Zhou Z, Kim S, et al. Process improvement of adeno-associated virus production. Frontiers in Chemical Engineering. 2022;**4**. Available from: https://www.frontiersin.org/articles/10.3389/fceng.2022.830421

[41] Hashimoto H, Mizushima T, Chijiwa T, Nakamura M, Suemizu H. Efficient production of recombinant adeno-associated viral vector, serotype DJ/8, carrying the GFP gene. Virus Research. 2017;**238**:63-68

[42] Martínez-Molina E, Chocarro-Wrona C, Martínez-Moreno D, Marchal JA, Boulaiz H. Large-scale production of lentiviral vectors: Current perspectives and challenges. Pharmaceutics. 2020;**12**(11):1051

[43] Witting SR, Li LH, Jasti A, Allen C, Cornetta K, Brady J, et al. Efficient large volume lentiviral vector production using flow electroporation. Human Gene Therapy. 2012;**23**(2):243-249

[44] Merten OW, Hebben M, Bovolenta C. Production of lentiviral vectors. Molecular Therapy Methods & Clinical Development. 2016;**3**:16017

[45] Bauler M, Roberts JK, Wu CC, Fan B, Ferrara F, Yip BH, et al. Production of lentiviral vectors using suspension cells grown in serum-free media. Molecular Therapy Methods & Clinical Development. 2020;**17**:58-68

[46] Meade O, Clark J, McCutchen M, Kerwin J. Exploring the design space of AAV transient-transfection in suspension cell lines. Methods in Enzymology. 2021;**660**:341-360

[47] Grieger JC, Soltys SM, Samulski RJ. Production of recombinant adeno-associated virus vectors using suspension HEK293 cells and continuous harvest of vector from the culture media for GMP FIX and FLT1 clinical vector. Molecular Therapy: The Journal of the American Society of Gene Therapy. 2016;**24**(2):287-297

[48] Gutiérrez-Granados S, Cervera L, Kamen AA, Gòdia F. Advancements in mammalian cell transient gene expression (TGE) technology for accelerated production of biologics. Critical Reviews in Biotechnology. 2018;**38**(6):918-940

[49] Celebi Torabfam G, Yetisgin AA, Erdem C, Cayli A, Kutlu O, Cetinel S. A feasibility study of different commercially available serum-free mediums to enhance lentivirus and adeno-associated virus production in HEK 293 suspension cells. Cytotechnology. 2022;**74**(6):635-655

[50] González-Domínguez I, Puente-Massaguer E, Lavado-García J, Cervera L, Gòdia F. Micrometric DNA/PEI polyplexes correlate with higher transient gene expression yields in HEK 293 cells. New Biotechnology. 2022;**68**:87-96

[51] Choosakoonkriang S, Lobo BA, Koe GS, Koe JG, Middaugh CR. Biophysical characterization of PEI/DNA complexes. Journal of Pharmaceutical Sciences. 2003;**92**(8):1710-1722

[52] Han X, Fang Q, Yao F, Wang X, Wang J, Yang S, et al. The heterogeneous nature of polyethylenimine-DNA complex formation affects transient gene expression. Cytotechnology. 2009;**60**(1-3):63

[53] Durocher Y, Pham PL, St-Laurent G, Jacob D, Cass B, Chahal P, et al. Scalable serum-free production of recombinant adeno-associated virus type 2 by transfection of 293 suspension cells. Journal of Virological Methods. 2007;**144**(1):32-40

[54] Zhao H, Lee KJ, Daris M, Lin Y, Wolfe T, Sheng J, et al. Creation of a high-yield AAV vector production platform in suspension cells using a design-of-experiment approach. Molecular Therapy Methods & Clinical Development. 2020;**18**:312-320

[55] Pezzoli D, Giupponi E, Mantovani D, Candiani G. Size matters for in vitro gene delivery: Investigating the relationships among complexation protocol, transfection medium, size and sedimentation. Scientific Reports. 2017;**7**:44134

[56] Sang Y, Xie K, Mu Y, Lei Y, Zhang B, Xiong S, et al. Salt ions and related parameters affect PEI-DNA particle size and transfection efficiency in Chinese hamster ovary cells. Cytotechnology. 2015;**67**(1):67-74

[57] Dekevic G, Tasto L, Czermak P, Salzig D. Statistical experimental designs to optimize the transient transfection of HEK 293T cells and determine a transfer criterion from adherent cells to larger-scale cell suspension cultures. Journal of Biotechnology. 2022;**346**:23-34

[58] Bono N, Ponti F, Mantovani D, Candiani G. Non-viral in vitro gene delivery: It is now time to set the Bar! Pharmaceutics. 2020;**12**(2):183

[59] Lavado-García J, Pérez-Rubio P, Cervera L, Gòdia F. The cell density effect in animal cell-based bioprocessing: Questions, insights and perspectives. Biotechnology Advances. 2022;**60**:108017

[60] Abaandou L, Quan D, Shiloach J. Affecting HEK293 cell growth and production performance by modifying the expression of specific genes. Cell. 2021;**10**(7):1667

[61] Broussau S, Jabbour N, Lachapelle G, Durocher Y, Tom R, Transfiguracion J, et al. Inducible packaging cells for large-scale production of lentiviral vectors in serum-free suspension culture. Molecular Therapy: The Journal of the American Society of Gene Therapy. 2008;**16**(3):500-507

[62] Broussau S, Lytvyn V, Simoneau M, Guilbault C, Leclerc M, Nazemi-Moghaddam N, et al. Packaging cells for lentiviral vectors generated using the cumate and coumermycin gene induction systems and nanowell single-cell cloning. Molecular Therapy Methods & Clinical Development. 2023;**29**:40-57

[63] Comisel R-M, Kara B, Fiesser FH, Farid SS. Gene therapy process change evaluation framework: Transient transfection and stable producer cell line comparison. Biochemical Engineering Journal. 2021;**176**:108202

[64] Pelascini LP, Janssen JM, Gonçalves MA. Histone deacetylase inhibition activates transgene expression from integration-defective lentiviral vectors in dividing and non-dividing cells. Human Gene Therapy. 2013;**24**(1):78-96

[65] Tran MY, Kamen AA. Production of lentiviral vectors using a HEK-293 producer cell line and advanced perfusion processing. Frontiers in Bioengineering and Biotechnology. 2022;**10**:887716

[66] Repligen. XCell ATF® Device Size Selection Guide. 2023. Available from: https://www.repligen.com/products/upstream-filtration/xcell-atf/xcell-atf-devices-and-controllers

[67] Repligen. KrosFlo® TFDF® Systems. 2023. Available from: https://www.repligen.com/products/upstream-filtration/krosflo-tfdf

[68] Sciences CL. Xcellerex™ Automated Perfusion System. 2023. Available from: https://www.cytivalifesciences.com/en/us/shop/cell-culture-and-fermentation/perfusion-systems/xcellerex-automated-perfusion-system-p-24623?

[69] Millipore-Sigma. Cellicon™ Perfusion Solution. 2023. Available from: https://www.emdmillipore.com/CA/en/product/Cellicon-Perfusion-Solution,MM_NF-C223762#specifications

[70] Biosystems A. VHU® Module. 2022. Available from: https://www.artemisbiosystems.com/wp-content/uploads/2022/01/VHU-Module-Datasheet-2022-01.pdf

[71] Sun Y, Huang L, Nie J, Feng K, Liu Y, Bai Z. Development of a perfusion process for serum-free adenovirus vector

herpes zoster vaccine production. AMB Express. 2022;**12**(1):1-11

[72] Mendes JP, Fernandes B, Pineda E, Kudugunti S, Bransby M, Gantier R, et al. AAV process intensification by perfusion bioreaction and integrated clarification. Frontiers in Bioengineering and Biotechnology. 2022;**10**:1020174

[73] Ansorge S, Cattaneo M, Rodenbrock A, Dorion-Thibaudeau J, Tremblay S, Manceur A, et al., editors. Towards Continuous Bioprocessing of Lentiviral Vectors. Integrated Continuous Biomanufacturing IV, ECI Symposium Series. Massachusetts, USA: Brewster (Cape Cod); 2019

[74] Tona RM, Shah R, Middaugh K, Steve J, Marques J, Roszell BR, et al. Process intensification for lentiviral vector manufacturing using tangential flow depth filtration. Molecular Therapy-Methods & Clinical Development. 2023;**29**:93-107

[75] Ghani K, Garnier A, Coelho H, Transfiguracion J, Trudel P, Kamen A. Retroviral vector production using suspension-adapted 293GPG cells in a 3L acoustic filter-based perfusion bioreactor. Biotechnology and Bioengineering. 2006;**95**(4):653-660

[76] Henry O, Dormond E, Perrier M, Kamen A. Insights into adenoviral vector production kinetics in acoustic filter-based perfusion cultures. Biotechnology and Bioengineering. 2004;**86**(7):765-774

[77] Getinge. Applikon BioSep. 2023. Available from: https://www.getinge.com/int/products/applikon-biosep/

[78] Blessing D, Vachey G, Pythoud C, Rey M, Padrun V, Wurm FM, et al. Scalable production of AAV vectors in orbitally shaken HEK293 cells. Molecular Therapy-Methods & Clinical Development. 2019;**13**:14-26

[79] Scheibenbogen K, Mikhed Y, McLaughlin K. Flexibility in Next Generation Facilities through Rocking Motion Bioreactors. 2022. Available from: https://www.sartorius.com/resource/blob/1256134/e0764443a93a591e150a6c1ca554a46d/rocking-motion-bioreactors-white-paper-en-b-sartorius-pdf-data.pdf

[80] Werner S, Hacker D, Baldi L, Tanner R, Kühner M, Broccard G, et al. Innovative, non-stirred bioreactors in scales from milliliters up to 1000 liters for suspension cultures of cells using disposable bags and containers—A Swiss contribution. Chimia. 2010;**64**(11):819

[81] Bauer I, Dreher T, Eibl D, Glöckler R, Husemann U, John G, et al. Recommendations for Process Engineering Characterisation of Single-use Bioreactors and Mixing Systems by Using Experimental Methods. Frankfurt am Main: DECHEMA Gesellschaft für Chemische Technik und Biotechnologie eV; 2020

[82] Nienow AW. Mass transfer and mixing across the scales in animal cell culture. In: Al-Rubeai M, editor. Animal Cell Culture. Cham: Springer International Publishing; 2015. pp. 137-167

[83] Dreher T, Husemann U, Adams T, de Wilde D, Greller G. Design space definition for a stirred single-use bioreactor family from 50 to 2000 L scale. Engineering in Life Sciences. 2014;**14**(3):304-310

[84] Tescione L, Lambropoulos J, Paranandi MR, Makagiansar H, Ryll T. Application of bioreactor design principles and multivariate analysis for development of cell culture scale down models. Biotechnology and Bioengineering. 2015;**112**(1):84-97

[85] Sanderson T, Erlandson T, Hazi N, MacIntyre A, Ingersoll BI, McLaughlin M, et al. Scalability comparison between 50 and 500 liter stirred tank bioreactor for production of rAAV viral vector. Cell and Gene Therapy Insights. 2021;**7**(9):1025-1033

[86] Arena TA, Chou B, Harms PD, Wong AW. An anti-apoptotic HEK293 cell line provides a robust and high titer platform for transient protein expression in bioreactors. MAbs. 2019;**11**(5):977-986

[87] Patel S, Fong E, George H. Considerations for Bioreactor Process Development and Scale-Up for Transient Transfection-Based Lentivirus Production in Suspension. 2020. Available from: https://www.emdmillipore.com/Web-US-Site/en_CA/-/USD/ShowDocument-Pronet?id=202008.048

[88] Chen P, Demirji J, Ivleva VB, Horwitz J, Schwartz R, Arnold F. The transient expression of CHIKV VLP in large stirred tank bioreactors. Cytotechnology. 2019;**71**(6):1079-1093

[89] Müller M, Lang C, Schmidberger T, Widmark H, Nascimento-Brooks L, Kahlig A. Next Level Scale-Up: How DOE and MVDA Improved Scale-Up Performance at a CDMO. 2021. Available from: https://www.sartorius.com/download/1156016/mvda-white-paper-en-b-sartorius-pdf-data.pdf

[90] Seidel S, Maschke RW, Mozaffari F, Eibl-Schindler R, Eibl D. Improvement of HEK293 cell growth by adapting hydrodynamic stress and predicting cell aggregate size distribution. Bioengineering. 2023;**10**(4):478

[91] Van't Riet K. Review of measuring methods and results in nonviscous gas-liquid mass transfer in stirred vessels. Industrial & Engineering Chemistry Process Design and Development. 1979;**18**(3):357-364

[92] Nienow AW. Reactor engineering in large scale animal cell culture. Cytotechnology. 2006;**50**(1-3):9

[93] Chalmers JJ. Mixing, aeration and cell damage, 30+ years later: What we learned, how it affected the cell culture industry and what we would like to know more about. Current Opinion in Chemical Engineering. 2015;**10**:94-102

[94] Schwarz H, Zhang Y, Zhan C, Malm M, Field R, Turner R, et al. Small-scale bioreactor supports high density HEK293 cell perfusion culture for the production of recombinant erythropoietin. Journal of Biotechnology. 2020;**309**:44-52

[95] Liu H, Liu XM, Wu BC, Ye LL, Ni XP, Huang PT, et al. Effects of hydrodynamic on aggregates formation, growth and metabolism of HEK293 cells in suspension culture. Sheng wu gong cheng xue bao = Chinese Journal of Biotechnology. 2006;**22**(1):101-106

[96] Nienow AW, editor Mass Transfer and Mixing Across the Scales in Animal Cell Culture. In: Al-Rubeai M, editor. Switzerland: Springer International Publishing; 2015

[97] Sharma R, Harrison STL, Tai SL. Advances in bioreactor systems for the production of biologicals in mammalian cells. ChemBioEng Reviews. 2022;**9**(1):42-62

[98] Matsunaga N, Kano K, Maki Y, Dobashi T. Culture scale-up studies as seen from the viewpoint of oxygen supply and dissolved carbon dioxide stripping. Journal of Bioscience and Bioengineering. 2009;**107**(4):412-418

[99] Sieblist C, Hägeholz O, Aehle M, Jenzsch M, Pohlscheidt M, Lübbert A.

Insights into large-scale cell-culture reactors: II. Gas-phase mixing and CO_2 stripping. Biotechnology Journal. 2011;**6**(12):1547-1556

[100] Xu S, Hoshan L, Jiang R, Gupta B, Brodean E, O'Neill K, et al. A practical approach in bioreactor scale-up and process transfer using a combination of constant P/V and vvm as the criterion. Biotechnology Progress. 2017;**33**(4):1146-1159

[101] Kelley B. Very large scale monoclonal antibody purification: The case for conventional unit operations. Biotechnology Progress. 2007;**23**(5):995-1008

[102] O'Mara P, Farrell A, Bones J, Twomey K. Staying alive! Sensors used for monitoring cell health in bioreactors. Talanta. 2018;**176**:130-139

[103] Frank GT. Transformation of biomanufacturing by single-use systems and technology. Current Opinion in Chemical Engineering. 2018;**22**:62-70

[104] Shukla AA, Gottschalk U. Single-use disposable technologies for biopharmaceutical manufacturing. Trends in Biotechnology. 2013;**31**(3):147-154

[105] Comisel R-M, Kara B, Fiesser FH, Farid SS. Lentiviral vector bioprocess economics for cell and gene therapy commercialization. Biochemical Engineering Journal. 2021;**167**:107868

[106] Lopes AG. Single-use in the biopharmaceutical industry: A review of current technology impact, challenges and limitations. Food and Bioproducts Processing. 2015;**93**:98-114

[107] Couper JR, Penney WR, Fair JR, Walas SM. 10 - mixing and agitation. In: Couper JR, Penney WR, Fair JR, Walas SM, editors. Chemical Process Equipment. 3rd ed. Boston: Butterworth-Heinemann; 2012. pp. 277-327

Section 6

Technological Implications of Cell Culture-Derived Brain Organoids in Biocomputing

Chapter 11

Organoid Intelligence: Bridging Artificial Intelligence for Biological Computing and Neurological Insights

Sangeeta Ballav, Amit Ranjan, Subhayan Sur and Soumya Basu

Abstract

Brain organoid implications have opened vast avenues in the realm of interdisciplinary research, particularly in the growing field of organoid intelligence (OI). A brain organoid is a three-dimensional (3D), lab-grown structure that mimics certain aspects of the human brain organization and function. The integration of organoid technology with computational methods to enhance the understanding of organoid behavior and to predict their responses to various stimuli is known as OI. The ability of brain organoids to adapt and memorize, is a key area of exploration. OI encapsulates the confluence of breakthroughs in stem cell technology, bioengineering, and artificial intelligence (AI). This chapter delves deep into the myriad potentials of OI, encompassing an enhanced understanding of human cognitive functions, and achieving significant biological computational proficiencies. Such advancements stand to offer a unique complementarity to conventional computing methods. The implications of brain organoids in the OI sphere signify a transformative stride towards a more intricate grasp of the human brain and its multifaceted intricacies. The intersection of biology and machine learning is a rapidly evolving field that is reshaping our understanding of life and health. This convergence is driving advancements in numerous areas, including genomics, drug discovery, personalized medicine, and synthetic biology.

Keywords: organoid intelligence, brain organoid, artificial intelligence, machine learning, biocomputing, neurogenesis, neurodegeneration

1. Introduction

The human brain is a remarkably intricate and complex organ that remains one of the greatest mysteries in scientific exploration. Despite significant progress in neurobiology, many aspects of brain development and disorders remain elusive. Brain organoids represent a revolutionary advancement in neuroscience research that serves as an unparalleled tool to study the development of the human brain and provides insights into the fundamental processes underlying neurogenesis, cell migration, and axon guidance [1]. These three-dimensional (3D), self-organized mini-brains hold the potential to transform our understanding of the brain's complexity, providing insights into early neurodevelopment, disease modeling, and drug testing [2]. Traditional two-dimensional (2D) cell cultures and animal models have limitations in fully capturing the complexity of human brain diseases. In contrast, brain organoids provide a more accurate model of the human brain's cellular and molecular environment (**Table 1**) [3].

In general, organoids are 3D structures grown from stem cells that mimic the architecture and functionality of real organs. They have been used to model various human tissues including the brain, liver, kidney, and gut. Organoids provide a more accurate representation of *in vivo* conditions compared to traditional 2D cell cultures, making them valuable tools for studying organ development, disease progression, and drug responses [4]. The advent of human stem cell-derived brain organoids has revolutionized the field of neuroscience, offering a promising avenue to replicate critical molecular and cellular aspects of learning, memory, and possibly aspects of cognition *in vitro*.

The term "organoid intelligence" (OI) has been coined to describe this emerging field, aiming to expand the definition of biocomputing towards brain-directed OI computing. The concept of organoid intelligence (OI) aims to harness the innate

Characteristic	2D cell culture	3D brain organoid models	Organoid intelligence
Dimensionality	Flat, single layer of cells	3D structure mimicking tissue organization	3D structure integrated with AI components
Cell interactions	Limited cell-cell and cell-matrix interactions	• Complex cell-cell and cell-matrix interactions • More representative of *in vivo* conditions	Complex interactions with added AI-driven feedback and analysis
Complexity	Simple, often a single cell type	Multiple cell types, mimics tissue architecture	Combines biological complexity with computational analysis
Functionality	Basic cellular functions	Advanced functions like tissue-specific activities, formation of neural networks	Advanced functions with AI-driven enhancements, data analysis, and predictions
Applications	• Basic research • Drug screening	• Disease modeling • Drug testing • Developmental biology	• Advanced research • Real-time analysis • Predictive modeling
Limitations	• Does not mimic *in vivo* conditions well • Lacks tissue context	• Still does not fully represent whole organs • Limited lifespan	• Technological challenges • Ethical considerations
Cost	Relatively low	Higher due to complexity	Highest due to integration of advanced technologies
Ethical concerns	Few, mostly related to source of cells	Concerns about creating "conscious" tissue, source of cells	Amplified concerns about consciousness, AI ethics, and potential misuse

Table 1.
Comparison of the characteristics of 2D cell culture, 3D brain organoid models, and organoid intelligence.

biological capabilities of brain organoids for biocomputing and synthetic intelligence by interfacing them with computer technology. The ability of brain organoids to adapt and memorize, is a key area of exploration. By leveraging the organoid's brain-like functionality, we can harness its capacity to process complex inputs, study the physiology of learning, and generate responses to control peripheral output devices (**Table 1**). This exploration of OI is a multidisciplinary endeavor that necessitates the engagement of both the scientific community and the public [5].

Biological computing refers to harnessing the power of biologically derived molecules to perform storage, retrieval, and processing. Human neurons are capable of incredible information processing to the microscopic size of their densely packed computational units with trillions of hyperdense synapses and high metabolic efficiency [6]. Moreover, these cells have the remarkable ability to regenerate and repair themselves continuously, similar to a computer system that operates without the need for external maintenance. In a study, an 8000-pound supercomputer claims to exceed the computational speed of the human adult brain which can only do so by consuming a million times more energy [2]. However, with new advancements in organoids, future machine learning could harness the computational efficiency and power of the brain. Biocomputers also have the potential to be more energy-efficient than traditional computers. Biological systems are capable of performing complex tasks with minimal energy consumption. In this aspect, the human brain can only perform complex cognitive tasks with a power consumption of around 20 watts, far less than that of a typical computer [7].

OI research aims to explore how a 3D brain cell culture can be made more computer-like. The many possible applications of this work include a new generation of biological and hybrid (biological-electronic) computing technologies. OI, thus, advances in the understanding of physiology of cognition, learning, and memory, and the pathophysiological effects of developmental and degenerative diseases, intoxication, and infection. OI also has the potential to unlock new neuromimetic artificial intelligence (AI) algorithms and aid the development of new brain-computer-interface technology. In this chapter, we have discussed the establishment and culture methods of brain organoids with clinical applications emphasizing their potential role in biocomputing and OI. For this, we have discussed current progress, challenges, and future perspectives. Further, we delve into the brain-computer interface (BCI), aiming to comprehend the nuances of brain connectivity, bridging the gap between brain organoids and computers, the intriguing aspect of training brain organoids with artificial intelligence (AI). This chapter offers a comprehensive insight into the convergence of biology and technology, paving the way for groundbreaking innovations in understanding and harnessing the power of the human brain. It also aims to explore the significance of brain organoids in revolutionizing neuroscience research, enabling a deeper comprehension of brain development and providing new avenues for understanding and treating brain disorders. The potential of OI to revolutionize computing, neurological research, and drug development has been also discussed extensively.

2. Brain organoids: a brief overview

Brain organoids are three-dimensional (3D) structures that serve as *in vitro* models to mimic the early stages of human brain development. These organoids are derived from pluripotent stem cells (PSCs), which can be obtained from different sources, such as induced pluripotent stem cells (iPSCs) reprogrammed from adult cells, or embryonic stem cells (ESCs) derived from early-stage embryos [8]. PSCs have the potential to develop into any cell type in the body, including the various cell types found in the brain (neurons and glial cells). The process of differentiation from a PSC into a specific type of cell is guided by a complex interplay of genetic and environmental factors. Various transcription factors are also used to activate or deactivate certain genes, or the cells are exposed to specific signaling molecules to mimic the conditions in a particular tissue [9]. By carefully controlling these factors, stem cells can be coaxed to differentiate into a wide variety of cell types, from neurons to heart cells. One of the most significant advantages of brain organoids is their ability to model the development of the human brain. Traditional animal models, such as mice, have significantly different brain structures and developmental processes, making them less than ideal for studying human neurodevelopment. In contrast, brain organoids can recapitulate key aspects of human brain development, including cell differentiation, migration, and interaction. This makes them a valuable tool for studying neurodevelopmental disorders, such as autism and schizophrenia [10].

2.1 Process of brain organoid formation

The process of generating brain organoids involves a technique called cerebral organoid differentiation that mimics the natural processes occurring during early brain development in the human embryo [11]. Firstly, PSCs are isolated from donated human tissues (for iPSCs) or derived from early-stage embryos (for ESCs). Once PSCs are obtained, they are directed to differentiate into neural cells using specialized growth factors such as fibroblast growth factors (FGFs), epidermal growth factor (EGF), epidermal growth factor (EGF) and, insulin-like growth factor (IGF), and signaling molecules such as bone morphogenetic proteins (BMPs), noggin, Wnt proteins and neurotrophins, to coax the cells into adopting a neural fate. This process is akin to the early stages of embryonic development when neural tissue begins to form [12]. Through the action of specific signaling pathways, the differentiating PSCs organize themselves into a structure called neuroectoderm. Neuroectoderm is a transient embryonic tissue that gives rise to the central nervous system, including the brain and spinal cord. Within the neuroectoderm, neural stem cells are formed, which are self-renewing and capable of generating various neural cell types [13]. After the formation of neural stem cells, they begin to self-organize and differentiate further. This process is driven by complex genetic and biochemical interactions which result in the formation of distinct brain regions that are akin to early brain development in embryos. These interactions include gene expression regulation, extracellular matrix (ECM) interactions, cell fate determination and differentiation of neural progenitor cells into specific cell types, formation of specific brain structures, chemical signaling between neurons and other cells, influencing synapse formation and neuronal function, mediate cell–cell interactions and guide the migration of cells to appropriate locations and contributing to the formation of specific brain regions. These brain regions include the structures resembling cerebral cortex, hippocampus, and other brain regions, each with specific neuronal types and functions [14]. As the self-organization and morphogenesis continue, the brain organoids recapitulate critical events observed during early human brain development, such as neurogenesis, gliogenesis, and the establishment of neural circuits (**Figure 1**). The resulting brain organoids display a certain level of complexity and functionality resembling early stages of the human brain. To maintain the vitality and growth of these organoids, it's essential to passage them periodically. Passaging involves carefully breaking down the larger organoids into smaller pieces, ensuring the core remains undamaged. These smaller

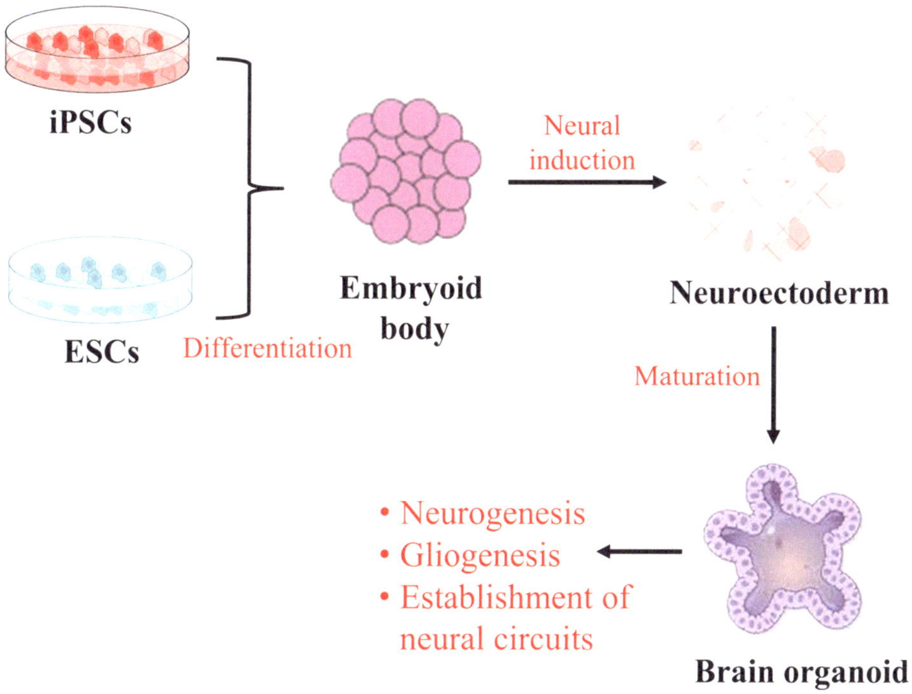

Figure 1.
Representation of process of generating brain organoids. PSCs isolated from iPSCs or ESCs, differentiate into an embryoid body. The differentiating PSCs organize themselves into a structure called neuroectoderm that subsequently matures into an organoid.

fragments are then reseeded in fresh culture media, providing them with the necessary nutrients and space for further growth. This process not only ensures the health of the organoids but also allows for the expansion of the culture, paving the way for multiple experiments and prolonged study durations.

The generation of brain organoids through cerebral organoid differentiation is a fascinating process that enables researchers to create 3D models of early human brain development in a laboratory setting. Many studies have established 3D culture system that allowed for the self-organization of neural progenitor cells into structures resembling different brain regions. They showed that these organoids could recapitulate key aspects of human brain development and be used as models to study microcephaly, a neurodevelopmental disorder characterized by reduced brain size [15, 16]. While some studies have generated neuromuscular organoids containing spinal cord and muscle tissue. The researchers showcased the potential for generating complex multi-tissue organoids using cerebral organoid differentiation techniques. In an instance, a study developed a model of human neuromuscular system, which comprised of both the nervous system and the muscles. To develop this model, the researchers used PSCs to generate 3D organoids that contained both neural and muscle tissue, similar to the human trunk's structure (from neck to abdomen). In these neuromuscular organoids, the neurons formed functional neuromuscular junctions with the muscle cells, closely mimicking the connections seen in the human body [17].

In contrast, there are reports highlighting the importance of cerebral organoids as a model system to study various infectious diseases. A study conducted by Cugola et al. focused on modeling Zika virus infection using cerebral organoids. The authors used human iPSCs to create cerebral organoids to model the early stages of brain development. They infected brain organoids with the Zika virus and observed the virus's effects on neural stem cells and brain development. The Zika virus infection resulted in cell death and disruption of organoid growth, closely resembling the microcephaly observed in human fetuses infected with the virus in utero. They also found that Zika virus infection altered the expression of genes involved in cell cycle regulation, DNA repair, and cell death, contributing to the observed detrimental effects on brain development [18]. These organoids offered a more accurate representation of human brain tissue than traditional 2D cultures, making them an ideal tool to investigate the impact of the Zika virus on brain development.

2.2 Application of brain organoids

The applications of brain organoids span across various fields, including disease modeling, drug testing, personalized medicine, and understanding human brain development (**Figure 2**).

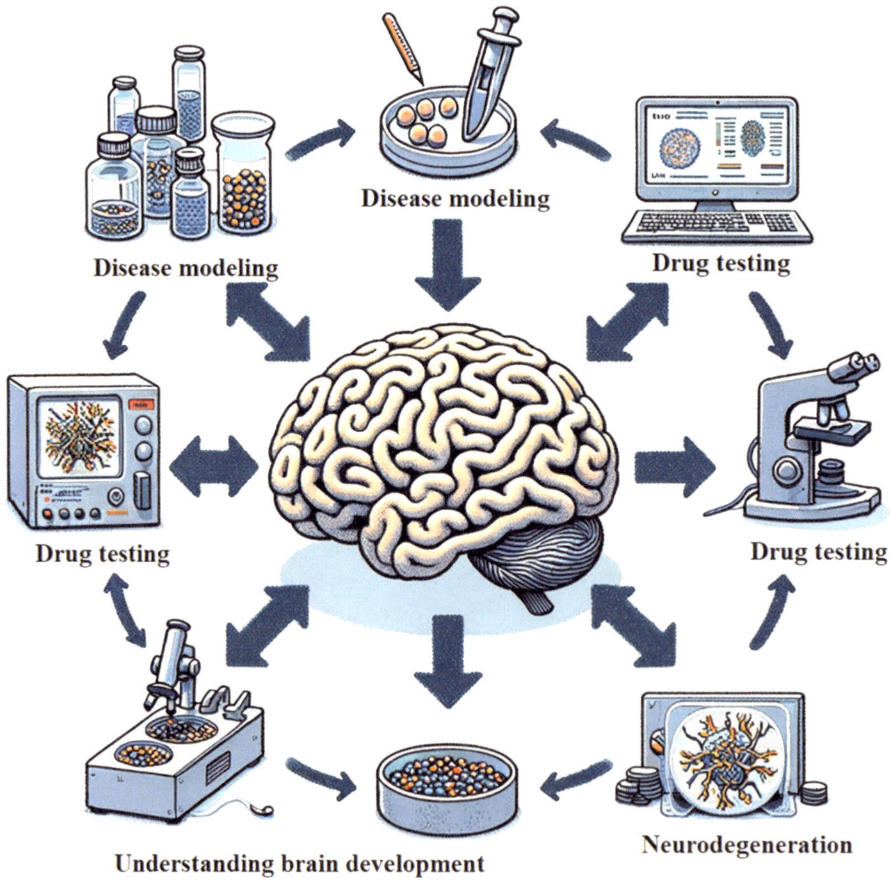

Figure 2.
Schematic flow chart illustrating the diverse applications and potential of organoid intelligence in the realm of neuroscience and medical research.

2.2.1 Disease modeling

The use of brain organoids in disease modeling has been a significant advancement in neuroscience. Brain organoids can be used to model various neurological and psychiatric diseases which are thought to arise during brain development. The advancements in methods for generating more complex brain organoids include the use of vascularized and mixed lineage tissue from PSCs. Moreover, synthetic biomaterials and microfluidic technology are boosting brain organoid development. The applications of brain organoids have also been seen in studying pre-term birth associated brain dysfunction, viral infections mediated neuroinflammation, neurodevelopmental, and neurodegenerative diseases [2]. Brain organoids can be used to model these early developmental processes, providing insights into how these disorders might originate. For example, researchers have used brain organoids to study the effects of genetic mutations associated with autism, revealing how these mutations can disrupt normal brain development. In another instance, brain organoids derived from patients with Alzheimer's disease can recapitulate key features of the disease, including the accumulation of amyloid-beta plaques and neurofibrillary tangles [19].

Similarly, brain organoids can be used to study disorders like autism, schizophrenia, and epilepsy, providing insights into disease mechanisms and potential therapeutic targets [20]. A study by Robles et al. reported that the *in vitro* models of the human brain can facilitate efficient development and clinical translation of therapeutic treatments of mental disorders. In their study, they addressed the limitations of current models of neural circuitry, which include animal models and post-mortem brain tissue. These models have limitations in assessing the functional alterations in short-range and long-range network connectivity between brain regions implicated in many mental disorders, such as schizophrenia and autism spectrum disorders.

To address these limitations, the researchers developed an *in vitro* model of the human brain that models the *in vivo* cerebral tract environment. They combined microfabrication and stem cell differentiation techniques to develop an *in vitro* cerebral tract model that anchors human-induced PSC-derived cerebral organoids and provides a scaffold to promote the formation of a functional connecting neuronal tract. Morphological and functional analyses revealed the expression of key neuronal markers as well as functional activity and signal propagation along with cerebral tracts connecting cerebral organoid pairs. The reported *in vitro* models enabled the investigation of critical neural circuitry involved in neurodevelopmental processes and had the potential to help devise personalized and targeted therapeutic strategies [21].

2.2.2 Drug testing

Traditional drug testing methods often rely on 2D cell cultures and animal models. Although these methods have provided valuable insights, they have limitations. 2D cell cultures lack the complex cellular interactions and architecture of the human brain. While animal models, although are more complex and do not fully replicate human physiology. In contrast, brain organoids offer a more accurate and physiologically relevant model of the human brain, making them an attractive platform for drug testing [22]. Researchers are now able to test the effects of potential drugs on these organoids by recreating the disease conditions and providing insights into the drug's efficacy and potential side-effects [23]. Brain organoids can also be used for personalized medicine. Many studies have created patient-specific organoids that carry the same genetic information as the patient [24, 25]. This allowed the testing of drugs on these patient-specific organoids and potentially predicted the patient's response to the drug. This approach has led to more personalized treatment strategies that improve patient outcomes. Rajan and colleagues illustrated an integrative genomic, *in vitro*, and *in vivo* functional treatment paradigm for high-grade glioma (GBM) using patient-derived organoids (PDOs). Their study suggested that rapidly implemented individualized drug response prediction models provide actionable information for the physician to combat recurrences or treatment resistances in GBM [24]. Similarly, another study suggested that veteran-derived human cerebral organoids not only can be used as an innovative human model to uncover the cellular responses to Gulf War toxicants but can also serve as a platform for developing personalized medicine approaches for the veterans [25]. Furthermore, brain organoids are used for high-throughput drug screening. By generating large numbers of organoids, researchers have tested the effects of many different drugs or drug combinations simultaneously. This accelerated the drug discovery process, potentially leading to the development of new treatments more quickly.

2.2.3 Understanding brain development

Brain organoids can be used to study the processes of human brain development and to understand how disruptions in these processes can lead to neurological and psychiatric disorders [18, 19]. Similarly, organoids derived from patients with autism have been shown to exhibit abnormal patterns of neuronal activity, providing insights into the underlying mechanisms of the disorder [26]. Human brain development is a complex process that involves the coordinated proliferation, migration, and differentiation of neural progenitor cells into a variety of specialized cell types, including neurons and glial cells. This process is tightly regulated by a network of genes and signaling pathways. Disruptions in these processes can lead to a range of neurological and psychiatric disorders, including autism, schizophrenia, and epilepsy [27]. Brain organoids provide a unique platform to study these processes in a controlled laboratory setting. One of the key advantages of using brain organoids to study brain development is their ability to recapitulate the 3D architecture of the brain. Unlike traditional 2D cell cultures, organoids can form complex structures that resemble those found in the developing human brain. This includes the formation of distinct brain regions, such as the cerebral cortex, and the development of intricate networks of neurons that are capable of transmitting electrical signals [26].

2.2.4 Neurodegeneration

Neurodegenerative diseases, including Alzheimer's, Parkinson's, Huntington's, and amyotrophic lateral sclerosis, are characterized by the progressive loss of structure or function of neurons. These diseases pose significant challenges to healthcare systems due to their chronic nature, lack of effective treatments, and the aging global population [28]. Brain organoids have emerged as a promising tool in the study of neurodegenerative diseases. They can be used to study the progression of these diseases and to test potential treatments. This is particularly important as these diseases often have a complex etiology involving both genetic and environmental factors. Brain organoids can be genetically manipulated to carry disease-specific mutations, allowing researchers to study the effects of these mutations on brain development and function [29]. Besides, various studies have also demonstrated the use of microglia-containing human brain organoids for studying brain development and pathology. They provide a new avenue to model brain development and pathology [30]. Microglia also play critical roles in neural development, synaptic formation and plasticity, and neural network maturation. Studies conducted over the past decades have implicated microglia dysfunction in multiple brain disorders, ranging from psychiatric disorders to neurodegenerative diseases [31].

3. Biocomputers: the intersection of biology and computing

Biocomputing involves the development and application of data storage, retrieval, and processing systems that use biological components. The concept is rooted in the idea that biological systems are inherently computational and can be harnessed to perform computing tasks. Biocomputers leverage the capabilities of biological molecules, particularly DNA and proteins, to perform computational calculations [32]. DNA has been used to create logic gates, the fundamental building blocks of digital circuits. These DNA-based logic gates can perform simple binary operations, such as AND, OR, and NOT functions, similar to their silicon-based counterparts in traditional computers [33]. One of the most significant advantages of biocomputers is their potential for parallel processing. Unlike traditional computers, which process information sequentially, biocomputers can perform many operations simultaneously. This is because biological reactions often occur in parallel, with many molecules interacting at the same time. This parallelism allows biocomputers to solve complex problems more efficiently than traditional computers [34]. Another advantage of biocomputers is their potential for miniaturization. Biological molecules are incredibly small, allowing for the development of ultra-compact computing systems. This is particularly relevant in the context of medical applications, where biocomputers could be used to develop smart drugs that can perform computations inside the body to diagnose or treat diseases [35].

3.1 History of biocomputers

The concept of biocomputing or using biological systems to perform computational tasks, has its roots in the fundamental understanding of life processes. The history of biocomputers is a fascinating journey that intertwines the fields of biology, computer science, and engineering (**Figure 3**). The idea of biological computation can be traced back to the mid-twentieth century when researchers began to understand the genetic code and the process of protein synthesis. The realization that DNA carries information and the cell translates this information into proteins, a process akin to reading and executing a program, laid the groundwork for the concept of biological computation [36]. However, the field of biocomputing truly began to take shape in the 1990s, with the advent of DNA computing. The pioneer of this field was Leonard Adleman, a computer scientist from the University of Southern California [37]. In 1994, Adleman published a groundbreaking paper in which he demonstrated that DNA could be used to solve a well-known mathematical problem, the seven-point Hamiltonian path problem, also known as the "traveling salesman" problem. Adleman's work showed that DNA molecules, with their ability to store information and undergo specific reactions, could be used to perform computations [38]. Following Adleman's work, the field of DNA computing rapidly expanded. Researchers developed methods to use DNA to perform logical operations and to solve more complex computational problems. DNA was used to create logic gates, the fundamental building blocks of digital circuits,

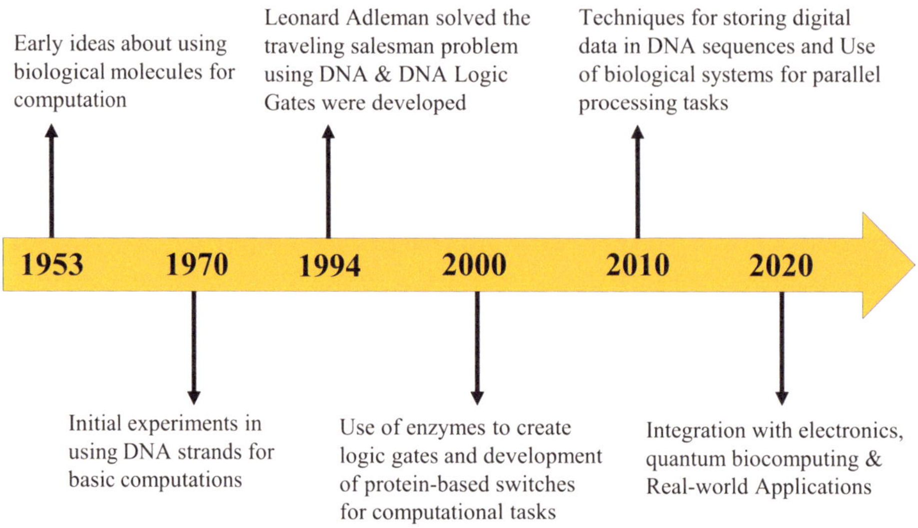

Figure 3.
Timeline for the development of biocomputers. A summary of key landmark studies leading to the fundamental understanding of biocomputers.

and to build DNA-based calculators and other devices [39]. In the 2000s, the field of biocomputing took another leap forward with the development of biological computers—living cells that are engineered to perform computational tasks. These biological computers use genetic circuits to process information and make decisions based on their inputs. Researchers have engineered bacteria that can "count" by flipping a genetic switch each time they divide [40].

3.2 Role of human brain cells in biocomputing

Biocomputing, at its core, is the use of biological materials or biological architectures for computational purposes. It's a field that merges biology, computer science, and information technology. Biocomputing has opened up exciting new possibilities in the field of neuroscience. One of the most intriguing developments is the use of human brain cells, or neurons, in biocomputing. Neurons have the potential to revolutionize the way we think about computing with their unique ability to process and transmit information. Neurons are the fundamental units of the brain, responsible for receiving sensory input from the external world, processing this information, and directing the body's response. They communicate with each other through specialized connections known as synapses, where electrical signals are converted into chemical signals and vice versa. This complex network of neurons and synapses forms the basis of the brain's computational power [41]. Just like transistors in an electronic circuit, neurons can switch on and off in response to certain inputs. However, unlike electronic transistors, neurons can process a wide range of inputs and can adapt their responses based on past experiences, a property known as plasticity [42]. Neurons can also be used to create biological neural networks, which can be trained to perform complex tasks. For example, researchers have grown networks of neurons on microelectrode arrays and trained these networks to control robotic devices. These biological neural networks have the potential to outperform traditional artificial neural networks in certain tasks, due to their inherent plasticity and adaptability [43].

One of the key applications of neurons in biocomputing is in the development of neuromorphic systems. Neuromorphic systems are designed to mimic the structure and function of the brain, with the aim of achieving the brain's computational efficiency and adaptability. These systems use artificial neurons and synapses to process information in a parallel and distributed manner, similar to the brain [41]. Biocomputing also allows for the creation of detailed computational models of neural systems. These models can simulate the behavior of neurons and neural networks, providing insights into how the brain processes information. This can help researchers understand normal brain function and the mechanisms underlying neurological disorders [44]. Besides, neuroscience research often generates large amounts of complex data. Biocomputing techniques, such as machine learning and data mining, can help analyze this data, uncovering patterns and relationships that might not be apparent through traditional analysis methods. Biocomputing plays a crucial role in the development of Brain-Computer Interfaces (BCIs), which are systems that enable direct communication between the brain and an external device. BCIs can be used to restore or augment lost neurological functions, such as movement in individuals with paralysis or communication in individuals with severe speech impairments [45]. By decoding the electrical activity of neurons, BCIs can allow individuals to control prosthetic limbs, computer cursors, or even drones with their thoughts. Another subfield of biocomputing is neuromorphic computing that involves designing computer systems based on the architecture of the brain. Neuromorphic computers mimic the brain's structure and function and could potentially perform complex tasks more efficiently than traditional computers [44].

3.3 The potential of biocomputers in medical research and treatment

One of the most promising applications of biocomputing is in disease diagnosis and monitoring. Researchers have developed biocomputers that can operate inside the human body, detect disease markers and even deliver targeted treatments. These biocomputers are designed to respond to specific biological signals, such as the presence of a particular protein associated with a disease [40]. In addition, biocomputing enhances the capabilities of medical imaging. By applying machine learning algorithms to imaging data, many researchers created models that can predict disease progression or response to treatment. This led to earlier and more accurate diagnoses, as well as more effective monitoring of disease progression [46]. Biocomputing techniques, such as machine learning and artificial intelligence (AI), can be used to analyze and interpret medical images. These techniques can identify patterns and features in images that may be difficult for humans to detect, leading to more accurate diagnoses. AI algorithms can be trained to identify tumors in medical images with a high degree of accuracy. Biocomputing algorithms can be used to reconstruct medical images, improving their quality and making them easier to interpret. This is particularly useful in modalities like computed tomography (CT) scan and magnetic resonance imaging (MRI), where raw data must be transformed into a visual format [47]. In medical imaging, segmentation is the process of identifying and delineating specific structures or regions of interest in an image. Biocomputing can automate this process, saving time and improving consistency. On the other hand, radiomics which is a field that extracts a large number of quantitative features from medical images using data-characterization algorithms, can be used to create predictive models for diagnosis, prognosis, or therapy response, which is particularly useful in personalized medicine. Biocomputing can also automate and improve the accuracy of the process image registration in which different sets of data are transformed into one coordinate system and is often necessary when multiple imaging modalities are used [48]. Biocomputing can be used to create 3D visualizations of medical images, providing a more comprehensive view of anatomical structures. This can be particularly useful for surgical planning and patient education.

Biocomputing also has significant potential in the field of drug discovery. Computational models of biological systems are being used to predict how potential drugs will interact with their targets in the body. This has speeded up the drug discovery process and reduced the need for animal testing [49]. Biocomputing allows for *in silico* screening of large databases of chemical compounds to identify potential drug candidates. These techniques help to predict how a compound will interact with a particular biological target based on its structure and properties. This has significantly speeded up the initial stages of drug discovery by reducing the number of compounds that need to be tested in the lab [50]. In addition, biocomputing is used to create predictive models of biological systems. These models simulate the interactions between potential drugs and their targets, predicting the efficacy and potential side effects of a drug before it is tested in the lab or in clinical trials. Once potential drug candidates are identified, biocomputing can be used to optimize these compounds. This can involve tweaking the structure of the compound to improve its efficacy, reduce side effects, or improve other properties such as solubility or stability [49]. Moreover, biocomputing has contributed to the development of personalized medicine. By using computational models to analyze a patient's genetic data, doctors could predict how the patient will respond to different treatments. This has led to more effective and personalized treatment plans [50].

4. Organoid intelligence (OI): the next frontier in biocomputing

OI refers to the integration of organoid technology with computational methods to enhance our understanding of organoid behavior and to predict their responses to various stimuli. High-throughput techniques such as single-cell RNA sequencing, proteomics, and imaging are used to generate comprehensive datasets from organoids. These datasets can provide detailed information about the organoid's cellular composition, structural organization, and functional dynamics [51]. Advanced computational methods, including machine learning and AI, are used to analyze the complex datasets generated from organoids. These analyses reveal patterns and relationships that are not readily apparent, providing insights into the organoid's behavior. The data and insights obtained from organoids are used to create predictive models. These models then simulate the organoid's response to various stimuli, such as drugs or genetic modifications. Thus, this guides the experimental design and accelerate the pace of research [52].

4.1 The concept of OI

The advent of organoid technology has revolutionized the field of biological research, particularly in the realm of neuroscience. Brain organoids provide a unique platform for studying neurodevelopmental processes, disease modeling, and drug testing. However, the intersection of organoid technology with AI has given rise to a new concept called organoid intelligence or OI. It is an emerging field that leverages the extraordinary biological processing power of brain organoids and the computational prowess of AI [53]. This fusion aims to create a new frontier in biocomputing and 'intelligence-in-a-dish', providing exceptional opportunities for understanding the human brain and developing novel therapeutic strategies. **Table 2** summarizes the

Aspect	Advantages	Disadvantages
Research potential	• Enables real-time analysis and feedback on organoid behavior. • Provides a platform for advanced research and predictive modeling.	• Still in its infancy, so the full potential and applications are not yet fully understood.
Data analysis	• AI can process and analyze vast amounts of data from organoids quickly. • Can identify patterns and insights that might be missed by human researchers.	• Risk of over-reliance on AI, potentially missing nuanced biological insights.
Personalized medicine	• Potential for real-time adaptation and personalized therapeutic strategies based on individual organoid responses.	• Ethical concerns about creating personalized brain models and potential misuse.
Technological integration	• Combines the strengths of biological and computational research. • Can lead to the development of novel technologies and methodologies.	• Technological challenges in seamlessly integrating organoids with AI systems.
Ethical concerns	• Can provide insights without the need for animal or human testing in some cases.	• Amplified concerns about creating "conscious" tissue. • Ethical dilemmas related to AI ethics, data privacy, and potential misuse.
Cost and accessibility	• Potential for scalable solutions in research and medicine with the integration of AI.	• High initial costs due to the integration of advanced technologies. • May not be accessible to all researchers or institutions.

Table 2.
Advantages and disadvantages of OI.

advantages and disadvantages of OI. Brain organoids are generated through a process that mimics the stages of human brain development. Starting from PSCs which have the potential to differentiate into any cell type in the body, researchers guide these cells to become neural progenitor cells. These progenitors then self-organize and differentiate further into various types of brain cells, forming organoids with structures similar to those found in the human brain [54]. The resulting organoids contain multiple types of neurons and supportive glial cells, and they exhibit complex activities such as neuron firing and network formation.

The concept of OI takes this a step further. By integrating AI methodologies, researchers analyzed and interpreted the complex patterns of activity within brain organoids. This provided insights into how neurons communicate and form networks, how these networks change over time, and how they respond to various stimuli. Furthermore, AI helped to optimize the process of organoid generation, improving the reproducibility and scalability of organoid production [52]. Recent research has highlighted more on the formation of OI using brain organoids. One of the study presented a novel approach in creating biologically relevant environment for the growth and development of human embryonic stem cell (hESC)-derived brain organoids. The authors used a decellularized adult porcine brain extracellular matrix (B-ECM) as a scaffold for the organoids [55]. The decellularization process involved removing cells from the ECM using chemical, physical, and enzymatic methods. This process resulted in a cell-free ECM scaffold with minimal interference from cellular antigens and metabolites. The B-ECM was then processed into a hydrogel, which was used as a scaffold for the organoids. The hydrogel formation process involved solubilizing the ECM proteins with acids and enzymes, specifically pepsin, which cleaves the non-helical protein regions outside of the triple helix protein structure of collagen. After neutralization to physiological pH, the hydrogel formation followed a collagen-based self-assembly process. The authors found that hESCs cultured in B-ECM hydrogels showed gene expression and differentiation outcomes similar to those grown in Matrigel [55]. This indicated that B-ECM hydrogels can be used as an alternative scaffold for human cerebral organoid formation. The use of B-ECM hydrogels as a scaffold for organoid formation is significant because it provides a more biologically relevant environment for the organoids, which could potentially enhance their development and functionality. This could have important implications for the study of neurodevelopment and disease, as well as for the development of OI (**Figure 4**) [56].

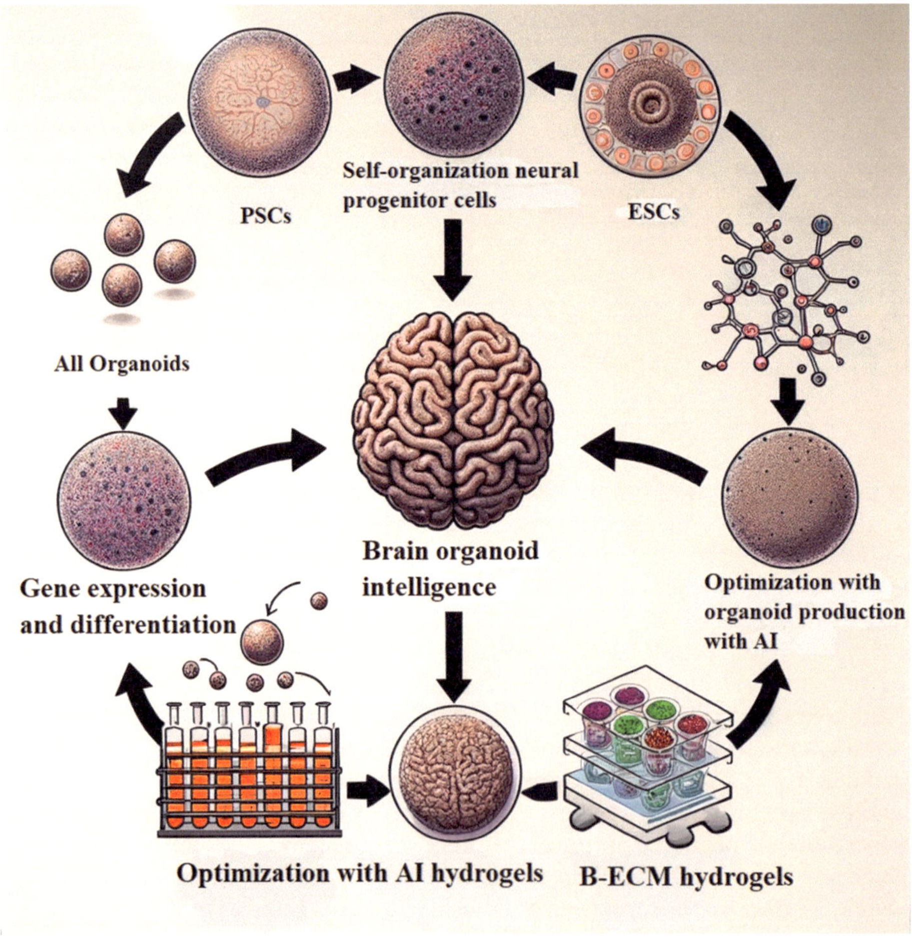

Figure 4.
Schematic representation of the development process of OI. The process starts with PSCs and progresses through differentiation into neural progenitor cells, self-organization into brain organoids, AI analysis of organoid activities, optimization of organoid production with AI, integration with biologically relevant scaffolds like B-ECM hydrogels, and culminates in the creation of 'intelligence-in-a-dish'.

4.2 Current research and advancements in the field

OI is an emerging interdisciplinary field that combines the biological processing capabilities of organoids with the computational power of AI. One of the significant advancements in the field of OI is the development of more biologically relevant environments for organoid growth. An exciting development is the integration of AI with organoid and organ-on-a-chip technologies. One of the studies discussed the convergence of AI and microfluidics in biomedicine, with a focus on organoids-on-chips [57]. This integration allowed for the rapid analysis of complex organoid data, potentially leading to the development of novel therapeutic strategies. Furthermore, advancements in the efficient culture of organoids derived from adult human stem cells have laid the foundation for PDO culture. This advancement was particularly significant for personalized medicine, as it allowed for the testing of drug responses on organoids derived from a patient's own cells [58].

Currently, researchers synthesize enhanced organoids that have the potential to mimic the computational and cognitive capabilities of an *in vivo* brain. A laboratory-grown brain organoid is characterized by high levels of myelination, a greater abundance of support cells, and a capacity for spontaneous neural activity, representing a substantial advancement from earlier iterations that lacked such physiological complexity [59]. With these synthesized mini-brains, few studies proposed using available 3D microelectrode arrays to perform external electrophysiological recordings of brain organoids to perform tasks such as learning, memory, and other cognitive functions [52]. Other studies have shown latent diffusion models that can reconstruct regional brain activation under fMRI, thereby decoding visual stimuli into high-fidelity images. Just as AI can be trained, OI could be taught to recognize and respond to specific environmental stimuli, potentially surpassing the performance of any modern computer. With the help of sensory units from organoids, OI may even be able to display decoded neural outputs on a digital screen, offering a novel way to understand complex biological processes. With the combined development of AI and OI, technology that leverages both systems could surpass the human brain as a superior information processing device [52]. Previous clinical trials for neurologic conditions such as Alzheimer's disease, autism, and schizophrenia have exhibited poor success rates, possibly attributed to the poor translation of research findings from animal models to human pathophysiology. Brain organoids created by reprogramming a small sample of a patient's skin cells, offer a non-invasive method to test brain tissue without the need for extracting tissue from living specimens [53]. Access to these recapitulated brains may allow for novel biomarker discovery using "omics" during initial disease stages, which could lead to innovative and personalized therapeutics.

4.3 The potential applications and implications of OI

OI marks the efficiency of the human brain in processing complex information compared to machines. While machines can perform tasks faster, the human brain is more efficient in terms of power consumption and data efficiency. This has led to high expectations for biological, brain-directed computing as an alternative to silicon-based computing, with the potential for the advances in computing speed, processing power, data efficiency, and storage capabilities, all with lower energy needs [51]. For instance, Smirnova et al. proposed biofeedback to systematically train organoids with increasingly complex sensory inputs and output opportunities. They envision interfacing the brain organoids with computers, sensors, and machine interfaces to facilitate supervised and unsupervised learning. The authors use the term "OI" for this approach to stress its complementarity to AI, where computers aim to perform tasks done by brains, often by modeling our understanding of learning [56]. Glioblastoma, a grade IV astrocytoma, is the most aggressive primary brain tumor, with an overall median survival of 16.0 months following standard treatment. Despite intensive treatment, the tumor almost invariably recurs. This poor prognosis is attributed to the initiation, propagation, and differentiation of cancer stem cells. Current *in vitro* models are limited at preserving the inter-and intra-tumoural heterogeneity of primary tumors. Therefore, brain cancer research models aims to recapitulate glioblastoma stem cell function, while remaining amenable for analysis [60]. Cerebral organoids are emerging as cutting-edge tools in glioblastoma research. The opportunity to generate cerebral organoids via iPSCs, and to perform co-cultures with patient-derived cancer stem cells (GLICO model), has enabled the analysis of cancer development in a context that better mimics brain tissue architecture.

Bioprinting technology is an advanced technique that can print various cells and biomaterials at desired locations, allowing for the creation of complex 3D biological structures. Evidence suggest that this technology can help standardize organoids and automate the fabrication process [61]. This technology has been increasingly used for the creation of brain organoids, which are miniaturized and simplified versions of an organ produced *in vitro* in 3D that show realistic micro-anatomy. In addition, the Blood-Brain Barrier (BBB) is a highly selective semipermeable border of endothelial cells that prevents solutes in the circulating blood from non-selectively crossing into the extracellular fluid of the central nervous system where neurons reside. The BBB is crucial for protecting the brain from harmful substances and maintaining homeostasis, but it also poses a significant challenge for drug delivery to the brain. Tan et al. developed 3D cellular and organoid culture for BBB in neuropharmacology. The authors used nanorobotics and AI to design, control, and target the BBB. Nanorobotics involved the use of microscopic robots for medical purposes, while AI was used to optimize these processes [58].

The implications of OI are vast, spanning from advancements in understanding brain development and diseases, to ethical considerations, and the potential for novel biocomputing models. Brain organoids offer unique opportunities for studying human brain development and neurological disorders. They allow scientists to model the complex interactions and processes that occur during brain development, which can lead to a better understanding of various neurological conditions. For instance, the use of brain organoids has already provided insights into the early stages of neurodevelopmental disorders, which are difficult to study in humans [62]. As we venture into creating entities that closely resemble human brains, ethical questions inevitably arise. One of the primary concerns is the potential for these organoids to develop some form of consciousness. If brain organoids were to reach a level of complexity where they could experience sensations or emotions, this would raise serious ethical questions about their use and treatment. There are also concerns about the potential misuse of this technology, such as in creating human-animal chimeras or the potential for unauthorized use in creating enhanced beings or biological weapons. The concept of OI introduces the possibility of using brain organoids as a form of biological computing. This could involve connecting brain organoids with real-world sensors and output devices, and even with each other, to create complex, networked interfaces. Such systems could allow for faster decision-making, continuous learning during tasks, and greater energy and data efficiency. This could revolutionize fields such as artificial intelligence, robotics, and data processing, opening up new avenues for exploration and innovation. Moreover, the implications of OI are profound, offers exciting possibilities for scientific discovery and technological innovation, while they also pose significant ethical challenges that need to be carefully navigated.

5. Brain-computer Interface (BCI): understanding brain connectivity

BCI is a rapidly evolving technology which revolutionizes our understanding of brain connectivity and has the potential to drastically alter the landscape of neuroscience, medicine, and human interaction with technology. BCIs enable direct communication between the brain and an external device, bypassing the traditional route of peripheral nerves and muscles. This technology has profound implications for understanding brain connectivity, with potential applications ranging from assisting individuals with neurodegenerative disorders to enhancing human cognition [61]. Understanding brain connectivity is a fundamental aspect of neuroscience. The human brain is a complex network of billions of neurons interconnected through trillions of synapses. These connections form intricate networks that facilitate the transmission and processing of information, leading to cognition, behavior, and consciousness. BCIs provide a unique window into these networks, allowing scientists to monitor and interpret the electrical activity of the brain in real-time. The intricate connections between different brain regions are essential for cognitive functions [63]. Brain organoids provide a unique opportunity to study neural connectivity and synapse formation. This enables researchers to investigate the impact of genetic mutations or environmental factors on neural network development and function, elucidating how disruptions in connectivity may contribute to neurodevelopmental disorders.

BCIs work by detecting changes in brain activity by typically using electroencephalography (EEG) or functional magnetic resonance imaging (fMRI). These changes are then translated into commands that control the external devices. For instance, a person might think about moving their right hand, which would generate a specific pattern of brain activity. The BCI can detect this pattern and translate it into a command to move a robotic arm [45]. One of the most promising applications of BCIs is in the field of neurorehabilitation. For individuals with neurodegenerative disorders or those who have suffered a stroke, BCIs can provide a means of communication and control that is independent of muscle control [64]. This can significantly improve the quality of life for these individuals, allowing them to interact with their environment and communicate with others despite their physical limitations. In addition to their therapeutic applications, BCIs also offer exciting possibilities for enhancing human cognition. By interfacing directly with the brain, BCIs potentially augment any cognitive abilities which can improve memory, attention, and even sensory perception. This could have far-reaching implications, from improving performance in everyday tasks to enhancing human capabilities in fields like science, engineering, and art [45].

6. Probes to connect brain organoids and computer

The integration of brain organoids and computers represents a cutting-edge frontier in the field of neuroscience and biotechnology. This integration is made possible through the use of organoids with computer systems. These probes serve as a bridge, enabling the transmission of information between the biological and digital realms [61]. The probe devices that connect brain organoids and computers are typically microelectrode arrays (MEAs) which are devices that contain multiple microscopic electrodes and can detect and record electrical activity from neurons. When these probes are inserted into a brain organoid, they pick up the electrical signals generated by the organoid's neurons. These signals are then amplified, digitized, and transmitted to a computer for analysis [65]. The use of MEAs in brain organoid research has several significant implications. Firstly, it allows for the real-time monitoring of organoid activity, providing insights into how neurons in the organoid communicate with each other. This can help researchers understand the fundamental processes of brain development and function. Secondly, MEAs can be used to stimulate the neurons in the organoid, allowing researchers to investigate how different types of stimulation affect neuronal activity. This could lead to the development of new treatments for neurological disorders which are characterized by abnormal neuronal activity. Thirdly, the integration of brain organoids and computers via MEAs opens up the possibility of creating biological computers or biocomputers. Biocomputers use biological materials, such as DNA or neurons, to perform computational operations. Brain organoids connected to computers could potentially be trained to perform specific tasks, such as pattern recognition or data processing, offering a novel approach to computing that leverages the unique capabilities of biological systems [55].

7. Training of brain organoid with AI

AI, particularly machine learning and deep learning, has the potential to transform the way one analyzes and interprets complex biological data. In the context of brain organoids, AI can be used to analyze large-scale data sets, identify patterns, and make predictions that would be impossible for humans to do manually. This can include everything from predicting the development of specific cell types to identifying the effects of different drugs or genetic modifications [23]. The first step in training brain organoids using AI is the creation of the organoids themselves. This is typically done using human PSCs which are cultured, and they differentiate into brain cells through the application of various growth factors. Over time, these cells self-organize into 3D that resemble aspects of the human brain. Once the organoids have been created, they are used to generate data for AI training. This typically involves using advanced imaging techniques to capture detailed images of the organoids at various stages of development. These images are then analyzed using machine learning algorithms that identify patterns and make predictions about the organoids' behavior. Training brain organoids with AI involves using machine learning algorithms to analyze the data generated by the organoids and make predictions about their behavior. This is done in a supervised or unsupervised manner. In supervised learning, the algorithm is trained on a set of labeled data, where the correct outcome is known. The algorithm then uses this training to make predictions on new, unlabeled data. In the context of brain organoids, this could involve training the algorithm to recognize specific cell types or developmental stages based on a set of labeled images [66]. On the other hand, unsupervised learning involves training the algorithm to identify patterns in the data without any pre-existing labels. This can be particularly useful in the context of brain organoids, where the complexity of the data can make it difficult to identify relevant patterns manually.

After the initial training, the model's performance is validated. This typically involves using the model to make predictions on the testing set and comparing these predictions to the actual outcomes. The model's parameters are adjusted and the model retrained multiple times to achieve optimal performance. Once the AI model has been trained and validated, it is used to guide the development of the brain organoids. This involve the use of the model to predict the effects of different growth factors or genetic modifications on the organoids' development, and then adjust these variables based on the model's predictions [51]. The development of the organoids should be continuously monitored, with the AI model's predictions being used to make ongoing adjustments. This iterative process of prediction, adjustment, and monitoring is key to guiding the organoids towards the development of organoid intelligence. Finally, the level of OI achieved is evaluated. This involve assessing the organoids' ability to perform basic information processing tasks, their short- and long-term memory capabilities, and other cognitive functions [67].

8. Challenges and future directions

Despite the exciting developments, there are also significant challenges in the field of biocomputing. One of the main challenges is the development of reliable and efficient methods for programming biological systems. Unlike traditional computers, which use a standard binary language, there is no universal language for programming biological systems. Additionally, biological systems are inherently noisy and variable, which can lead to errors in computation [68]. Another challenge is the integration of biocomputers with traditional electronic devices. While there has been progress in developing interfaces between biological and electronic systems, there is still much work to be done to achieve seamless integration. Besides culturing neurons and maintaining them in a healthy state is a complex task that requires precise control over the cellular environment. Furthermore, interfacing neurons with electronic devices is a significant technical challenge, due to the differences in scale and material properties between biological and electronic systems. However, ongoing research and technological advancements are helping to overcome these challenges. As our understanding of biology and our computing capabilities continue to grow, the potential of biocomputing in medical research and treatment will only increase.

In contrast, the concept of OI is not without its challenges and ethical considerations. As brain organoids become increasingly complex and start to exhibit higher-order functions, questions arise about their moral and legal status. Some researchers have raised concerns about the potential for organoids to develop consciousness, although this is currently purely speculative and far from the capabilities of current organoid technology. Moreover, the use of AI in organoid research brings its own set of challenges. AI algorithms require large amounts of data for training, and the data generated by organoid research is both complex and high-dimensional. Ensuring the accuracy and interpretability of AI models is crucial for their effective application in organoid research [62]. Along with creating these intelligent, human-like brains, some consider the possibility of organoids autonomously developing sentience, awareness, or pain perception. The ethical issue of OI includes the possibility of being fed biased or harmful information as well as concerns regarding cost, privacy and data mining, equity among users, intellectual property protections, and disruption of human-occupied jobs. To overcome these relevant issues surrounding OI, researchers suggest performing a comprehensive ethical analysis that includes a review from diverse groups to foster responsibility, accountability, and trust with the stakeholders. To ensure the successful integration of this technology, unmet limitations will need to be addressed first.

Currently, the main limitation of OI and brain organoids is the discrepancy in complexity compared to actual human brains. Today, brain organoids are below 500 micrometers in diameter and have less than 100,000 cells. These relatively non-complex models fail to show the developmental asymmetry nor the predictable anatomy that is needed to supplement preclinical trials of neurological conditions. As a result, decades of work will likely be necessary to develop organoids advanced enough to replace modern AI. Moreover, additional time is needed to develop the technology required for fully transmitting the computations of the organoid to a digital interface before it can practically convey clinically relevant information. Despite these hurdles, the proposition of OI "intelligence-in-a-dish" is an exciting new field that will surely attract more research. Complex brain organoids pose beneficial applications for understanding the development and treatment of neurological diseases. A long road of rigorous testing and embedded ethical mediation will be required before OI and brain organoids reach their potential.

9. Conclusion

Despite various challenges, the potential benefits of OI are immense. By providing a platform for studying the human brain in detail, brain organoids could revolutionize the understanding of neurodevelopmental and neurodegenerative disorders. The integration of AI could accelerate the research, enabling rapid analysis of complex organoid data and potentially leading to the development of novel therapeutic strategies. In conclusion, the intersection of brain organoid technology and AI in the form of OI represents a promising new frontier in neuroscience and biocomputing. As this field continues to evolve, it holds the potential to significantly advance the understanding of human brain and pave the way for novel treatments for a range of neurological disorders. OI represent a groundbreaking technology that has the potential to revolutionize our understanding of human brain development and disorders. As this field progresses, it is essential to strike a balance between scientific advancement and ethical considerations. With continued research and innovation, OI hold the promise of shaping the future of neuroscience, providing novel insights into brain complexity and unlocking new therapeutic opportunities for brain disorders.

Acknowledgements

The authors are thankful to Dr. D. Y. Patil Biotechnology and Bioinformatics Institute, Dr. D. Y. Patil Vidyapeeth, Pune for the physical infrastructure. Sangeeta Ballav is thankful to Dr. D. Y. Patil Vidyapeeth, Pune for Junior Research Fellowship (DPU/291/2021). This work is supported by the Ramalingaswami Re-entry fellowship, Department of Biotechnology, Govt. of India to S. Sur [BT/RLF/Re-entry/47/2021] and Intramural Grants, Dr. D. Y. Patil Vidyapeeth (DPU), Pimpri, Pune, India to S. Basu [DPU/644-43/2021].

Conflict of interest

There are no conflicts of interest.

Author details

Sangeeta Ballav*, Amit Ranjan, Subhayan Sur and Soumya Basu*
Cancer and Translational Research Centre, Dr. D.Y. Patil Biotechnology and Bioinformatics Institute, Dr. D. Y. Patil Vidyapeeth, Pune, Maharashtra, India

*Address all correspondence to: sangeetaballav@gmail.com and soumya.bs@gmail.com

References

[1] Zhao Z, Chen X, Dowbaj AM, et al. Organoids. Nature Reviews Methods Primers. 2022;**2**:94. DOI: 10.1038/s43586-022-00174-y

[2] Adlakha YK. Human 3D brain organoids: Steering the demolecularization of brain and neurological diseases. Cell Death Discovery. 2023;**9**(1):221. DOI: 10.1038/s41420-023-01523-w

[3] Ballav S, Deshmukh AJ, Siddiqui S, Aich J, Basu S. Two-Dimensional and Three-Dimensional Cell Culture and Their Applications. London, UK: Intechopen; 2021. DOI: 10.5772/intechopen.100382. Available from: https://books.google.co.in/books?hl=en&lr=&id=_Pt2EAAAQBAJ&oi=fnd&pg=PA45&dq=info:wLF994UpmZoJ:scholar.google.com&ots=x6GuPYO_xd&sig=vZb80d2-4q394jageXpHXdZPlwg&redir_esc=y#v=onepage&q&f=false

[4] Chen JH, Chu XP, Zhang JT, et al. Genomic characteristics and drug screening among organoids derived from non-small cell lung cancer patients. Thoracic Cancer. 2020;**11**(8):2279-2290. DOI: 10.1111/1759-7714.13542

[5] Morales Pantoja IE, Smirnova L, Muotri AR, et al. First organoid intelligence (OI) workshop to form an OI community. Frontiers in Artificial Intelligence. 2023;**6**:1116870. DOI: 10.3389/frai.2023.1116870

[6] Goñi-Moreno A, Nikel PI. High-performance biocomputing in synthetic biology-integrated transcriptional and metabolic circuits. Frontiers in Bioengineering and Biotechnology. 2019;**7**:40. DOI: 10.3389/fbioe.2019.00040

[7] Mehonic A, Kenyon AJ. Brain-inspired computing needs a master plan. Nature. 2022;**604**(7905):255-260. DOI: 10.1038/s41586-021-04362-w

[8] Agboola OS, Hu X, Shan Z, Wu Y, Lei L. Brain organoid: A 3D technology for investigating cellular composition and interactions in human neurological development and disease models in vitro. Stem Cell Research & Therapy. 2021;**12**(1):430. DOI: 10.1186/s13287-021-02369-8

[9] Romito A, Cobellis G. Pluripotent stem cells: Current understanding and future directions. Stem Cells International. 2016;**2016**:9451492. DOI: 10.1155/2016/9451492

[10] Andrews MG, Nowakowski TJ. Human brain development through the lens of cerebral organoid models. Brain Research. 2019;**1725**:146470. DOI: 10.1016/j.brainres.2019.146470

[11] Lancaster MA, Knoblich JA. Generation of cerebral organoids from human pluripotent stem cells. Nature Protocols. 2014;**9**(10):2329-2340. DOI: 10.1038/nprot.2014.158

[12] Chang EA, Jin SW, Nam MH, Kim SD. Human induced pluripotent stem cells : Clinical significance and applications in neurologic diseases. Journal of Korean Neurosurgical Association. 2019;**62**(5):493-501. DOI: 10.3340/jkns.2018.0222

[13] Xue X, Sun Y, Resto-Irizarry AM, et al. Mechanics-guided embryonic patterning of neuroectoderm tissue from human pluripotent stem cells. Nature Materials. 2018;**17**(7):633-641. DOI: 10.1038/s41563-018-0082-9

[14] Di Lullo E, Kriegstein AR. The use of brain organoids to investigate neural development and disease. Nature Reviews. Neuroscience. 2017;**18**(10):573-584. DOI: 10.1038/nrn.2017.107

[15] Lancaster MA, Renner M, Martin CA, et al. Cerebral organoids model human brain development and microcephaly. Nature. 2013;**501**(7467):373-379. DOI: 10.1038/nature12517

[16] Stachowiak EK, Benson CA, Narla ST, et al. Cerebral organoids reveal early cortical maldevelopment in schizophrenia-computational anatomy and genomics, role of FGFR1. Translational Psychiatry. 2017;**7**(11):6. DOI: 10.1038/s41398-017-0054-x

[17] Faustino Martins JM, Fischer C, Urzi A, et al. Self-organizing 3D human trunk neuromuscular organoids. Cell Stem Cell. 2020;**27**(3):498. DOI: 10.1016/j.stem.2020.08.011

[18] Salick MR, Wells MF, Eggan K, Kaykas A. Modelling zika virus infection of the developing human brain In vitro using stem cell derived cerebral organoids. Journal of Visualized Experiments. 2017;**127**:56404. DOI: 10.3791/56404

[19] Das S, Li Z, Wachter A, et al. Distinct transcriptomic responses to Aβ plaques, neurofibrillary tangles, and APOE in Alzheimer's disease. Alzheimer's & Dementia. 2023. DOI: 10.1002/alz.13387

[20] Angulo Salavarria MM, Dell'Amico C, D'Agostino A, Conti L, Onorati M. Cortico-thalamic development and disease: From cells, to circuits, to schizophrenia. Frontiers in Neuroanatomy. 2023;**17**:1130797. DOI: 10.3389/fnana.2023.1130797

[21] Robles DA, Boreland AJ, Pang ZP, Zahn JD. A cerebral organoid connectivity apparatus to model neuronal tract circuitry. Micromachines (Basel). 2021;**12**(12):1574. DOI: 10.3390/mi12121574

[22] Zhou JQ, Zeng LH, Li CT, et al. Brain organoids are new tool for drug screening of neurological diseases. Neural Regeneration Research. 2023;**18**(9):1884-1889. DOI: 10.4103/1673-5374.367983

[23] Shala AL, Arduino I, Salihu MB, Denora N. Quercetin and its Nano-formulations for brain tumor therapy-current developments and future perspectives for paediatric studies. Pharmaceutics. 2023;**15**(3):963. DOI: 10.3390/pharmaceutics15030963

[24] Rajan RG, Fernandez-Vega V, Sperry J, et al. In vitro and In vivo drug-response profiling using patient-derived high-grade glioma. Cancers (Basel). 2023;**15**(13):3289. DOI: 10.3390/cancers15133289

[25] Yates PL, Case K, Sun X, Sullivan K, Baas PW, Qiang L. Veteran-derived cerebral organoids display multifaceted pathological defects in studies on gulf war illness. Frontiers in Cellular Neuroscience. 2022;**16**:979652. DOI: 10.3389/fncel.2022.979652

[26] Ortiz-Cruz CA, Marquez EJ, Linares-García CI, Perera-Murcia GR, Ramiro-Cortés Y. Haploinsufficiency of Shank3 increases the orientation selectivity of V1 neurons. Scientific Reports. 2022;**12**(1):22230. DOI: 10.1038/s41598-022-26402-9

[27] El Shami M, Savani MR, Gattie LC, et al. Human plasma-like medium facilitates metabolic tracing and enables upregulation of immune signaling pathways in glioblastoma explants. Preprint. bioRxiv. 2023. DOI: 10.1101/2023.05.29.542774

[28] Logan A, Belli A, Di Pietro V, et al. The mechanism of action of a novel neuroprotective low molecular weight dextran sulphate: New platform therapy for neurodegenerative diseases like amyotrophic lateral sclerosis. Frontiers in Pharmacology. 2022;**13**:983853. DOI: 10.3389/fphar.2022.983853

[29] Castiglione H, Vigneron PA, Baquerre C, Yates F, Rontard J, Honegger T. Human brain organoids-on-Chip: Advances, challenges, and perspectives for preclinical applications. Pharmaceutics. 2022;**14**(11):2301. DOI: 10.3390/pharmaceutics14112301

[30] Hong Y, Dong X, Chang L, et al. Microglia-containing cerebral organoids derived from induced pluripotent stem cells for the study of neurological diseases. iScience. 2023;**26**(3):106267. DOI: 10.1016/j.isci.2023.106267

[31] Schafer ST, Mansour AA, Schlachetzki JCM, et al. An in vivo neuroimmune organoid model to study human microglia phenotypes. Cell. 2023;**186**(10):2111-2126.e20. DOI: 10.1016/j.cell.2023.04.022

[32] Benenson Y. Biocomputers: From test tubes to live cells. Molecular BioSystems. 2009;**5**(7):675-685. DOI: 10.1039/b902484k

[33] Fan D, Wang J, Wang E, Dong S. Propelling DNA computing with Materials' power: Recent advancements in innovative DNA logic computing systems and smart bio-applications. Advanced Science. 2020;**7**(24):2001766. DOI: 10.1002/advs.202001766

[34] Gearheart CM, Arazi B, Rouchka EC. DNA-based random number generation in security circuitry. Bio Systems. 2010;**100**(3):208-214. DOI: 10.1016/j.biosystems.2010.03.005

[35] Grozinger L, Amos M, Gorochowski TE, et al. Pathways to cellular supremacy in biocomputing. Nature Communications. 2019;**10**(1):5250. DOI: 10.1038/s41467-019-13232-z

[36] Tamura K. The genetic code: Francis Crick's legacy and beyond. Life (Basel). 2016;**6**(3):36. DOI: 10.3390/life6030036

[37] Wang Z, Wu X, Wu T. A parallel DNA algorithm for solving the quota traveling salesman problem based on biocomputing model. Computational Intelligence and Neuroscience. 2022;**2022**:1450756. DOI: 10.1155/2022/1450756

[38] Andreev VP, Gillespie BW, Helfand BT, Merion RM. Misclassification errors in unsupervised classification methods. Comparison based on the simulation of targeted proteomics data. Journal of Proteomics & Bioinformatics. 2016;**Suppl 14**:005. DOI: 10.4172/jpb.S14-005

[39] Su X, Smith LM. Demonstration of a universal surface DNA computer. Nucleic Acids Research. 2004;**32**(10):3115-3123. DOI: 10.1093/nar/gkh635

[40] Zhu W, Han Y, Zhou Q. Performing CTL model checking via DNA computing. Soft Computing. 2019;**23**:3945-3963. DOI: doi.org/10.1007/s00500-018-3314-7

[41] Fleck JS, Jansen SMJ, Wollny D, et al. Inferring and perturbing cell fate regulomes in human brain organoids. Nature. 2023;**621**(7978):365-372. DOI: 10.1038/s41586-022-05279-8

[42] Qiao N, Mostafa H, Corradi F, et al. A reconfigurable on-line learning spiking neuromorphic processor comprising 256 neurons and 128K synapses. Frontiers in Neuroscience. 2015;**9**:141

[43] Jeon I, Kim T. Distinctive properties of biological neural networks and recent

advances in bottom-up approaches toward a better biologically plausible neural network. Frontiers in Computational Neuroscience. 2023;**17**:1092185. DOI: 10.3389/fncom.2023.1092185

[44] Kotaleski JH, Blackwell KT. Modelling the molecular mechanisms of synaptic plasticity using systems biology approaches. Nature Reviews. Neuroscience. 2010;**11**(4):239-251. DOI: 10.1038/nrn2807

[45] Shih JJ, Krusienski DJ, Wolpaw JR. Brain-computer interfaces in medicine. Mayo Clinic Proceedings. 2012;**87**(3):268-279. DOI: 10.1016/j.mayocp.2011.12.008

[46] Evans AC, Thadani NN, Suh J. Biocomputing nanoplatforms as therapeutics and diagnostics. Journal of Controlled Release. 2016;**240**:387-393. DOI: 10.1016/j.jconrel.2016.01.045

[47] Yaqub M, Jinchao F, Arshid K, et al. Deep learning-based image reconstruction for different medical imaging modalities. Computational and Mathematical Methods in Medicine. 2022;**2022**:8750648. DOI: 10.1155/2022/8750648

[48] Balluff B, Heeren RMA, Race AM. An overview of image registration for aligning mass spectrometry imaging with clinically relevant imaging modalities. Journal of Mass Spectrometry and Advances in the Clinical Lab. 2021;**23**:26-38. DOI: 10.1016/j.jmsacl.2021.12.006

[49] Paul D, Sanap G, Shenoy S, Kalyane D, Kalia K, Tekade RK. Artificial intelligence in drug discovery and development. Drug Discovery Today. 2021;**26**(1):80-93. DOI: 10.1016/j.drudis.2020.10.010

[50] Sliwoski G, Kothiwale S, Meiler J, Lowe EW Jr. Computational methods in drug discovery. Pharmacological Reviews. 2013;**66**(1):334-395. DOI: 10.1124/pr.112.007336

[51] Smirnova L, Morales Pantoja IE, Hartung T. Organoid intelligence (OI)—The ultimate functionality of a brain microphysiological system. ALTEX. 2023;**40**(2):191-203. DOI: 10.14573/altex.2303261

[52] Kofman S, Mohan N, Sun X, Ibric L, Piermarini E, Qiang L. Human mini brains and spinal cords in a dish: Modeling strategies, current challenges, and prospective advances. Journal of Tissue Engineering. 2022;**13**:20417314221113391. DOI: 10.1177/20417314221113391

[53] Matsui TK, Tsuru Y, Kuwako KI. Challenges in Modeling human neural circuit formation via brain organoid technology. Frontiers in Cellular Neuroscience. 2020;**14**:607399. Published 2020 Dec 3. DOI: 10.3389/fncel.2020.607399

[54] Setia H, Muotri AR. Brain organoids as a model system for human neurodevelopment and disease. Seminars in Cell & Developmental Biology. 2019;**95**:93-97. DOI: 10.1016/j.semcdb.2019.03.002

[55] Simsa R, Rothenbucher T, Gürbüz H, et al. Brain organoid formation on decellularized porcine brain ECM hydrogels. PLoS One. 2021;**16**(1):e0245685. DOI: 10.1371/journal.pone.0245685

[56] Li Z, Yue M, Liu Y, et al. Advances of engineered hydrogel organoids within the stem cell field: A systematic review. Gels. 2022;**8**(6):379. DOI: 10.3390/gels8060379

[57] Polini A, Moroni L. The convergence of high-tech emerging technologies

into the next stage of organ-on-a-chips. Biomaterials and Biosystems. 2021;**1**:100012. Published 2021 Feb 20. DOI: 10.1016/j.bbiosy.2021.100012

[58] Pamarthy S, Sabaawy HE. Patient derived organoids in prostate cancer: Improving therapeutic efficacy in precision medicine. Molecular Cancer. 2021;**20**(1):125. Published 2021 Sep 29. DOI: 10.1186/s12943-021-01426-3

[59] Qian X, Song H, Ming GL. Brain organoids: Advances, applications and challenges. Development. 2019;**146**(8):dev166074. Published 2019 Apr 16. DOI: 10.1242/dev.166074

[60] Weth FR, Peng L, Paterson E, Tan ST, Gray C. Utility of the cerebral organoid glioma 'GLICO' model for screening applications. Cells. 2022;**12**(1):153. Published 2022 Dec 30. DOI: 10.3390/cells12010153

[61] Lee H. Engineering In vitro models: Bioprinting of organoids with artificial intelligence. Cyborg and Bionic Systems. 2023;**4**:0018. DOI: 10.34133/cbsystems.0018

[62] Trujillo CA, Muotri AR. Brain organoids and the study of neurodevelopment. Trends in Molecular Medicine. 2018;**24**(12):982-990. DOI: 10.1016/j.molmed.2018.09.005

[63] Sui Y, Yu H, Zhang C, Chen Y, Jiang C, Li L. Deep brain-machine interfaces: Sensing and modulating the human deep brain. National Science Review. 2022;**9**(10):nwac212. Published 2022 Oct 7. DOI: 10.1093/nsr/nwac212

[64] McFarland DJ, Wolpaw JR. Brain-computer interfaces for communication and control. Communications of the ACM. 2011;**54**(5):60-66. DOI: 10.1145/1941487.1941506

[65] Huang Q, Tang B, Romero JC, et al. Shell microelectrode arrays (MEAs) for brain organoids. Science Advances. 2022;**8**(33):eabq5031. DOI: 10.1126/sciadv.abq5031

[66] Esmaeili M, Vettukattil R, Banitalebi H, Krogh NR, Geitung JT. Explainable artificial intelligence for human-machine interaction in brain tumor localization. Journal of Personalized Medicine. 2021;**11**(11):1213. Published 2021 Nov 16. DOI: 10.3390/jpm11111213

[67] Yang H, Lam KY, Xiao L, et al. Author correction: Lead federated neuromorphic learning for wireless edge artificial intelligence. Nature Communications. 2022;**13**(1):4838. Published 2022 Aug 17. DOI: 10.1038/s41467-022-32602-8

[68] Bourret R, Martinez E, Vialla F, Giquel C, Thonnat-Marin A, De Vos J. Human-animal chimeras: Ethical issues about farming chimeric animals bearing human organs. Stem Cell Research & Therapy. 2016;**7**(1):87. Published 2016 Jun 29. DOI: 10.1186/s13287-016-0345-9